T0320356

# Maintenance Parts Management Excellence

Most successful organizations recognize Maintenance Parts and Procurement as a critical success factor to Asset Management Excellence and their fundamental supply chain value proposition. This book works as a guide to all the stakeholders that influence the success of their Maintenance Parts Operation and their enterprise's bottom line.

*Maintenance Parts Management Excellence: A Holistic Anatomy* defines the Maintenance Parts Management's role in Asset Management Excellence and expands on the importance of the Parts Inventory Planner role in an organization. It discusses how to create a unique Maintenance Parts Management Strategy for an organization and offers insights into the multiple strategies needed to create and maintain a Maintenance Parts inventory policy. The book also provides an organized overall approach to creating Maintenance Parts Management Excellence in an enterprise.

Executives with an organization responsible for the construction, management, and disposal of all asset classes (plant, equipment, IT assets), consultants responsible for assignments associated with optimizing lifecycle decisions for clients, maintenance, and reliability professionals within an organization, will benefit from this professional plus book. Upper-level undergraduate engineering students, as well as graduate students of management who focus on operations management and engineering graduate students addressing issues of maintenance and reliability engineering, may also be interested in this book.

# Maintenance Parts Management Excellence

## A Holistic Anatomy

Don M. Barry

CRC Press is an imprint of the
Taylor & Francis Group, an **informa** business

Designed cover image: Don M. Barry

First edition published 2023
by CRC Press
6000 Broken Sound Parkway NW, Suite 300, Boca Raton, FL 33487-2742

and by CRC Press
4 Park Square, Milton Park, Abingdon, Oxon, OX14 4RN

*CRC Press is an imprint of Taylor & Francis Group, LLC*



ISBN: 978-1-032-38295-1 (hbk)
ISBN: 978-1-032-38369-9 (pbk)
ISBN: 978-1-003-34467-4 (ebk)

DOI: 10.1201/9781003344674

Typeset in Times
by SPi Technologies India Pvt Ltd (Straive)

# Contents

# Preface

The book is organized in the same fashion as the second edition of *Asset Management Excellence – Optimizing Equipment Life-cycle Decisions*, published in late 2010, CRC Press. These books holistically address their topics, revealing critical issues paramount to leadership driving excellence in Asset Management and Parts Management.

The *Maintenance Parts Management Excellence: A Holistic Anatomy* focuses more on the management dynamics of Maintenance Parts. It holistically looks at the many elements suggested in this book. Its approach includes strategy, organization, metrics, store room logistics, systems, parts sourcing and refurbishment, inventory planning and optimization, asset lifecycle alignment, and process re-thinking. In addition, MPE includes a checklist to help consider where management should prioritize their efforts for the short and longer-term.

Maintenance Parts Management Excellence is a topic that most organizations poorly understand. Getting to Maintenance Parts Management Excellence takes a corporate community, a committed group of stakeholders, and a unique culture. Maintenance Parts supports Asset Management. Asset Management supports Operational Excellence, and successful operations create a happy customer experience.

Don Barry was the country Parts Manager of IBM Canada's Field Maintenance organization for almost eight years. His total direct parts experience exceeds 20+ years. His combined experience with parts surpasses 35 years when you add his direct consulting to the equation. While consulting, he taught a Maintenance Parts Management Excellence course privately and in the University of Toronto's Maintenance Masters program. He has since expanded his training and provides consulting and training to corporations and individuals serious about understanding leading practices in Maintenance Parts Management.

The primary source for this book is Don's direct experience from when he was the owner of one of the most effective parts operations globally and mentoring and guiding his consulting clients.

In this book, Don shares the elements that he believes allowed his parts operation to become one of the best globally. Then, he breaks the Maintenance Parts business down, leveraging different models to help the reader understand how these elements fit together and apply the same principles to their situation.

Don explains the influences of Asset Lifecycle alignment, Organizational reporting, and Inventory planning. Asset Risk and Reliability integration is an essential component of a leading Maintenance Parts stocking policy elements. He suggests that it does not matter where Maintenance Parts reports within the enterprise as long as the corporate behavior is correct and aligned to drive Parts Management Excellence.

# Acknowledgments

Thanks to all who have supported my personal experience in Asset Management, Maintenance Parts Management, and this book.

# About the Author

**Don Barry** is a principal consultant and thought leader for Asset Acumen Consulting Inc – supporting Risk and Reliability Strategies, Strategy consulting, Enterprise Asset Management solutions, IT and IoT in Asset Management training, and Asset Performance Management solutions. Don also supports his clients in mentoring, consulting, and teaching leading maintenance Parts management practices. Previously leading IBM Canada's Supply Chain and Enterprise Asset Management Practice (for 15 years), as well as the Global Lead for IBM's Asset Management Center of Competency and as an Associate Partner, Don is an expert at creating distribution process improvements and developing inventory reduction strategies with increased service levels. He specializes in solving asset management, distribution, and asset re-utilization business problems. His in-depth experience includes world-class inventory optimization, re-utilization techniques, and leading the coordination of many service business mergers and divestitures. He has over 44 years in service delivery support systems and application development, including three years in field service management and 15 years directly in business process development and supply chain management.

Mr. Barry's direct client list has included industries such as Upstream Oil and Gas, Pipelines, Power Generation and T&D Utilities, Mining, Airlines, Computer Electronics Manufacturers, Steel Manufacturers, and Federal and Provincial Governments.

Mr. Barry's noted areas of experience include:

- Supply Chain Operations
- Maintenance Parts Management
- Enterprise Asset Management systems program executive
- Merger and Acquisition
- Change Management
- Reliability-Centered Maintenance (RCM) Practitioner
- Instructor of the "Maintenance Parts Excellence Program"
- Instructor of the University of Toronto's "Physical Asset Management Program"
- Instructor of Quality and Management transformation at IBM
- Prime contributor to the second edition of *Asset Management Excellence – Optimizing Equipment Life-cycle Decisions*, published in late 2010, CRC Press.
- Worked as the Asset Management Lead for the Calgary-based IBM Natural Resources Solution Centre

# References

The following publications contributed to the research and knowledge of the author in support of this book:


- *Asset Management Excellence Optimizing Equipment Life-cycle Decisions* 2nd Edition, Edited by Campbell, Jardine, McGlynn, CRC Press, 2011.
- *Uptime, Strategies of Excellence in Maintenance Management* – John Dixon Campbell, Productivity Press, 1995.
- *Reliability-centered Maintenance II, 2nd edition*, John Moubray, Industrial Press Inc., 1997.
- *MRO (Maintenance, Repair, and Overhaul) Handbook*, John Campbell, Coppers and Lybrand, Plant, Engineering, and Maintenance (PEM) Magazine, Vol. 22, Issue 2.
- *International Standard ISO55001 Asset management* – Management systems – Requirements ISO55001:2014 (E).
- *The Memory Jogger, A Pocket Guide of Tools for Continuous Improvement*, GOAL/QPC 1988.


The following training, teaching exercises, and general experience contributed to the research and knowledge of the author in support of this book:


- Maintenance Parts Management Excellence Training, program lecture notes, Don Barry, Asset Acumen Consulting Inc., 2020.
- Physical Asset Management Program, Lecture notes – UofT, Impact of Change Management – Don Barry, Asset Acumen Consulting Inc., 2019.
- IBM Parts Logistics Quality Program notes, Don Barry, 1991.
- Quality archived notes in IBM Parts Logistics Services, Don Barry, 1991.
- Survey of extensive Asset Management related projects and Change Management, D. Barry, 2006.
- Project notes – KPI development workshops, D Barry, Asset Acumen Consulting Inc., 2019.
- RCM2 Practitioner Training, The Aladon Network, 2006.

# 1 Introduction
## *Why Maintenance Parts Management Is Important*

### AFTER-PRODUCT SALES SERVICE PARTS OR PLANT MAINTENANCE PARTS

Maintenance parts (spare parts or service parts) require a focus across all industries, where replacing a part is needed to maintain the expected function or performance of a piece of equipment or asset. The equipment (or asset) can be:

- A product that a company has manufactured and sold to a customer, for example, a part of a train, aircraft, computer, copier, washing machine, elevator, or electrical transformer or
- A piece of equipment that a company uses in its organization (an apparatus for manufacturing, a power generator, an electricity transmission line, a telephone switching system, vehicle, aircraft, and railroad tracks)

Keeping this in mind, the maintenance parts management portion of an enterprise can be:

- A contributing business unit to the services or after-sales business or
- A straight maintenance, repair, and overhaul (MRO) business (Figure 1.1)

The after-sales maintenance parts business supports manufactured products sold to their enterprise's customers. Many of these assets are sold with a recommended parts list that should be in stock if the buyer expects to maintain the investment aligning to the original equipment manufacturer (OEM) recommended maintenance program. If this is the case, the buyer will manage these parts the same way an MRO would. If a third party maintains the asset, a supply chain needs to be in place to provide maintenance parts to the third-party maintenance and repair service providers.

After-sales maintenance parts can be straight parts sales. Still, it often means a maintenance part is considered a product maintenance activity delivered (sold as a portion of the maintenance service) through the OEM service team. Industries that perform this type of parts service can include:

- Automotive
- Aerospace and defense
- Engineering and construction
- High tech (computers, smartphones, etc.)

DOI: 10.1201/9781003344674-1

**FIGURE 1.1** What is maintenance parts management. (Adapted from copyright 101 MPE Introduction ©2020 from Maintenance Parts Management Excellence Training edited by Barry. Reproduced by permission of Asset Acumen Consulting Inc.)

Many OEMs maintain their products and manage the demand for these service parts. For example, automotive dealerships often make more profit maintaining vehicles than selling them.

The MRO business is more about managing spare parts to support equipment maintenance used within their enterprise. It entails supplying parts to the maintenance department to keep the business's critical assets operating to meet the product quality, and level of service promised to customers and to maintain its reputation for on-time delivery. Industries that perform this type of parts service can include:

- Mining, oil, and gas
- Telephone infrastructure enterprises
- Transportation companies
- Utilities
- Other product-producing companies (including supporting discreet and process manufacturers)

Procuring assets; managing the asset lifecycle; maintaining the asset performance and cost; and supporting asset maintenance, analytics, and parts processes are included in asset management.

For consistency throughout this book, the term "maintenance parts management" will represent both types of maintenance parts missions, MRO and after-sales business support, unless otherwise suggested. Maintenance parts management is a critical element in the pursuit of supporting and planning for the sustainment of critical operating assets performing their desired function.

A key goal of asset management is ensuring that the asset performs its desired function. Maintenance planning and execution and parts management are critical elements in achieving this goal. Even if the maintenance of the assets is outsourced to a third party, the responsibility for the maintenance program and planning, and by extension, the parts management planning should remain with the company's asset management and maintenance departments.

## THE ROLE OF MAINTENANCE PARTS MANAGEMENT IN SUPPORT OF ASSET MANAGEMENT

Asset management is a growing concern for many organizations. The asset management strategic activities an enterprise engages in to improve its financial return can offer exciting new initiatives. The range of supporting activities can include conforming to the new ISO55000 standards. Asset management can also mean adding the Internet of Things (IoT) technologies with sensors and analytics. The sensors would enable early insights into what assets may be starting to fail functionally and provide an early warning. With an early warning, a maintenance action can mitigate the operating asset's operational risks (reduce downtime or poor-quality output), safety, or environmental risks. A dynamic early warning can help reduce failure impacts, reduce costs, and improve customer satisfaction through improved service levels.

Executing asset management maintenance is fundamentally a parts and labor business. It is most often difficult to do the maintenance of mechanical assets efficiently without the necessary parts. *Typically executing maintenance is 40% labor costs, 40% parts-related costs (usage and management), and 20% other support costs such as infrastructure and technical overhead.* Most parts maintenance activities are disbursed or tracked through the enterprise's local store location and supporting systems.

Leading asset management organizations have implemented standards, such as ISO55000, and support a reliability-centered maintenance (RCM) or a total productive maintenance (TPM) culture. Yet, these best intentions of implementing a standard or discipline culture will be undermined if the communication and support are not in place to provide the *"right part* (with the tools & equipment instructions and schematics), *with the right quantity, to the right place, in the committed timeline"* when needed.

Someone in the enterprise needs the proper stewardship to ensure that the correct part is stored close to the asset's forecasted need and managed so that it does not arrive broken or corroded (defective) when needed (Figure 1.2).

The figure describes a typical maintenance repair transaction for urgent work. It would consist of the operator identifying the asset or area from their install base that is no longer working and submitting a call for maintenance. After confirming the type of asset/defect to be addressed and approved for a response action, a technician would be assigned to travel to the asset site. Often, a part will be identified as needed to diagnose an issue and repair the defect. This is where the maintenance parts business units' delivery mission is called upon to source the part from their closest stock room.

If the part is available in the local stock room, it is picked and provided to the maintenance tech, either as an over-the-counter delivery or through the help of a delivery solution that could include a truck or even an aircraft vendor. The advanced work required to prepare for the urgent parts order and support the order after it is delivered is the maintenance parts business unit's mission. Successfully selecting the right part to be stocked in the most effective stock room in advance of the urgent parts need requires insight and communications from many areas of the enterprise and a risk-balanced understanding of the financial implications (to stock or not to stock the part). It is critical to track and manage all the parts to the technician and the new (or used) parts that could be returned. The ability to facilitate maintenance parts stewardship, including the repair or disposition of these used defective parts that align with operations and business needs, is also critical and part of the maintenance parts business unit's responsibility. This stewardship is just part of the cycle of responsibility a

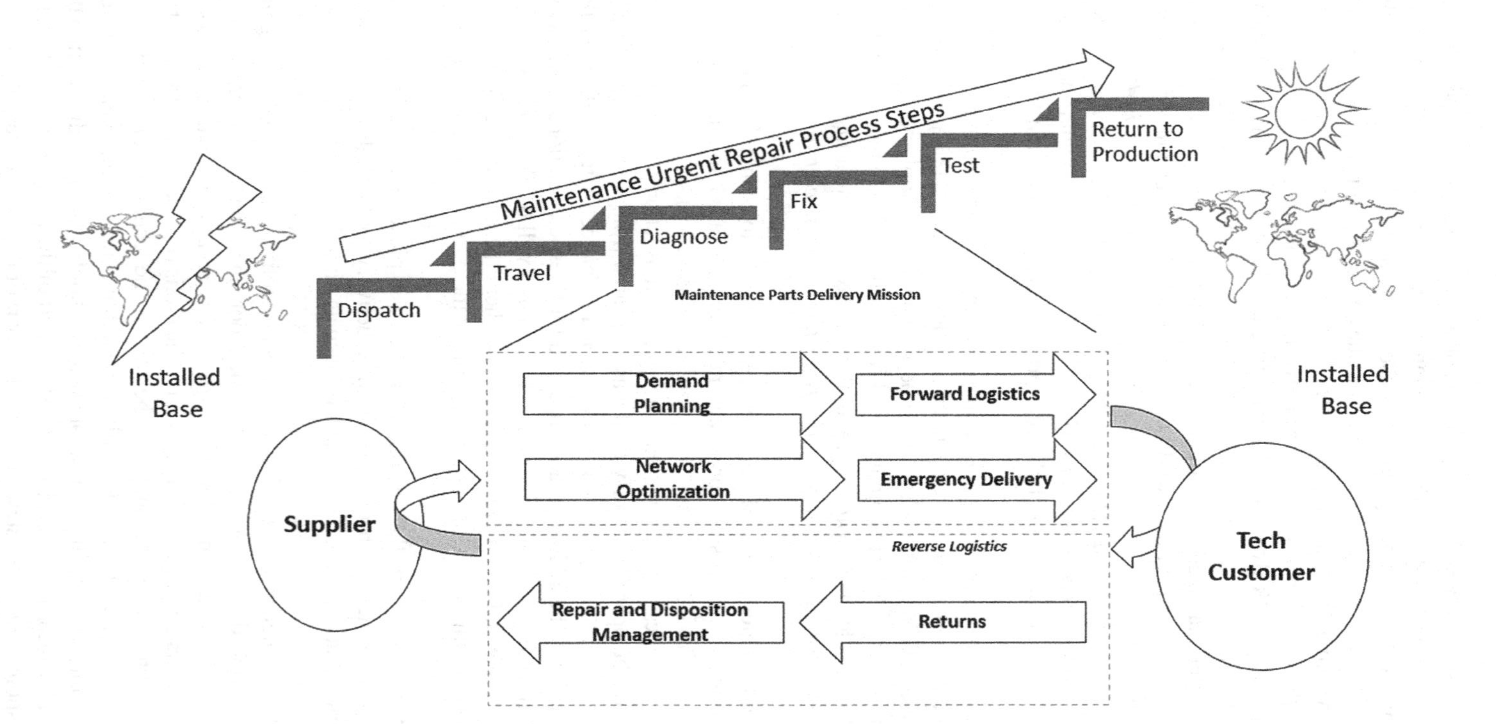


**FIGURE 1.2** Simple steps to repair urgent work. (Adapted from copyright 101 MPE Introduction ©2020 from Maintenance Parts Management Excellence Training edited by Barry. Reproduced by permission of Asset Acumen Consulting Inc.)

leading maintenance parts operation will assume to drive success into the enterprise. Understanding the asset maintenance program, the parts failure rate, the early warning tracking of potential failure, and the consequence of part failure are all elements that help a parts manager facilitate having the necessary parts when needed. It is not a culture that breeds success if the maintenance technician just goes to the counter, asks for a part, and complains when it may not be available, as many organizations seem to find themselves experiencing.

Most enterprising organizations exist to provide value by creating manufactured products or delivering services that generate revenue and profit. The maintenance organization plays a crucial role in ensuring that the essential assets that contribute to its operation (e.g., manufacturing or service assets) create "value" in its supply chain and are maintained to deliver the targeted, sustained value.

While operating the asset to create value in a supply chain is the operator's role, they cannot succeed in creating that value without occasionally needing maintenance. Likewise, maintenance cannot do its role without receiving the support of the parts component of a parts and labor maintenance action. Maintenance parts management cannot do their role without the support of procurement and the invoices paid by their accounts payable group to maintain the support relationship with the parts suppliers.

Achieving maintenance parts excellence requires that the enterprise empower the supporting maintenance parts organization to support the "value" mission of the enterprise, which, in turn, helps both maintenance and operations.

*The maintenance parts organization's mission should be to provide the required service levels of parts support* (accuracy, proximity, timeliness, and availability) *at optimal cost.*

Cost management does play a role in maintenance parts management. Cost is a necessary element of the real mission of an enterprise to create supply chain value. Value is created through an enterprise's employees and their associated assets to generate the revenue and profit expected by the business's shareholders.

## WHO SHOULD OWN MAINTENANCE PARTS MANAGEMENT?

When establishing a new enterprise's foundational business units and responsibilities (finance, sales, operations, and maintenance), some organizations consider maintenance parts management an afterthought.

The maintenance part management organization will report to finance, procurement, maintenance, operations, or HR in various scenarios. The reporting structure for maintenance parts is essential. However, the reporting structure is less important than ensuring *optimal behavior of the various stakeholders* who *create* and support *maintenance parts excellence*. The key for each listed business unit and design engineering is that they all have a role to play to ensure that they work in a well communicated and executed parts support set of processes and infrastructure.

Key stakeholders should understand the enterprise business objectives and be responsive to the requirements and expectations of the business, including making the right decisions to support enterprise sales, operations, maintenance, and parts support. For example, parts planning requires input from the sales plan, the operations plan, the procurement plan, and the maintenance plan. Individual and team communication, corporate and cultural behaviors play as much a role in the success of an excellent parts operation as the people directly charged with parts management. A successful enterprise will *establish a set of criteria for asset management decision-making* that includes the foundational steps to ensure that the required parts support is in place and well-executed.

The enterprise operations and asset management objectives must be understood and served when establishing a maintenance parts excellence culture. Likewise, the relationship between the maintenance parts business unit and other areas such as operations, maintenance, design engineering, finance, legal, and HR must be understood.

How do these business units support each other with policy and process, and how can they be supported through information technology (IT)? All must help each other with the best intentions to expect process and enterprise success.

Thought will need to be given regarding how "parts" will be accounted for on the corporate books (as an asset until used or expensed when procured). Parts accounting is often left to the finance organization to sort out. Company capital asset declaration is far from the mindset of the operations and maintenance teams. Maintenance parts is often perceived as a place to position employees that did not fit in other roles – still an afterthought. Yet maintenance parts management can be as complex as maintenance or operations and needs to be as sophisticated in its collected support from all the stakeholders and IT. Materials handling disciplines need to be in place for stock rooms and warehouses. Experienced inventory planners must influence the targeted service levels and execute the most effective inventory stocking policies.

## BENEFITS OF DOING MAINTENANCE PARTS MANAGEMENT WELL

The cost of "not" executing maintenance parts management well can be debilitating to an enterprise. The obvious notion is that if the part is unavailable, the critical asset (the needed value chain component) cannot produce value for the enterprise. A part is in stock to mitigate the critical asset's unexpected lost function/production. Suppose a vital asset is out of production, and one is "waiting for a part." In that case, the missing production value can estimate the part-related impact cost until the part is in the hand of the maintenance technician. The operational impact of a failed asset due to parts can include lost production time, downstream lost operator time, customer dissatisfaction, contract delivery penalties, and damage to the company delivery reputation if it happens frequently.

Maintenance technician costs are also incurred by not having the part close at hand when needed. If the technician must wait for the part, the enterprise also experiences lost technician time, which is a cost to the business. If the maintenance technician needs to "put tools down" and leave the asset location until the part arrives and then "tool up" again, it compounds the technician-related maintenance costs. The *lost production costs, technician costs, and parts expediting charges, all contribute to the impact of a part not being close and available* to mitigate the unexpected lost production of the asset.

The benefits can be significant when a proper enterprise-wide parts management focus is well-executed.

A high-level benefits list can include:

- Increased asset availability
- Saved maintenance technician time
- Equipment purchases extended
- Warranties claimed
- Inventories (one time) and associated inventory carrying costs (annual) reduced
- Maintenance parts expediting costs reduced (management time and travel)
- Maintenance parts unit costs are reduced through an effective procurement program, consolidating vendors, and parts repair programs, where appropriate

Focusing on maintenance parts management and its related parts inventories can create multiple views of potential savings for an organization. This book will reveal many ways in which the benefits can be estimated. The estimates will differ depending on the business context and dynamics and the willingness of the affected stakeholders to commit to the potential initiatives.

Based on the current inventory state and the existing management disciplines, a potential view of inventory reduction opportunities could be available, as shown in Table 1.1.

Parts will play a role in many of the asset management opportunities. Think how parts would contribute to just a 1% benefit of many asset management process initiatives.

Imagine the impact of a 1% benefit to the enterprise if you could:

- Improve asset availability (Are you operating 7/24?)
- Improve operator/trades utilization and efficiency (Are you labor-intensive?)
- Extend asset life (Is your enterprise capital intensive?)
- Improve response time (How dispersed are your facilities?) (Can you predict parts needed for likely failures?)

**TABLE 1.1**
**Potential Benefits of a Maintenance Parts Excellence Focus**

| Business Scenarios | Potential Inventory Benefit Percentage Points |
|---|---|
| Production availability | Up 2–11% |
| Labor utilization | Up 8–22% |
| Equipment purchases | Down 2–7% |
| Warranty recoveries | Up 5–50% |
| Inventory needs | Down 20–35% |
| Inventory carrying costs | Down 5–25% |
| Material costs | Reduced 10–40% |
| Material transportation costs | Reduced 10–20% |
| Purchasing labor | Reduced 10–40% |
| Parts unit costs | Reduced 10–40% |
| Parts expediting costs | Reduced 10–50% |
| Parts dispositioning tax benefit timelines | Brought forward by years |

- Increase spare parts inventory level/turns (Are you carrying the correct spares to avoid catastrophic downtime?)
- Improve inventory service level (Are parts there when we need them?)
- Leverage a supplier's pooled inventory as part of your service level agreements to quickly access costly and critical parts
- Reduce commissioning time and surplus for new build (Have you adequately identified emerging asset spare parts support?)
- Increase the quality of asset support output by reducing rejects, rework, returns, parts loss, giveaway, shrinkage, waste, or improving repair yield (What is the cost of poor quality?)

It is often difficult to identify the potential savings from any initiative. If you can quantify a 1% benefit, it is often a matter of determining how many "1%" we can accumulate. This exercise can be helpful as Parts is often not funded for process improvements.

Not all savings are considered equal to an organization's bottom line. An example of potential savings categories and the initiatives considered are shown in Table 1.2.

Figure 1.3 shows an example of the potential benefits of a focus on maintenance parts management has created opportunities for *reduced inventory expense, inventory carrying costs, and improved parts service levels*. It has also *created productivity benefits for maintenance planners and technicians and, in turn, reduced maintenance costs*.

**TABLE 1.2**
**Example of Benefits from a Maintenance Parts Excellence Focus**

| Initiative Category | Potential Initiative Benefits | Direct Savings Range (%) | Contribution to Total Savings Range (%) |
|---|---|---|---|
| Reduced inventory expense | • Optimal stocking lists and levels<br>• Improved procurement options<br>• Reduced obsolescence | 15–35 | 50–65 |
| Increased planner productivity | • Automated tasks and alerts<br>• Better performance against plan<br>• Improved proactive maintenance performance<br>• Reduced break/fix | 5–30 | 4–8 |
| Improved service levels | • Better scheduling<br>• Service level performance<br>• Client loyalty | 2–20 | 4–8 |
| Increased craft productivity | • Productive utilization<br>• Defect resolution | 5–15 | 4–8 |
| Reduced repair costs | • Reduced delays<br>• Reduced tool-down due to missing parts<br>• More recoverable warranty | 20–30 | 2–5 |
| Reduced expediting costs | • Lower freight costs<br>• Fewer losses | 10–25 | 2–5 |

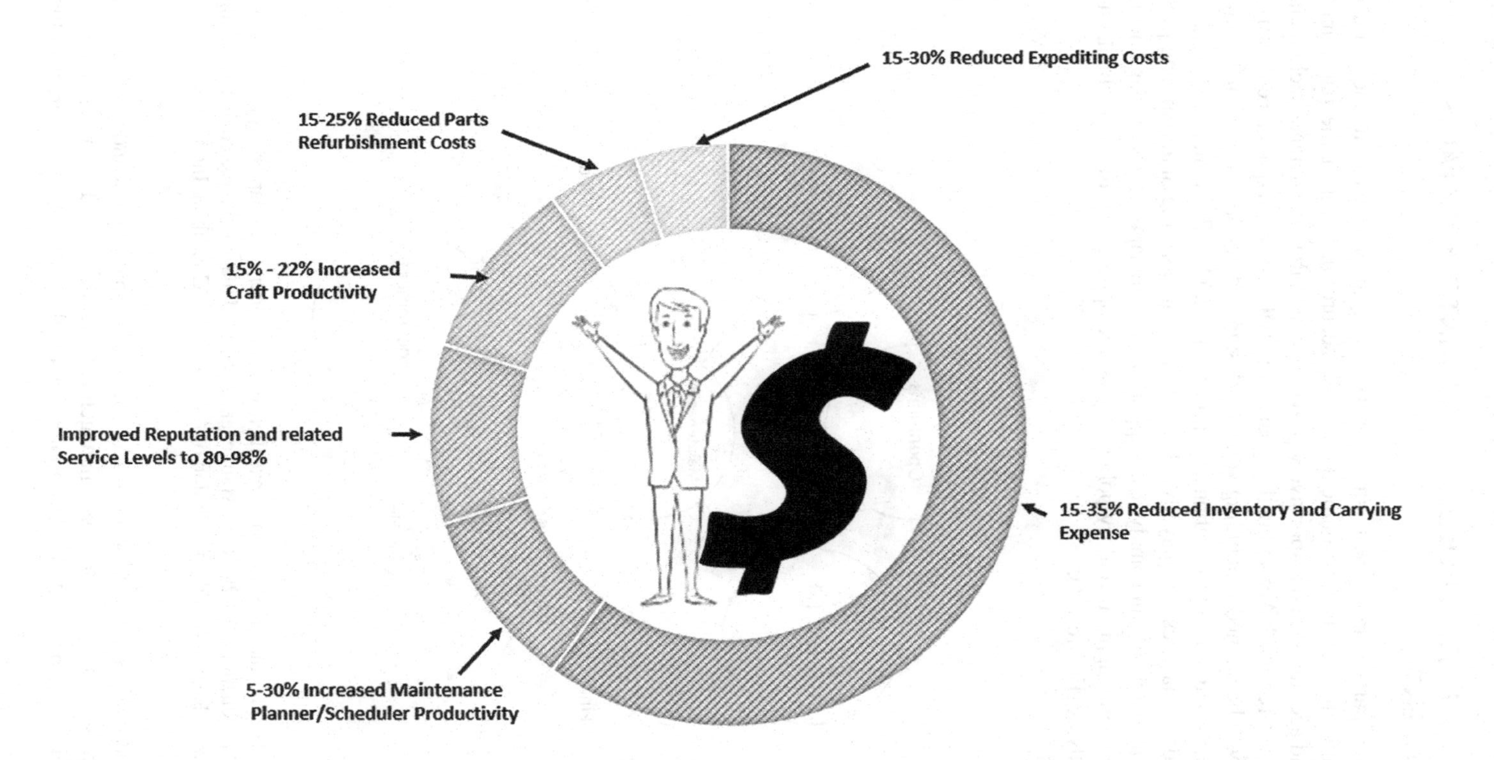


**FIGURE 1.3** Example of potential benefits found from a focus on maintenance parts inventory. (Adapted from copyright 101 MPE Introduction ©2020 from Maintenance Parts Management Excellence Training edited by Barry. Reproduced by permission of Asset Acumen Consulting Inc.)

## WHO ARE THE STAKEHOLDERS FOR MAINTENANCE PARTS MANAGEMENT?

Maintenance parts management aims to have the right parts close to the critical assets, maintained in an environment that will secure and protect the parts until needed, and accurately and responsively provide them to the maintenance technician when urgently required. Achieving this goal at optimal cost will require an enterprise community to be aligned - including vendors, design engineering, operations, maintenance, parts planning and warehousing, procurement, IT, HR, finance, and legal.

IT and data analytics can play a vital stakeholder support role in automating processes, replenishment, and analytics, supporting maintenance parts management. The basics of managing inventory policy rely on the opportunity to do practical statistical analysis (Figure 1.4).

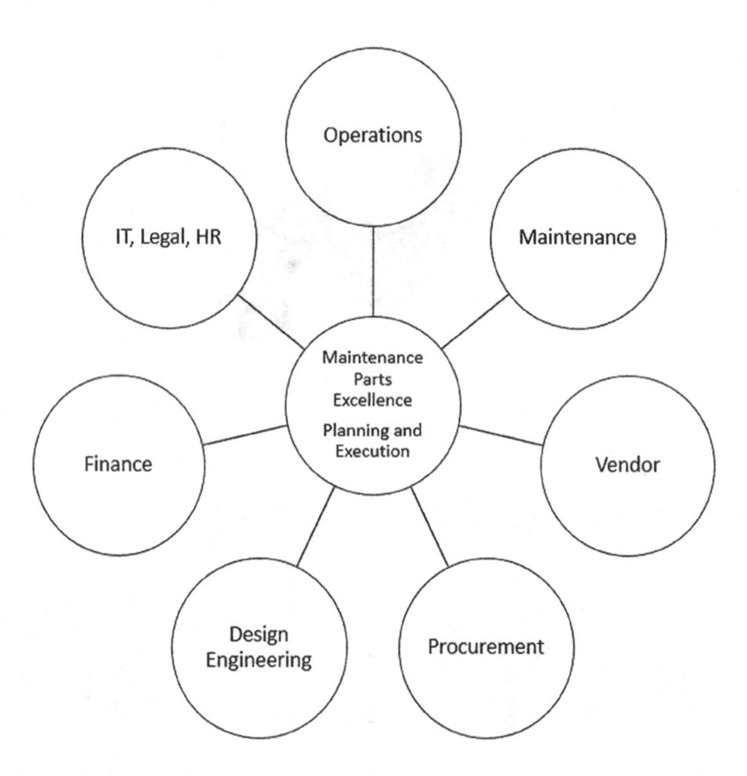


**FIGURE 1.4** The maintenance parts excellence stakeholders. (Adapted from copyright 101 MPE Introduction ©2020 from Maintenance Parts Management Excellence Training edited by Barry. Reproduced by permission of Asset Acumen Consulting Inc.)

Most successful organizations now recognize maintenance parts and procurement as a critical success factor to asset management excellence and their fundamental supply chain value proposition. This book is intended to guide the listed areas of the enterprise.

# 2 Maintenance Parts Excellence Influences

It is interesting to be introduced to a new maintenance and parts organization and learn their prioritized issues specific to maintenance parts management. For example, their latest perception may be that:

- The parts warehouse team should improve parts availability
- Cycle time for urgent parts orders should be improved
- Health and safety issues should be corrected in a timely fashion
- Parts quality is an issue in the warehouse
- The right parts are seldom stocked for urgent repairs when needed
- Somebody needs to get a handle on the parts inventory stocking policy for new assets and new construction
- The relationship between maintenance, operations, design engineering, maintenance parts, and finance needs to be better defined and exercised to support maintenance parts excellence
- A better understanding of what is a leading practice in maintenance parts management needs to be baselined, and a prioritized list of actions needs to be created, communicated, and executed
- Inventory efficiencies are not well-established or managed
- Inventory dispositioning (scrap) is challenging to get approved
- Not all parts activity is tracked in the maintenance/parts systems
- Finance restricts the number of needed parts that may be stocked
- Parts need to be hidden from the inventory system by maintenance technicians to support future urgent needs

The above is a partial list, and almost any asset-intensive organization could develop a more comprehensive one, perhaps not too different from the above. For example, when the maintenance parts management is prioritized, issues are tabled across a mix of stakeholders: that stakeholder's group would typically be the warehouse manager, inventory planner, material handler, maintenance staff and management, operations staff and management, design engineering staff and management, information technology (IT), human relations (HR), and finance (Figure 2.1).

A maintenance excellence pyramid framework has been used hundreds of times to facilitate specific issues an enterprise may encounter to achieve asset management excellence. This framework, in turn, was used to help them prioritize the actions they should plan to take to improve their asset management execution and management.

DOI: 10.1201/9781003344674-2

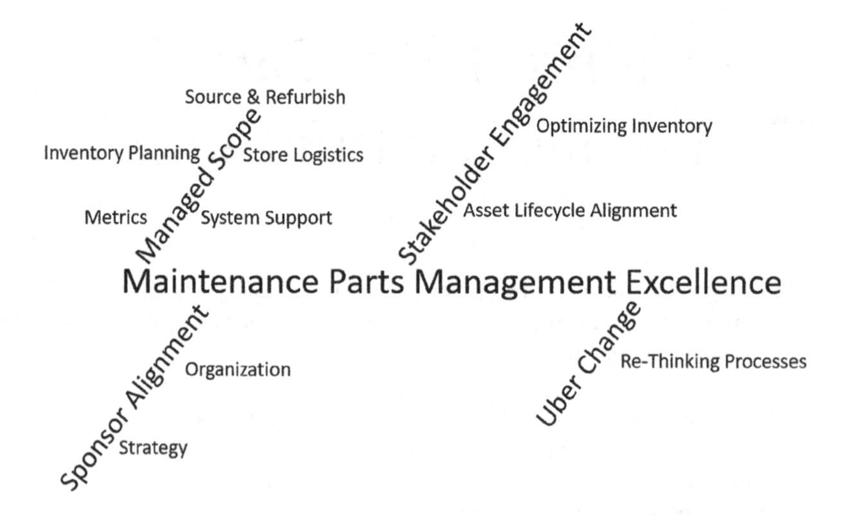


**FIGURE 2.1** Maintenance parts excellence topics. (Copyright MPE ©2022 from Maintenance Parts Management Excellence Training edited by Barry. Reproduced by permission of Asset Acumen Consulting Inc.)

New standards such as PAS55 and ISO55000 to ISO55002 have supplemented and complemented this approach and other emerging frameworks developed to guide and engage stakeholders to higher levels of asset uptime and asset management.

## ORIGINS OF THE MAINTENANCE PARTS EXCELLENCE PYRAMID

Maintenance parts management is one of the ten elements from the original maintenance excellence pyramid. (See *Asset Management Excellence Optimizing Equipment Life-cycle Decisions*, 2nd Edition).

Typically, four initial areas drive return on asset (ROA) – driving costs down while driving up production, safety, environmental, and regulatory compliance. Having conducted numerous assessments: experience has taught me that maintenance parts is one of the three to four primary areas of focus that can help an organization handle its asset management costs.

These focus areas include:

- Efficient execution of work through planning, scheduling, and approvals management
- Efficient and effective support of maintenance execution with spares and support materials management activity
- Proactive definition of what maintenance should be done to create "effective" management of risk and reliability for reasonably likely failures
- "Optimization" of the above processes while leveraging critical data and technology (Industry 4.0, Enterprise Asset Management (EAM) tools, IoT, analytics, Artificial Intelligence (AI), Blockchain)

Experience has also taught me that these four areas require a collaborative culture and approach to influence success, including:

- The stakeholder community's "mindset" needs to evolve from traditional thinking to scientific, business-based thinking
- An inclusive culture that supports a notion that operations, maintenance planning, and inventory planning work as one is in place and executing well
- A culture that proactively eliminates barriers to creating a cooperative approach among production, operations, and engineering (including leveraging technology, where needed)
- Executing a strategically planned and managed change program, which is established and led by management while improving cross-functional processes with technological advances
- Leveraging local and outside knowledge to understand leading practices in planning, scheduling, proactively identifying, predicting, and prescribing maintenance requirements (Figure 2.2)

The framework approach to maintenance parts excellence has been structured through the influence of many asset management-related business management pyramids. It is different in content and purpose. However, it tries to align with the method used in John Dixon Campbell's book *Uptime* (Productivity Press) and reinforced in *Asset Management Excellence, Optimizing Equipment Life-cycle Decisions* (2nd Edition Edited by Campbell, Jardine, and McGlynn).

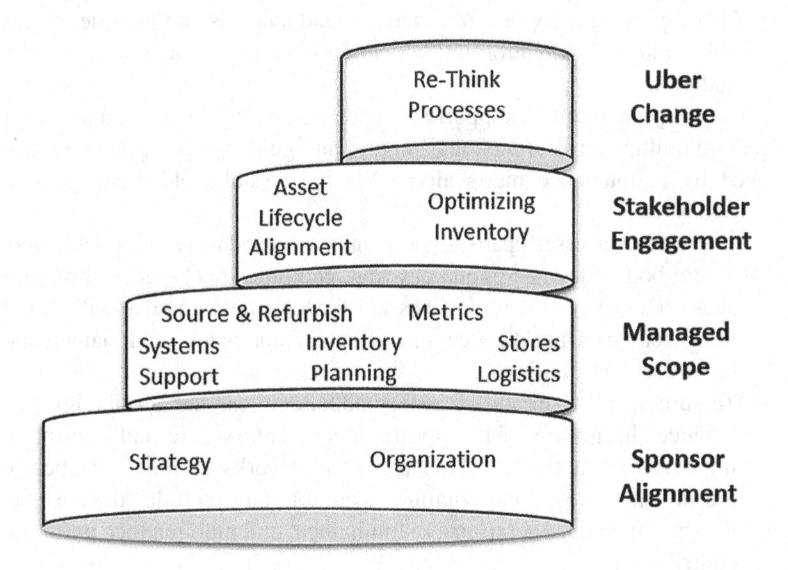


**FIGURE 2.2** Maintenance parts management excellence pyramid. (Copyright MPE ©2022 from Maintenance Parts Management Excellence Training edited by Barry. Reproduced by permission of Asset Acumen Consulting Inc.)

Maintenance parts excellence is likewise structured in four parts:

- Beginning with visible and effective "sponsor alignment":
  - To establish the foundational guidance needed to align with the enterprise business goals, maintenance goals, and the related supporting maintenance parts management goals
  - To organize the maintenance parts management goals aligned to support the enterprise goals so all stakeholders in the enterprise understand their role and contribution to the program's success
  - To ensure that leadership is "seen" leading and sponsoring the parts operation to support asset management excellence
  - Understand how change management initiatives may need to be injected into the foundational culture to empower staff and enable leaders to lead and be successful
- Then controlling the "managed scope" of the functions assigned explicitly to maintenance parts management:
  - To establish the face of maintenance parts to the maintenance technician, stores logistics, including shipping, receiving, order fulfillment, inventory stores stewardship
  - To measure success with effective metrics that work across all Business dimensions (levels) in support of maintenance parts excellence
  - To establish how parts are acquired into the enterprise's maintenance parts network, we have a focus on parts source and refurbishment
  - To create, manage, and execute inventory stocking policy across the enterprise set of storerooms, we have a focus on inventory planning
  - To support the activities, transactions, and analysis and provide a workable audit trail with information systems having a role at this level in data and systems support
- Some key activities supporting and advancing maintenance parts require multiple cross-functional support and guidance that cannot be managed by maintenance marts alone. We have "stakeholder engagement" activities:
  - To establish an asset's parts support philosophy and inventory stocking policy for both initial spares and ongoing recommended "spares" through to an asset's end-of-life, we have asset lifecycle alignment that will include design engineering, dinance, operations, maintenance, and maintenance parts staff contributions
  - To support all item and inventory adds, deletes, and service levels to balance the notion of the optimizing inventory role and continuous improvement activities. This focus area works to make the best of the inventory resource dynamics available and includes design engineering, finance, operations, maintenance, and maintenance parts staff contributions
- Ultimately the highest level is called "uber change" to influence the need for significant change to meet future needs and lead the charge in establishing

the enterprise as a leader in its industry. Foundationally uber change can include:

- Process re-engineering element to allow for new methods, systems, or tools in achieving maintenance parts excellence
- The ability to predict and bridge the technology chasm with new algorithms, associated automation
- Ensure that all maintenance parts processes align with standard legal, financial, and business regulatory requirements
- Re-thinking the current processes to prepare for the future

## THE MAINTENANCE PARTS MANAGEMENT COMMUNITY INFLUENCERS

The maintenance technician is often the source of basic perceptions of how well a maintenance parts operation performs. They usually have experiences influencing their belief about what inventory should be stocked and where it should be stocked. This opinion typically comes from operational fears and experience when parts were previously unavailable when needed.

Curiously, it often seems that the parts personnel are:

- "Forgotten" when the service levels desired are consistently well met
- "Remembered" when an essential part is unavailable and must be expedited

Who owns inventory management is an interesting question depending on whether you are working to identify the valid owner or why a part is unavailable when urgently required.

A complete list of influencers on maintenance parts management may include:

- Stores warehouse leadership and material handlers
- Inventory planner
- Operations
- Maintenance leadership and technicians
- Design engineering
- Vendors
- Finance
- Health and safety
- Information technology (IT)
- Human relations (HR)
- Legal

The challenge is that each of these business units may have conflicting priorities. Ultimately you may suggest that you want the highest service level at optimal cost, but at what service level, and what is an acceptable or even optimal cost?

Finance will want inventory levels and controls balanced with production costs and outputs while achieving corporate health and safety targets. Many other business units may suggest that they can rally around such a principle, but how will they feel when:

- A new critical asset is being installed into the enterprise. Who decides the inventory policy for this new asset?
- A maintenance technician wants to hide parts away in a remote location – just in case – to be used when urgent repairs are required. Who confirms if and how these parts should be stocked and accounted for in the organizations accounting books?
- A maintenance technician removes a part from stock after hours and forgets to account for this transaction in the maintenance parts inventory systems. Who is accountable for the inventory miss on the following inventory audit or when the missing part is requested for an urgent parts order?
- A set of parts have been identified for scrap. Who decides if the risk versus financial benefit is appropriate to remove these assets from the books?
- Purchasing sees an opportunity to bulk buy some parts at a better cost but must acquire four years of usage. Who is accountable for the lower inventory turnover rate against the inventory effectiveness metric?
- 40% of parts items used are identified as repairable at a favorable cost with an expected quality at or better than new. Who determines the process, business risk, and queue algorithm for this alternative sourcing of parts?

The genesis of a new investment into a new set of parts is when a new asset starts its lifecycle and is determined to be added to the enterprise. If we assume this new asset is a crucial production asset, then the design engineering team (perhaps with the influence of the OEM vendor) will establish a list of related spare parts or initial spare parts (ISP). The ISP would be reviewed to confirm whether each piece should be stocked or not. This ISP "list" typically would be the foundational parts needed and would complement the maintenance parts management "inventory policy," which could also have an algorithm that would include recent activities and expected local overrides. The overall "inventory policy" for supporting this new asset would be reviewed and signed off by business elements, including maintenance, operations, engineering, maintenance parts, and finance. The success of the inventory scheme for this specific asset is highly dependent on the participating stakeholders' foresight, discipline, and teamwork.

Communications are essential to the success of maintenance parts excellence.

- It begins when a potential parts item need is identified
- It ends with the technician receiving a part within an agreed-to timeline

Communications failed: if a part is ordered at a parts counter and the technician expects the Part to be there when needed without any prior support and forecast.

The maintenance parts organization will often report to finance, procurement, maintenance, operations, or HR. As suggested earlier, the reporting structure for maintenance parts is important but less important than ensuring that the *behavior of the various stakeholders* that *create* and support *maintenance parts excellence is aligned, supportive, and driven to be optimal.* The key for each listed business unit and design engineering is that they all have a role in maintaining a well-communicated and executed parts support infrastructure. These key stakeholders should understand the enterprise's business objectives and be responsive to the requirements and expectations of the business, including making the right decisions to support sales, operations, maintenance, costs, and parts support.

It takes a community of cross-functional business units to make a leading maintenance parts operation successful and achieve maintenance parts excellence.

## STARTING A MAINTENANCE PARTS EXCELLENCE CULTURE

Many organizations do not know where to start when reviewing their maintenance parts operation. They typically recognize that including all the stakeholders in such an initiative is essential.

Some of the considered approaches to improving their parts operation may include:

- Reviewing any new program changes recently made in the maintenance parts operation
- Reviewing what a leading maintenance parts operation could include as potential enhancements in their organization
- Facilitating a "high-level" assessment of their new maintenance parts management policies, practices, and programs
- Having an independent expert review and assess their current organization to facilitate identifying gaps in leading practices
- Complementing the third-party review by assessing their current organization and identifying gaps in leading practices
- Targeting known or perceived gaps as part of the process (e.g., repairable parts, warehouse flow, tool crib processes, cycle counts, initial spare parts process, inventory policy)
- Assessing the ability of the current set of IT systems to support current and future needs
- Developing input to current and future budgets specific to systems, tools, and training enhancements for the mutually agreed changes

## MAINTENANCE PARTS MANAGEMENT ASSESSMENT APPROACH

It is best to have representatives from each of the potentially impacted stakeholder groups, as listed earlier in this chapter, participate in each step of the review process and any related workshops.

Inclusiveness will help buy into any of the findings and recommendations from such an initiative.

Figure 2.3 shows a maintenance parts management assessment approach that has been successfully deployed in the past to:

- Collect current landscape and pain points by business unit:
  - Leverage a questionnaire with leading practice suggestions representing the ten elements of the maintenance parts excellence pyramid
  - Interview key stakeholders to confirm landscape and pain points by Business Unit
- Review "as is" processes and interview those impacted
- Analyze and present results to the involved stakeholders with management
- Review leading practices in maintenance parts excellence and prioritize what is best for your organization/enterprise
- From the prioritized initiatives, draft a "high-level" plan to implement prioritized initiatives
- Define success criteria for each of the Initiatives
  - Design principles
  - Design points
- Define a "high-level" implementation "roadmap" to implement the initiatives
- Implement the initiatives
  - Governance model, prioritized initiatives, policy changes, collected defined benefits, high-level implementation plan, budget by initiative
  - Commitment and buy-in from the stakeholders and management

Insight into the level two steps of this process is shown in the following table (Table 2.1).

This approach can benefit the enterprise by getting the key stakeholders together and identifying the current perceived and actual issues. To provoke thought and discussion, the questionnaire results and interviews will provide an early understanding of where they are internally performing against their knowledge of leading practices or best-in-class. It will help them recognize the role they likely need to play to ensure the success of identified initiatives. It will provide a tangible record of how the stakeholders came together and what was agreed to in prioritized actions (Figure 2.4).

Some typical outcomes of a maintenance parts operational review can include:

- Training materials on leading practices in inventory management
- A comprehensive evaluation report of the proposed new concept inventory planning strategy with recommendations on where to improve techniques or possible tactics
- A summary of existing measurable maintenance parts management KPIs
- A comprehensive evaluation report of current or proposed maintenance parts management policy with:
  - In-depth assessment of the current or proposed maintenance parts management policy
  - Recommendations of areas of improvement
- A prioritized list of needs that requires systems or training enhancements
- A benefits and cost estimate to implement prioritized needs

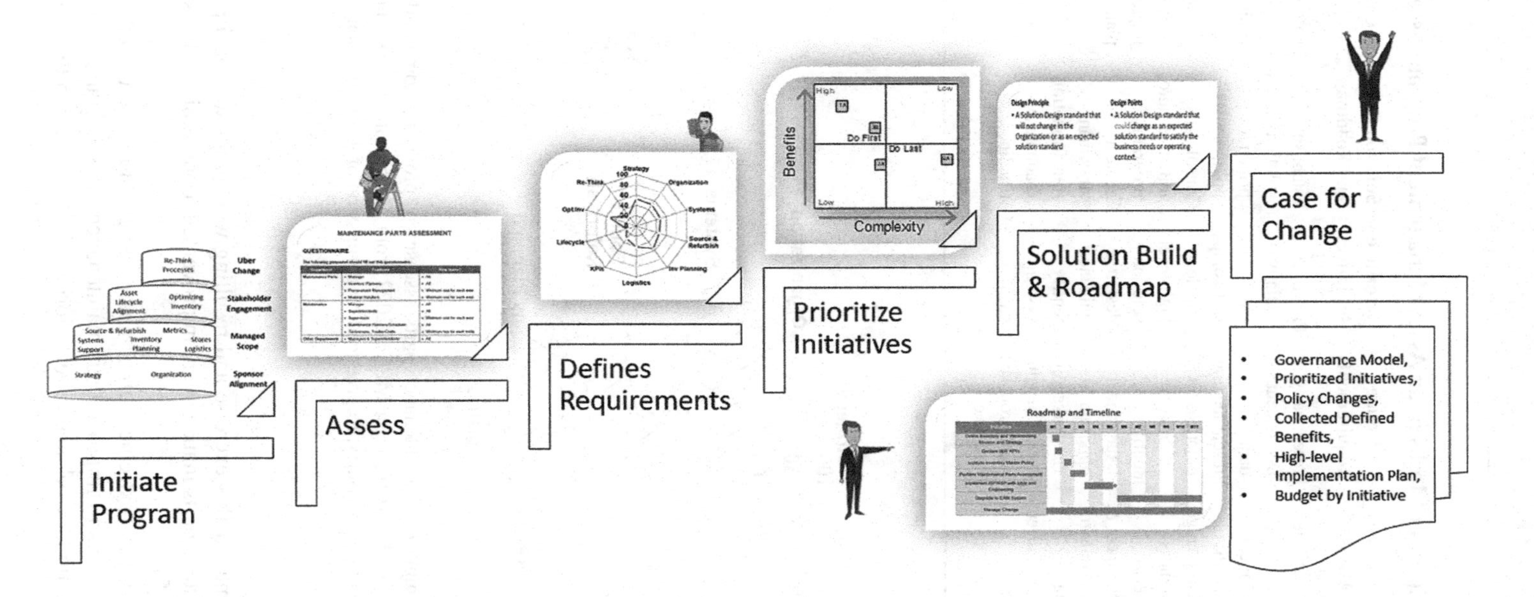


**FIGURE 2.3** Example of an asset management program assessment, strategy development, and business case process flow. (Copyright MPE ©2022 from Maintenance Parts Management Excellence Training edited by Barry. Reproduced by permission of Asset Acumen Consulting Inc.)

**TABLE 2.1**
**Level Two Description of Tasks in an Asset Management Program Assessment**

| Initiate Program | Assess | Define Requirements | Prioritize Initiatives | Solution Build and Roadmap | Case for Change |
|---|---|---|---|---|---|
| Kick-off project<br>Establish governance<br>Establish roles<br>Distribute questionnaires | Tour facilities<br>Review "as is" documents and KPIs<br>Assess change readiness<br>Collect assessment & questionnaire data<br>Review current systems support | Present early assessment data<br>Present leading practices calibration<br>Identify issues and opportunities via workshops<br>Identify benefits from proposed initiatives | Prioritize initiatives via workshops<br>Provide high-level solution design guidance<br>Identify system gaps | Establish solution design principles/ points<br>Consolidate initiatives into programs<br>Create a strawman transformation roadmap and timeline | Draft a transformation business case & case for change<br>Document a high-level implementation plan |

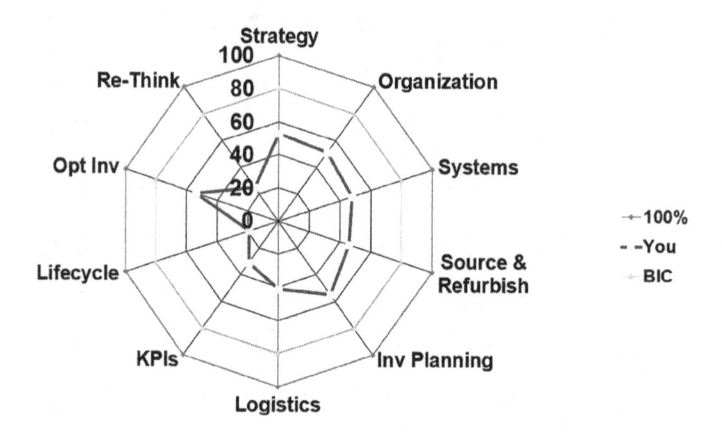


**FIGURE 2.4** Example of a maintenance parts assessment providing some early gaps to perceived best in class 1. (Copyright MPE ©2022 from Maintenance Parts Management Excellence Training edited by Barry. Reproduced by permission of Asset Acumen Consulting Inc.)

One of the outcomes of this exercise and resulting workshops will be a prioritized matrix of maintenance parts management improvement opportunities that should be pursued (Figure 2.5).

The matrix identifies "extremely high" financial benefit and perceived low effort as prioritized initiatives to improve their maintenance parts operation. These early steps can fund additional work and lower-priority initiatives later in the program.

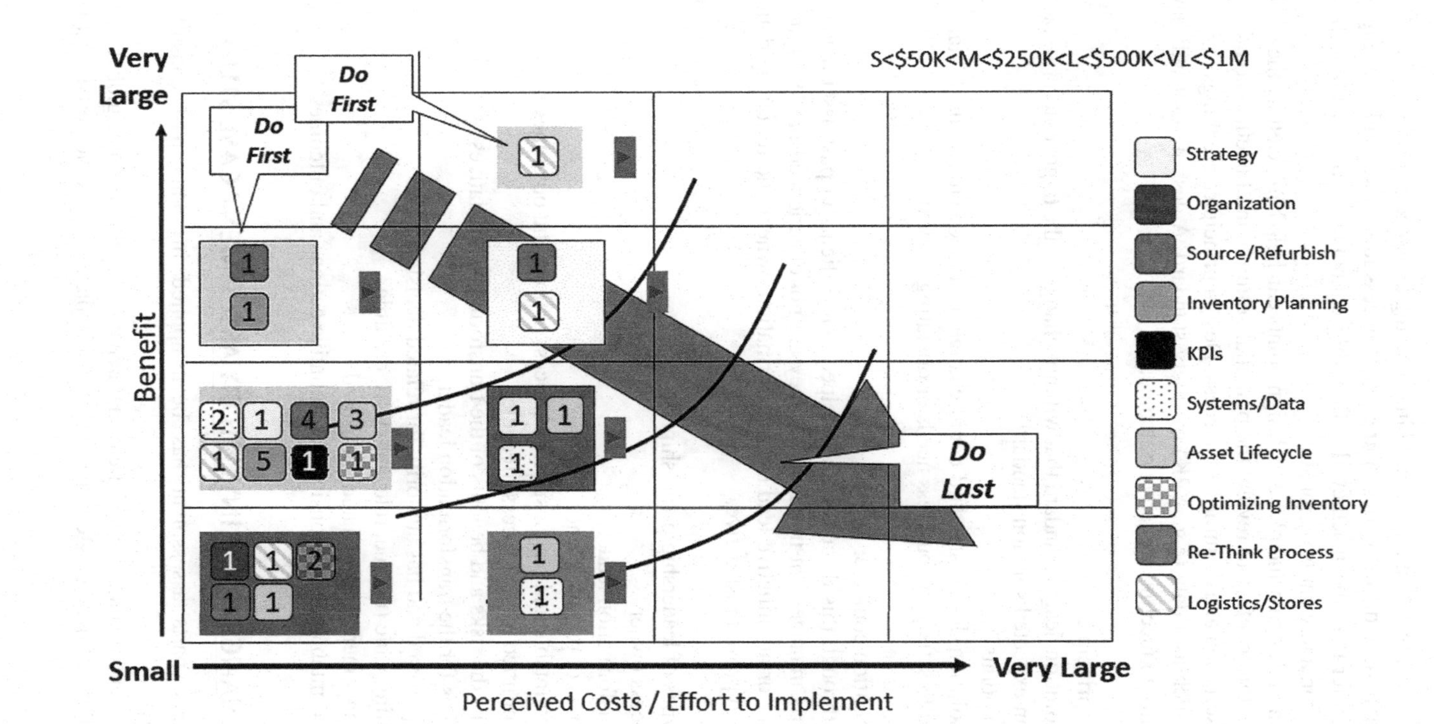


**FIGURE 2.5** Example of a maintenance parts activity matrix. (Adapted from copyright 101 MPE Introduction ©2020 from Maintenance Parts Management Excellence Training edited by Barry. Reproduced by permission of Asset Acumen Consulting Inc.)

## DESIGN PRINCIPLES AND DESIGN POINTS

Often an organization is not experienced in executing a change in its enterprise. As a result, they may have difficulty describing what success looks like while waiting for it or perhaps after it has been achieved. Often, they cannot agree on what must be in the final solution and what is optional.

The notion of design principles and design points can help here. Getting the stakeholder leadership together to review the potential solution and benefits should also include a description and definition of success. Design principles are things that will not change, such as a date (e.g., Y2K) or safety standard. A design point will have more flexibility in meeting the target outcome.

- Design principle
  - A solution design standard that will not change in the target enterprise or as an expected solution standard
- Design points
  - A solution design standard could change as an expected solution standard to satisfy the business needs or operating context.

A formal definition could be:

Working through this discussion has helped many clients in past asset management and parts management initiatives. If they are not experienced in identifying a process owner or an initiative lead (or transformation owner), the following definitions have also proven instructive.

- Two types of leadership/ownership:
  - Process owner
  - Transformation leader
- Attributes of process owner
  - Accountable and responsible for process actions and outcomes
  - Committed to the process lifecycle
  - Owns the risks and benefits of the transformation initiatives
- Attributes for the transformation leader:
  - Empowered and has authority for related decisions
  - Willing, able (unique, relevant skill), credible
  - Management assigned and sponsored
  - Accountable and responsible for initiative actions and outcomes

## A MAINTENANCE PARTS INVENTORY ASSESSMENT CASE STUDY

A maintenance parts assessment has been applied in many past consulting engagements.

For one extensive international field maintenance operation, their global headquarters challenged their single large country to implement specific new initiatives from other countries.

A maintenance parts inventory assessment was implemented for the target local country. It was found that the target country had already executed the prescribed inventory initiatives, and additional inventory risk would impact the business services levels, profit, and reputation.

A report outlining all the findings was sent to their international headquarters. As a result, some identified and already implemented initiatives described in the target country became the new standard for other countries in that enterprise.

## SMALL CASE STUDY RESULT SET

A maintenance parts management assessment (using the process shown in Figure 2.3) was facilitated with a cross-functional stakeholder group like the list shown in Figure 1.6 for a small power and water utility. Saving opportunities suggested that over five years, $7M to $14M of operational savings were identified. This initiative represented an approximate reduction of the current inventory of more than 20% and meant an increase in maintenance parts management service levels and hours of operation.

Other initiatives identified included opportunities by:

- Ensuring that all parts transactions in that enterprise would be in their parts item master so that they could track parts requests in future analytics and consider all past needs in ongoing replenishment algorithms
- Creating a re-designed initial spare parts/recommended spare parts (ISP/RSP) process and owner to:
  - Ensure that ISPs are provided by design engineering and ownership is transferred to Maintenance, who will own the RSP list throughout the asset lifecycle
  - Drive a reliability-centered maintenance (RCM) approach to be used to support ISP where practical
- Enhancing the inventory planner role, who will be responsible to consolidate RSP/ISP with demand management responsibilities
- Assigning surplus inventories to the inventory planner to be reviewed monthly for potential redistribution, sale, or scrap based on value, past activity, and RSP settings
- Agreeing that one territorial stock room is now policy (versus small stock rooms on one campus or one small city)
- Collecting parts transaction data so that it can be identified as unplanned and planned demand (have an "exceptional demand" field will be added to the parts system)
- Ensuring that proper weekly parts demand is calculated and applied to the accurate min/max policies and algorithms
- Adding a "where used" field will be added to the parts system to help with ISP and surplus decisions
- Review "inventory provision" and "inventory turnover" calculations with finance

The above is a partial but representative list of initiatives from one maintenance parts assessment described in Figure 2.3 and leveraging a questionnaire like the one shown in this book's Appendix. One of the key benefits of using this approach (along with the identified savings and prioritized actions) was that it included the key stakeholders that support the maintenance parts management operation. The overall assessment approach helped significantly bridge issues that may have arisen after the workshops in recognizing each stakeholder's role in making the maintenance parts management operation excellent.

The following chapters will describe the ten elements introduced in the maintenance parts excellence pyramid (Figure 2.2) and other important influences to consider in executing maintenance parts excellence.

# 3 Maintenance Parts Strategy Development

## MAINTENANCE PARTS VERSUS PRODUCTION WAREHOUSING OPERATIONS

Perhaps it is evident that some organizations jump to the expectation that maintenance parts management and product supply chain management (as in a manufacturing environment) have a lot in common. While it is true that both need committed and experienced inventory, warehousing, and materials handling skilled resources, the two franchise's, missions and processes are pretty distinct. Product supply chains address the finished goods manufacturing process with planned production schedules and distribution networks. Maintenance parts must focus on a random and volatile, less predictable urgent order fulfillment mission with little in common with a well-executed manufacturing process (Figure 3.1).

Up to the start of this century, supply chain management (SCM) was a high-focus area for business process savings, and SCM evolved to leverage IT automation for their forecasting.

However, in most enterprise scenarios, maintenance parts management has been the poor cousin of SCM, with very little IT investment, finance focus, or process investment, for that matter. Compared to SCM, maintenance parts management has been subtly enhanced in commercially available IT solutions to help them achieve optimal inventory or process excellence. If SCM and maintenance parts management were in the same building, *"what they may have most in common are washrooms and parking lots."*

Product supply chain management is, at a high level, primarily linear in its manufacturing process. Its focus is on the efficiency and productivity of the process and the cost and quality of throughput. On the other hand, maintenance parts management is "all about the installed asset base value." The maintenance parts management stakeholder is focused on the cost to support and service levels of the installed asset and must be tenacious about keeping that asset operating at its planned potential.

To ensure their competitiveness in the rapidly changing global marketplace or to provide value in their area, mining sites, manufacturers, utilities, and transportation companies, to name a few, are globally seeking ways to radically reduce their asset and operational costs while increasing their value. Maintenance parts is a critical contributor to this value proposition. Maintenance parts spending, including expenditures for services and spare parts, will directly impact services costs. However, the actual percentage will depend on the industry and how they execute well across their enterprise.

Optimal maintenance parts placement, which includes supporting asset productivity throughout its lifecycle, will directly impact the productive asset service level the parts are supporting.

DOI: 10.1201/9781003344674-3

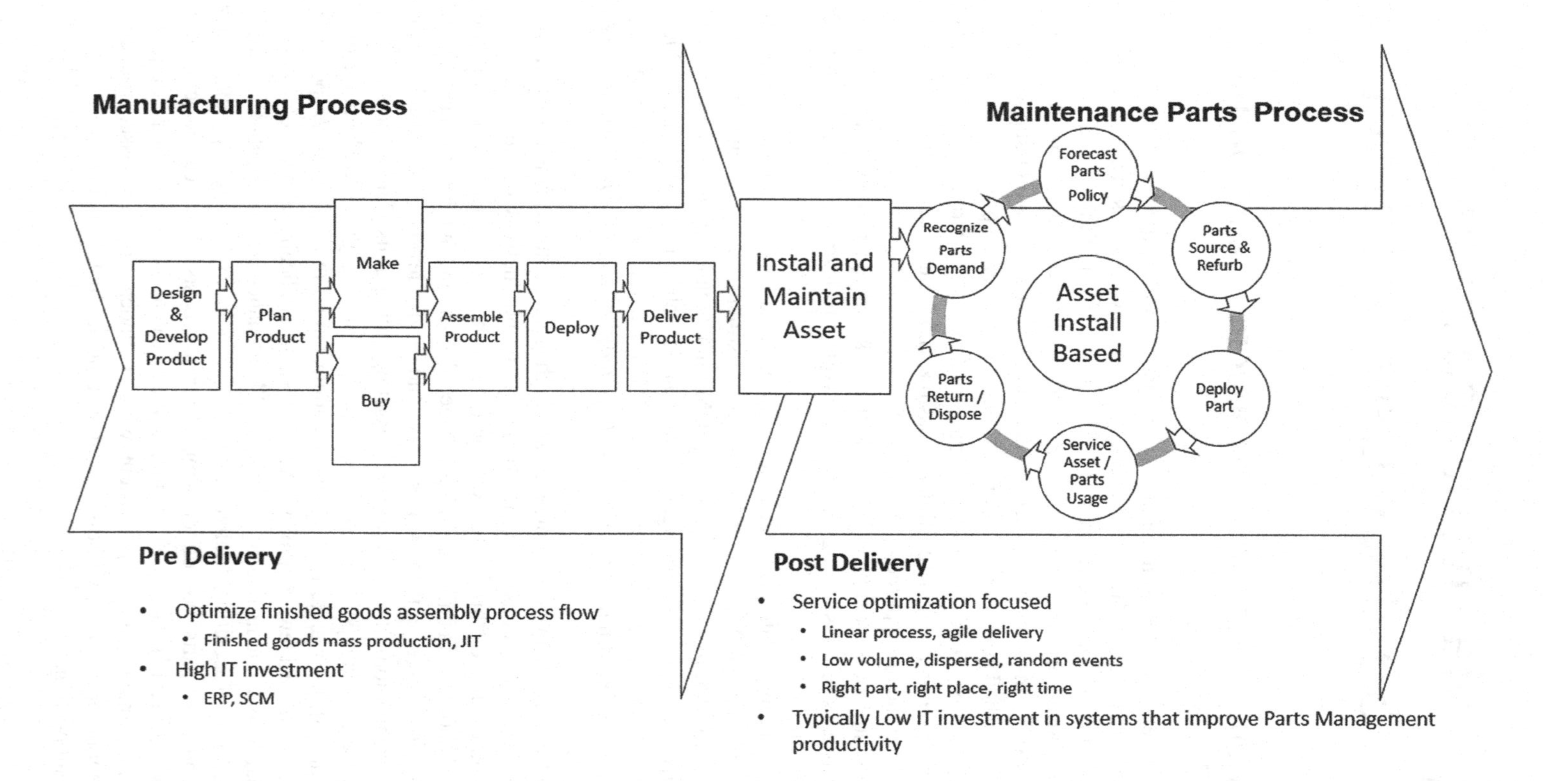


**FIGURE 3.1** Generic overall strategy development process. (Adapted from copyright 102 MPE Org and Metrics ©2020 from Maintenance Parts Management Excellence Training edited by Barry. Reproduced by permission of Asset Acumen Consulting Inc.)

Ineffectively managed maintenance parts processes can result in:

- Unnecessarily high costs
- Low production
- Low worker productivity
- Poor product quality
- Poor brand reputation
- Poor inventory accuracy and stewardship
- Inflated inventory-holding costs

Ineffective management of maintenance parts policy and processes can impact an enterprise's bottom line (Table 3.1).

Addressing the above issues and making the payback optimal drive service level and cost benefits that will enhance the company's financial and reputational position.

**TABLE 3.1**
**Potential Benefits to Optimal Maintenance Parts Management Excellence**

| Business Scenarios | | Benefit Points |
|---|---|---|
| Production availability | Having the Part available when called upon for an urgent repair can reduce downtime and lost production time of critical asset costs by 3 to 9 times | Up 2–11% |
| Labor utilization | Having the Part available when called upon for an urgent repair can reduce maintenance technician costs by 3 to 9 times – not including parts expediting costs | Up 8–22% |
| Equipment purchases | Having the Part available can enhance maintenance tasks and help to extend asset life | Down 2–7% |
| Warranty recoveries | Having a good asset and parts warranty tracking system can increase the percentage of workable warranty claims back to a supplier | Up 5–50% |
| Inventory needs | Having an effective inventory policy that can account for Initial Spare Parts and forecast demand activity can help to optimize inventory levels and identify surplus parts for timely and effective disposition | Down 20–35% |
| Inventory carrying costs | An optimized inventory well executed will minimize inventory and scrap and effectively reduce inventory carrying costs | Down 5–25% |
| Material costs | A well-thought-through procurement, refurbishment, and parts costing discipline | Reduced 10–40% |
| Material transportation costs | Consolidating shipments, eliminating surplus, and replenishment church between stocking sites | Reduced 10–20% |
| Purchasing labor | Consolidating purchases through a vendor analysis and requirements forecast will reduce purchasing workload and transportation costs for both regular and expedited orders | Reduced 10–40% |
| Parts expediting costs | Having the right Part close to the forecasted need will reduce the expediting costs for urgent and immediate needs | Reduced 10–50% |
| Parts dispositioning tax benefit timelines | An effectively defined inventory policy and surplus parts program will allow for timely and effective disposition – finance's tax-related benefits are realized years earlier | Brought forward by years |

## PRODUCTIVITY AND PRODUCTION ISSUES DUE TO POOR MAINTENANCE PARTS MANAGEMENT

Supporting maintenance and operations with the right part (when needed) can significantly impact operations availability resulting from prolonged equipment breakdowns in a production process. For the maintenance technician, the wait time for parts can be significant, particularly if they have to "tool down, leave the site," and then "return and tool up" when the part arrives.

Not managing the many influences on maintenance parts costs can create high "hidden costs" of low productivity. Enterprises that lack the policies and processes to promote rigorous spare-part inventory management run an increased risk of production and productivity losses. The increased risk is genuine when a part is critical to production in a down scenario.

The simple way to look at this is if the part is available when the maintenance technician requires the part. The time to repair the asset is "X." If a part is needed but not available to repair an asset in an urgent scenario, then the maintenance technician's time to repair the asset can be 3–9 times "X." The ability to coordinate the right part to the right place within the committed time can add up quickly, and while this adds up, *the asset is not producing, so the "lost production" cost can also add up!*

These parts-related costs can be significant for asset-intensive industries, such as mining sites, manufacturers, utilities, transportation companies, and others who report the highest maintenance parts costs as a percentage of revenues.

A process for tracking equipment productivity should be enabled to help trace downtime back to stockouts of spare parts and confirm if the necessary parts were in or out of stocking policy.

Often spare parts become surplus to actual requirements over an asset's lifecycle. Surplus parts can accumulate due to early and mid-life purchasing decisions. They are acquired as surplus to the asset's install bill of materials and end up in the "spare parts inventory" or simply get ignored as an asset becomes less critical to the business or is discontinued from service, and the maintenance parts operations is not informed. The accumulation of parts results in parts that pile up in warehouses long after needing them. Excess inventories may result from not having a rigorous part purchasing planning process that includes setting stock levels (policy) that trigger reordering. Without central coordination, a company's multiple plants scenario can duplicate parts purchases and create unnecessary inventory stocks.

To achieve maintenance parts excellence, an enterprise should consider the many influences that create an optimal inventory management policy and make this integral to developing their maintenance parts strategy.

## CREATING A MAINTENANCE PARTS EXCELLENCE STRATEGY

Creating a strategy does not mean getting together to ultimately implement a software solution or enhancement for maintenance parts management, but it can be!

Strategy often means understanding where we are (as is) and where we want to be (to be). Then understanding the "gap" between the "as is" and the "to be," develop a business benefits case if that "gap" is mitigated and select a solution or set of initiatives to execute to realize the benefit (Figure 3.2).

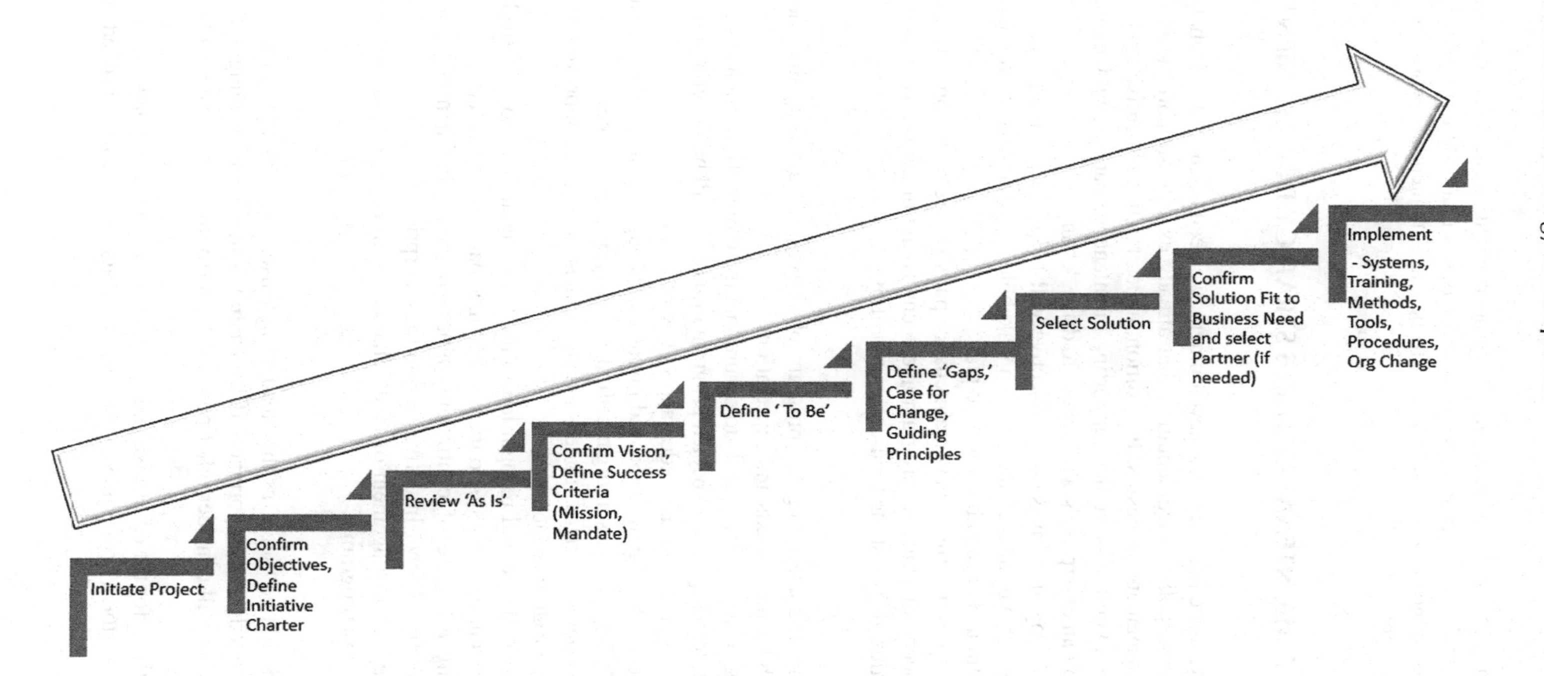


**FIGURE 3.2** Generic overall strategy development process. (Adapted from copyright 101 MPE Introduction ©2020 from Maintenance Parts Management Excellence Training edited by Barry. Reproduced by permission of Asset Acumen Consulting Inc.)

Creating a strategy also means getting the proper management focus and getting the correct stakeholders to agree on the prioritized issues that need to be addressed. From there, determine what initiatives need to be implemented and when to drive benefits to the enterprise.

## INITIATING A MAINTENANCE PARTS STRATEGY DEVELOPMENT PROGRAM

To start a maintenance parts strategy development, we need to understand what process and corporate policy documentation that already exist to support the maintenance parts management processes. In addition, understanding executive leadership culture and current culture-specifics supporting maintenance parts is fundamental to get a baseline on an enterprise's willingness to drive "change."

Defining the program's mission or objective will help manage the scope of the potential change. It will also confirm which stakeholders need to be represented to drive a successful and executable strategic plan.

The leadership sponsoring a new maintenance parts strategy development should expect to be "seen" driving this program and seen driving the initiatives out of the plan. Leaders must demonstrate their commitment by:

- Reviewing and approving the program objectives and ensuring that they align with business needs and expectations
- Lobbying across business units to ensure that the required stakeholders are represented and that the expected program initiatives will be integrated into the business processes as needed across business units
- Being "seen" driving the approved initiatives through all areas of the business, including changes in organization or policy, if appropriate
- Getting systems changes funded and implemented that are approved and key to the program's success
- Providing guidance and recognition to the team members who effectively drive the identified benefits to the maintenance parts strategy plan
- Supporting a culture of continuous improvement for maintenance parts management and other initiatives across the enterprise
- Supporting the management of "risk" that is identified in the maintenance parts strategy program

Management should be "seen" as the "lead and sponsor" of such a program so that staff can acknowledge its importance. If leadership is not "seen" leading a critical program, the team will consider other perceived priority initiatives that should consume their time.

Existing policies for the enterprise and specifically for the maintenance parts area need to be understood, maintained, or changed if agreed to and approved when working through the strategy development. The maintenance parts strategy should confirm that it supports a policy that is:

- Considering the size, scope, and operating context of the enterprise and business environment
- Considering the organizational and process gaps that may need to be addressed in this program
- Aligned with the enterprise policies in terms of service and financial objectives
- Aligned with the organizational plan and with enterprise organizational rules provided by HR

It is vital to get the stakeholders to agree on the scope of what they want to influence and improve. For example, will the maintenance parts strategy plan address a prioritized list of activities within a supported timeline?

A mission statement can help align the scope of the maintenance parts program. An example of a working mission statement can be:

- Clearly understand all our client's logistical needs
- Effectively and expediently re-design/upgrade our processes, tools, and facilities
- Fully meet and comply with all our client's maintenance parts logistics necessities
- Drive for completing this mission by 20XX (5 years in the future)

Add some guiding principles for the stakeholders in the workshops, and the process can begin (Figure 3.3).

**FIGURE 3.3** Example of workshop mission icon and motto. (Adapted from copyright 101 MPE Introduction ©2020 from Maintenance Parts Management Excellence Training edited by Barry. Reproduced by permission of Asset Acumen Consulting Inc.)

Example of guidance that could be used:

- Driving to continue to support the corporate enterprise mission
- Continue to support day-to-day logistics execution
- Policy design leveraging stakeholder participation
- Leveraging the policy to navigate the process
- Top enterprise executive support will be evident as a priority

The stakeholders must ensure that the enterprise objectives are being served while addressing the maintenance parts program objectives. For example, maintenance parts must consider the needs beyond just the maintenance parts management team or the maintenance technical management and staff. The parts' needs will also include requirements from operations staff and management, design engineering staff and management, HR, legal, and finance.

The maintenance parts strategy program objectives should:

- Be cost-justified against the investment opportunities and priorities of the enterprise
- Be managed and tracked to confirm both the cost and benefit of an approved initiative
- Include all prioritized requirements
- Consider the change impacts and actions needed to change culture, roles, processes, or policies
- Be transparent in communicating the considered opportunities, prioritized initiatives, and progress on the approved initiatives and timelines
- Be modified and updated – when appropriate and approved by management

Specific to a maintenance parts program, the objectives can include:

- Establishing and updating the maintenance parts program strategy, including related mission, guidance, and objectives
- Ensuring that the maintenance parts related systems can support the delivery of the maintenance parts program strategy
- Ensuring the suitability, adequacy, and effectiveness of the maintenance parts program
- Establishing and updating the maintenance parts plan and changed policies
- Reporting on the performance of the maintenance parts program strategy initiatives to top management

The strategy should define the scope of a maintenance parts excellence outcome. For example, the maintenance parts manager may have a different expected outcome definition than the maintenance technician, the design engineer, or the finance manager.

When a cross-functional group of stakeholders comes together to improve their business outcome through a maintenance parts strategy workshop, each likely has a diverse expected outcome from defining an "improved" maintenance parts management. This list may include (but is not limited to):

- Improved asset production and overall equipment effectiveness (OEE)
- Minimize inventory values
- Maximize asset-created value to inventory ratios
- Maximize inventory turnover
- Minimize inventory provisions
- Maximize parts availability
- Minimize parts procurement time
- Ensure asset lifecycle support
- Minimize the parts that are stocked for planned and scheduled maintenance
- Maximize the work orders with kitted parts
- Maximize parts sales and profit, if sales are applicable

A new strategy program should challenge if all the stakeholders are represented and confirm if they can support the expected programs and initiatives that are likely to come. You should challenge whether a stakeholder:

- Is relevant to the maintenance parts management process or solution
- Understands what is required of them concerning maintenance parts management
- Can decide to represent a decision from their business unit concerning maintenance parts management
- Can define and own the metric that may be applied to their business unit, to support the expected benefits to the enterprise

## EXECUTING THE MAINTENANCE PARTS PROGRAM STRATEGY INITIATIVES

Once the maintenance strategy plan has been approved, assign a program lead to drive the initiatives. Initiatives typically can be grouped, and where practical, it would also be effective to assign a lead or champion to each group of the approved initiatives. With the top management sponsor's guidance, stewardship, and accountability, this program team will have a support system recognized as key to the enterprise's success and get the necessary mindset to facilitate change. This team will communicate the agreed-to program and timeline and declare how the initiatives and related benefits will be achieved. They will promote the desired outcomes - steering from the undesired results - and support a culture of change and post-initiative implementation of continuous improvement.

Successful maintenance parts strategic plans will support, document, and display:

- What initiatives will be done
- Who will be assigned/responsible per initiative
- An audit trail on how the initiatives were selected and prioritized

- How these initiatives fit into the plan, how they will be implemented and the timeline for implementation and expected benefits (tangible, intangible)
- Identified risks and a risk management plan (contingency)
- A defined period for the Plan (e.g., annual or five-year plan)

Approach options to developing a maintenance parts strategic plan can include many hybrid techniques.

Some options to consider are:

- Leveraging the maintenance parts assessment program, as shown in Chapter 2
- "Design thinking" workshops (discussed later in this chapter)
- Leverage traditional business process re-engineering methods, including root cause analysis, parrado, mission, vision, and goals documentation (discussed in a later chapter) (Figure 3.4)

Leveraging the questionnaire provided in this book (Appendix A1 Maintenance Parts Excellence Questionnaire) and following the approach shown in Chapter 2 (Figure 2.2 Example of an asset management program strategy and business case process flow) has been a popular method for this author over the past decade. This approach will:

- Get cross-function teams together
- Review existing processes and policies
- Provide an understanding of where the enterprise is in terms of maintenance parts excellence by leveraging the provided questionnaire
- Through interviews, collect insights for key stakeholders across the enterprise as input to their current progress, issues, and opportunities
- Review existing culture and their experience with change
- Take the time to learn and calibrate maintenance parts terms and related leading practices
- Collect known or existing desired changes to the enterprise in support of maintenance parts excellence outcomes
- Invite the team to consider additional insights on leading practices in maintenance parts management
- Create a list of potential opportunities to improve the enterprise maintenance parts program
- Establish high-level benefits and effort quantification against each initiative
- Leverage the initiative quantifications to drive the initiative prioritization
- Look for ways to group high-value initiatives into sub-programs (e.g., strategy, systems, inventory planning, inventory lifecycle)
- Create a draft of the initiatives/sub-program implementation timeline and related cost benefits
- Review with sponsor and management for approval risks review and changes to the policy if needed.

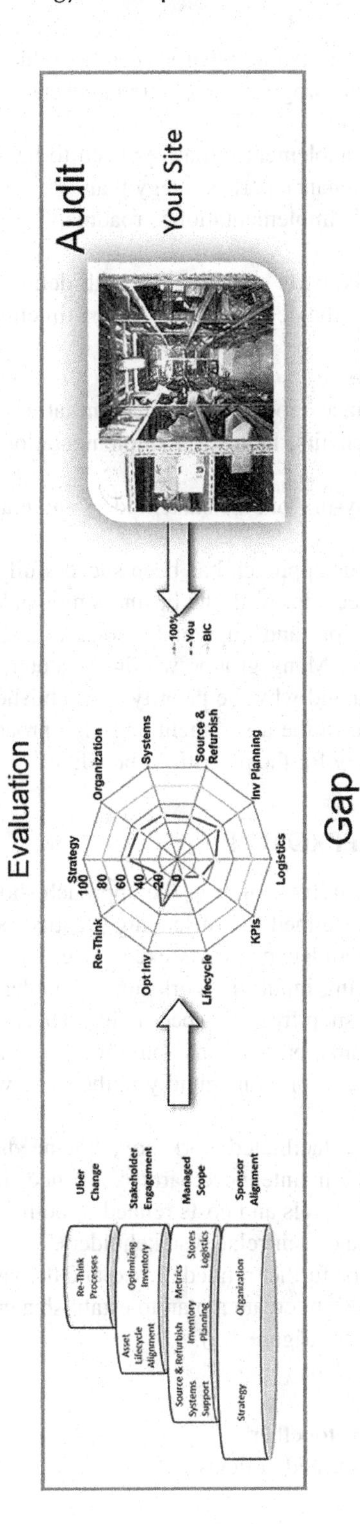


**FIGURE 3.4** Leveraging the Maintenance Parts Excellence Questionnaire. (Adapted from copyright 101 MPE Introduction ©2020 from Maintenance Parts Management Excellence Training edited by Barry. Reproduced by permission of Asset Acumen Consulting Inc.)

- Work with management to define what success should look like for each approved initiative or sub-program (leveraging a discussion on design principles and design points)
- Confirm what change enablement initiatives need to be added to invest in the success of the maintenance parts strategy plan
- Define a "high-level" implementation "roadmap" to implement the initiatives
- Finalize the maintenance parts strategy plan with definitions documented, including costs benefits, the definition of success, timeline, and stakeholders involved
- Implement the initiatives
  - Establish a governance model, prioritized initiatives, policy changes, collected defined benefits, high-level implementation plan, budget by initiative
  - Commitment and buy-in from the stakeholders and management

Over the past few decades, this approach has been successfully used with business stakeholder participation. Once approved, the business unit stakeholders participate in the plan's input, prioritization, and implementation. As a result, their new plan becomes closer to acceptance. Many groups within the enterprise have now contributed to the change agenda and why the priority is established the way it is. The stakeholders directly involved in the assessment and plan process can advocate for the expected changes and lobby for the prioritized benefits.

## A DESIGN THINKING APPROACH

A design thinking approach can be similar to the approach shown earlier, except it does not start by reviewing a defined list of leading practices specific to the target business area, in this case, maintenance parts excellence. Instead, design thinking facilitates an approach to taking input via workshops to understand the end user's (stakeholder) needs in a workshop free of a specific list. This approach seeks empathy with the process participants, brainstorms solutions in participative workshops, and looks to develop conceptual solutions rapidly in the same workshop investment of participants.

For example, from an actual facilitated workshop, "the needs" may be defined as "an intent to design and build a maintenance parts excellence culture and set of processes to improve the services levels and costs related to both the enterprise and the maintenance parts operation and their related stakeholders."

The workshop object may be further refined to "expand the vision, align the stakeholders, and identify initiatives" to create a plan to establish a culture and discipline of maintenance parts excellence (Figure 3.5).

This approach will:

- Get cross-function teams together
- Review existing processes and policies

Sample 'Hill' Statements

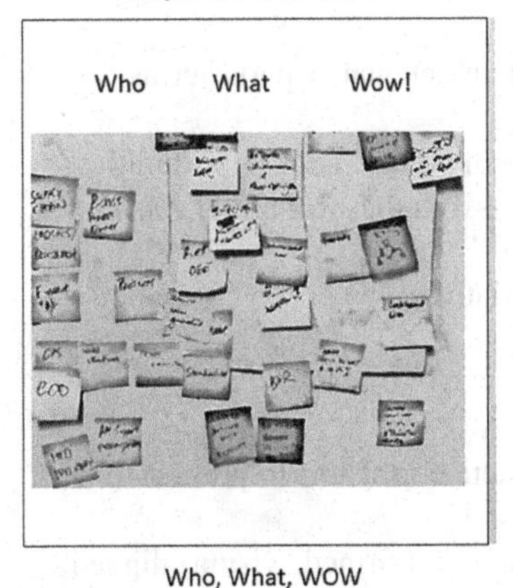


Who, What, WOW

Sample Empathy Map

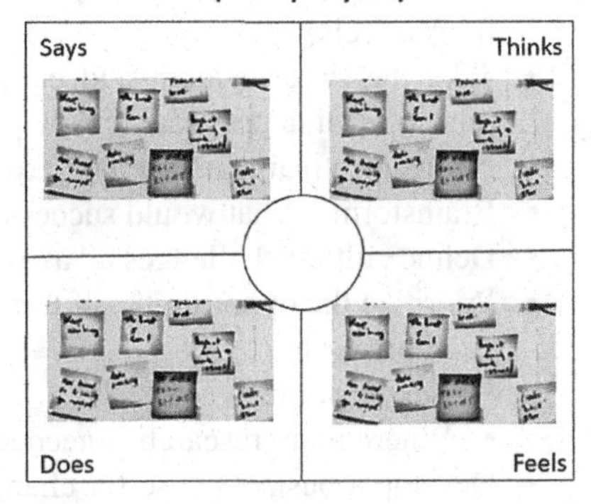


What a Stakeholder – Says, Does, Thinks, Feels

**FIGURE 3.5** Example of a design thinking workshop board.

- Provide an understanding of where the enterprise is in terms of maintenance parts excellence by leveraging a presentation by key stakeholders
- Collect known or existing desired changes to the enterprise in support of maintenance parts excellence outcomes
- Review existing culture and their experience with change
- Review how other organizations are working toward or meeting similar desired goals
- Define design thinking workshop goals such as:
  - Confirming the vision for the solution
  - Confirm who the stakeholders and target client are for maintenance parts excellence
  - Further defining opportunities and initiatives to drive cost and service level benefits
  - Define how staff will be impacted and discuss change enablement options
  - Understand and agree to the business case for the prioritized actions
- Define target culture and process through "hill" statements (targeted future culture):
  - Examples:
  - Wait time for parts will be nominal for all urgent requests
  - Initial spare parts data will always predict a spare part's optimal stock level based on risks like the asset's lifecycle usage and ability to be supported by the vendor
- Brainstorm the who, what, and how statement that would support the hill statements
- Confirm through brainstorming who the stakeholders should be for maintenance parts Eecellence

- Brainstorm an "empathy map" from each stakeholder identified (says, does, thinks, feels)
- Take the time to learn and calibrate what maintenance parts terms and related leading practices are
- Brainstorm maintenance parts excellence "pain points" and "opportunities"
- Brainstorm "What would success look like" – if in the *Wall Street Journal*
- Define culture challenges – "as is" and "to be"
- Prioritize the opportunities if turned into initiatives:
    - Short-term
    - Longer-term
    - Where more research is needed
- Develop a business case for change (can practice in teams to present to an investor)
- Create a plan to implement with the mission confirmed, vision aligned, business drivers defined, impacted stakeholders identified, opportunities and initiatives prioritized

Leveraging traditional business process re-engineering methods, including root cause analysis, parrado, mission, vision, and goals documentation, will be expanded in Preparing for Uber Change in Maintenance Parts Management (Chapter 13) later in the book.

# 4 Organization Management in Maintenance Parts

## ORGANIZATION INTRODUCTION

The maintenance parts organization often reports to finance, procurement, maintenance, operations, or even HR. The reporting structure for maintenance parts is important but less critical than ensuring that the *behaviors of the various stakeholders* that *create* and support *maintenance parts excellence* are motivated to work to an optimal parts operation. The key for each listed business unit and design engineering and others is that they all have a role in ensuring a well-communicated and executed parts support infrastructure. These key stakeholders will understand the enterprise's business objectives and be responsive to the requirements and expectations of the business, including making the right decisions to support sales, operations, maintenance, and parts support.

It takes a community of cross-functional business units to make a leading maintenance parts operation and achieve maintenance parts excellence.

The method or hierarchy an enterprise may elect to staff its maintenance parts management operation can be influenced by the size, culture, history, management mix, and workload or regulatory positioning among, no doubt, many other mindsets. It would be challenging to suggest that one hierarchical solution is the single definition of "leading practice" for maintenance parts management.

This chapter will suggest practical guidelines that can influence the best workable solution for an enterprise considering an organization assessment. We will review the foundation of a suitable materials management structure to drive behavior and the affected stakeholders. We will also review and understand how they can succeed personally and for the enterprise in pursuing maintenance parts excellence (Table 4.1).

From a maintenance parts excellence perspective, the inventory manager's goal may be to provide high parts service levels at optimal costs and confirm that parts are procured and maintained to sustain their quality. Others in the parts management group may focus on material handling activities, inventory integrity, internal relationship service levels, purchasing costs, etc. Others, such as design engineering, may need to understand how to close out a design, implementation,

DOI: 10.1201/9781003344674-4

**TABLE 4.1**
**Diverse Goals Influencing Maintenance Parts Organization**

| Stakeholder | Parts Role | Parts Goal |
|---|---|---|
| Inventory manager | Management | • Highest parts service levels at optimal inventory levels and cost<br>• Quality parts |
| Warehouse manager | Management | • Perceived service levels<br>• Shipping/Receiving accuracy,<br>• Inventory integrity at near 100% |
| Procurement manager | Management | • Consolidated transactions,<br>• Low parts unit cost, low purchasing costs |
| Maintenance technician | Management | • High parts fulfillment levels – on demand<br>• Quality parts |
| Operations manager | Influencer | • High asset availability, throughput and production quality,<br>• Minimal downtime |
| Design engineer | Influencer | • Quality and low-cost commissioning process |
| Finance manager | Influencer | • Low capital costs<br>• Minimal inventory, inventory reserves<br>• High returns from production<br>• Accurate financial asset reporting |

and commissioning process to hand the asset reins to operations and maintenance and move on to the next project pending within the enterprise.

Across many organizations, the maintenance parts management operation has reported to a diverse set of business units. The pros and cons of any of them should be considered (after all), as they may have other influences that drive their behavior beyond a classical maintenance parts focus (Table 4.2).

The business pendulum may influence how a parts operation could be organized. For example: if you are a new business with a greenfield site, maintenance parts management may be closely organized by Finance or operations as maintenance may not yet exist. If you are a small operation with just a couple of staff managing parts, it could fit into finance, maintenance, finance, or HR. If you are shrinking or downsizing, finance may be where it is assigned. In any event, the behavior of all the stakeholders is more important than where it reports.

It is vital to confirm the high-level responsibilities for maintenance parts management that will arise across the supported asset's lifecycle. Finance will be looking to manage all aspects of the cost of an asset across its lifecycle, including maintenance parts that will need to be purchased and stored. Depending on how an enterprise is set up and its overall management policies: The design engineer will be concerned about the asset from an early analysis phase through the conceptual design, functional design, and detailed design phases, and overseeing the construction phase and working with maintenance and operations in the commissioning phase of an asset lifecycle. The design engineer's goal is to complete the target asset's design (functional and operating) and hand this over to operations and maintenance at the end of

**TABLE 4.2**
**Ownership Considerations for Maintenance Parts Management Reporting Structure**

| Business Unit Owner | Pros | Cons |
|---|---|---|
| Maintenance/ Operations | Quick to accommodate changes to stock as they are closer to the source of the problems and the results of decisions | Tend to overstock if the planning function is weak – they will stock all items on a bill of material instead of only the critical items to ensure that parts are available to start a job instead of pre-planning the job and buying only what is needed just-in-time |
| Inventory Management/ Purchasing | Concentrate more closely on inventory policy, analyzing and implementing cost-effective replenishment, and maximizing the vendor relationship | Not always quick to respond to new maintenance demands because of the need for input from Design Engineering. Discussion, collaboration, and subsequent negotiation for new setups or changes in materials can be a constant workload. Conservative stocking policy for items and tight control levels (e.g., minimum safety stocks) and cheap versus meet specs |
| IT/HR | Work to facilitate all business requirements – tend to lead more through a Quality focus or transformation projects support automation tools. Can be effective as a facilitator of inventory management support across an asset lifecycle | Not inherently any portion of the parts business processes except supporting the staff and implementing process and technology solutions |
| General Manager, Finance | Inventory/stores decisions are based on what's best for the company – not for the department | Stores now have three interests to mediate and balance, which can complicate decision-making and responsiveness |

the asset's commissioning phase. To do this, the asset must be working and accepted. The supporting data (including the suggested initial spare parts – ISP) must be provided to the maintenance parts management group and the asset formally accepted by operations and maintenance.

For maintenance parts, it is likely that the ISP list would be given to the inventory planner (if applicable) and lobbied for the ISP list approvals from maintenance, operations, and finance.

If the asset has a stringent set of asset controls (e.g., aviation assets), then parts usage and inventory support levels would need to be tightly tracked through the asset's operation through to sale and decommissioning (Figure 4.1).

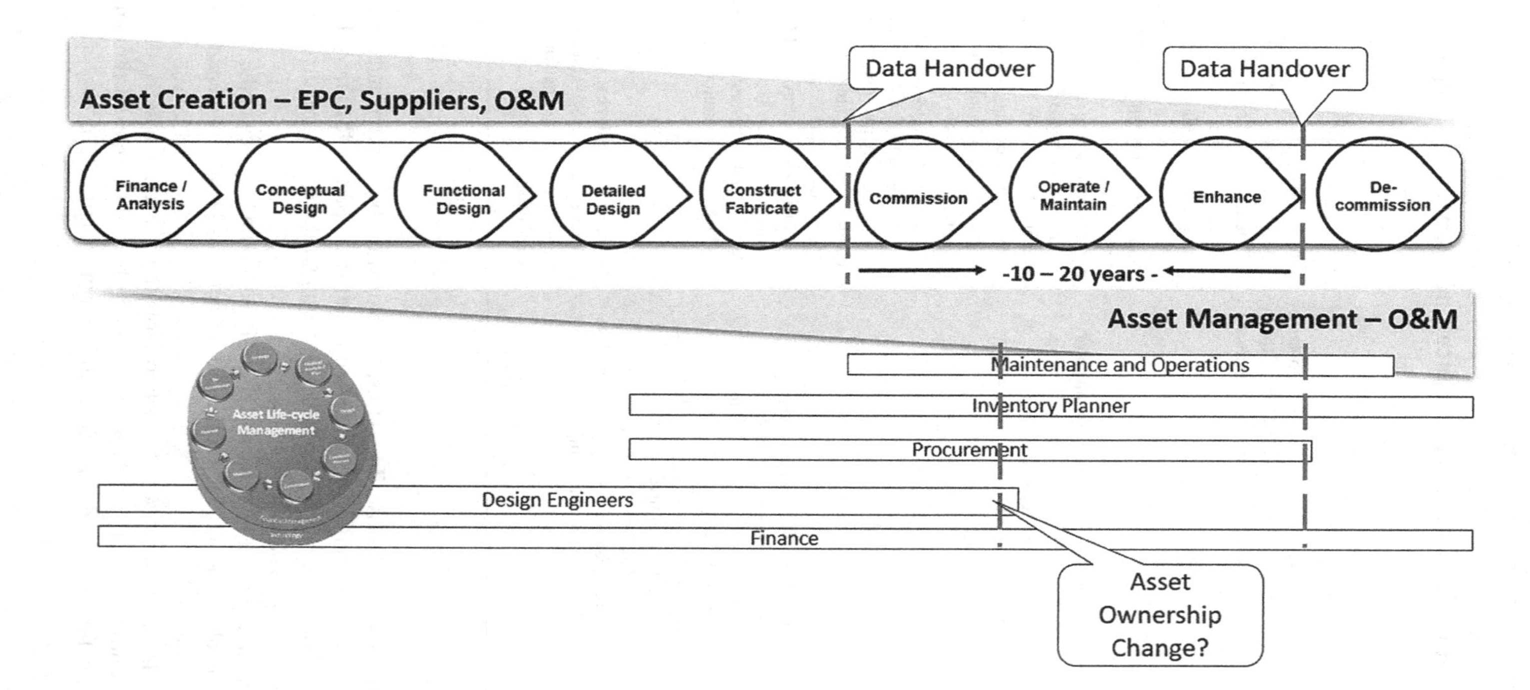


**FIGURE 4.1** Asset lifecycle timeline and business unit responsibility alignment for maintenance parts. (Adapted from copyright 101 MPE Introduction ©2020 from Maintenance Parts Management Excellence Training edited by Barry. Reproduced by permission of Asset Acumen Consulting Inc.)

## CONSIDERATIONS FOR THE MAINTENANCE PARTS MANAGEMENT GROUP

The basic maintenance parts management set includes material handling, shipping/receiving asset parts stewardship, and related health and safety activities. If ordered parts are delivered, then vehicle management may also be involved. Field inventory (beyond a prime stock room-room location) often needs to be managed within a remote site or in a maintenance technician's assigned service vehicle. For scoping this out, we will call this "logistics." The logistics block executes parts issues, receiving and storage, perform inventory controls and internal audits, and manages related material handling health and safety considerations (Figure 4.2).

### Issuing: Issues, Delivery, Pick Lists

These experienced personnel are on the front lines supporting the maintenance technician with his planned and unplanned parts requests. They often help the technician identify the actual part needed if the part identification is a bit ambiguous or incomplete. The maintenance technician should be responsible for making the correct part request the first time.

### Receiving and Storing: Receipts, Item Verification, Invoice Validation, and Item Storing

Parts receiving is a vital part of the inventory integrity of the operation. The receiver sees the part first in a given stock room and must compare what was "ordered" to "received." If you receive a part incorrectly, your inventory is incorrect from the start. Unfortunately, inventory inaccuracies are often difficult to identify before the part is requested from stock.

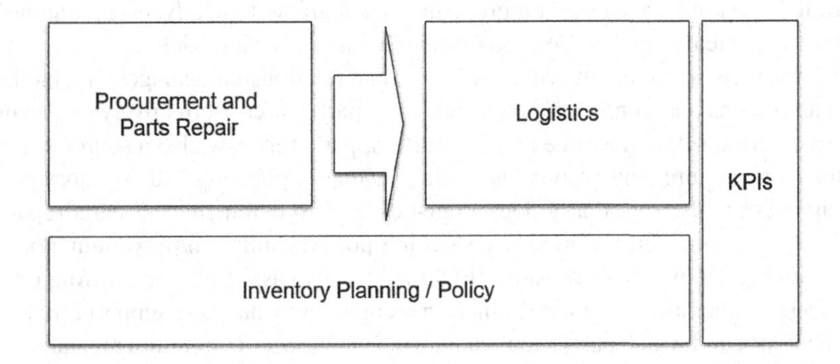


**FIGURE 4.2** High-level maintenance parts management block diagram. (Copyright Figure 6.1 ©2011, adapted from *Asset Management Excellence, Optimizing Equipment Life-cycle Decisions* edited by Campbell, Jardine, McGlynn. Reproduced by permission of Taylor & Francis Group, LLC, a division of Informa plc.)

### Inventory Control, Locating, Counting, Tagging

The Inventory integrity role is one where the most diligence must be exhibited. Correct identification and location of parts are crucial to effective service and inventory management. Well-trained and motivated staff are required for this comprehensive set of activities.

### Audit and Analysis, Item Setup, Disbursement Analysis, Projects, Accuracy Control

While many would suggest that this role is a luxury, you should expect that this work will need to be done. This role is needed to facilitate activity analysis and change to improve service levels, inventory accuracy, pick accuracy, inventory value, effective replenishment algorithms, or continuous improvement. In many operations, the warehouse manager facilitates this role.

### Health and Safety Considerations

Warehousing is typically where maintenance parts management focuses on health and safety. Health and safety disciplines are required in maintenance parts because the material handlers reside and work in a closed environment with racking and moving fork trucks around them. Documentation for handling and shipping dangerous goods (e.g., chemicals), dealing with bulk racking that could cause injury if tipped, and dealing with pressured gases. The warehouse is also where box cutters are used to assist with receiving and picking so accidents can happen. Typically, the warehouse manager and another designate are assigned to audit this area. Other considerations for fire wardens and emergency preparedness should be taken in an urgent threat (natural or otherwise).

Generally, multi-skilling is helpful in Logistics. Multi-skilling allows for operational flexibility, given the dynamic demand of when shipments may arrive, urgent orders come in, or a maintenance technicians service van needs restocking. Many operations do not have a cast of more than a few staff, so typically, every warehouse team member learns how to do each of the roles listed for logistics.

Procurement has a role in vendor selection and relationship management, facilitating transactions and confirming payments. If a part can cost-effectively be repaired and returned to stock in a "like new" condition, parts repair is also a source channel for the parts procurement group of activities. Inventory planning and inventory policy management is the role that guides Logistics and procurement and parts repair to work in terms of a formal Inventory stocking policy, minimum/maximum stocking levels, and economic order quantity (EOQ). They also assist with input from design engineering, operations, or maintenance on recognizing what parts support is required for each asset and which parts should have a designation as a minimum recommended spare part (RSP) (also could be an initial spare part – ISP from engineering).

High-level KPIs are typically developed to drive an empowered ability to autonomously serve maintenance-related parts management and tracking processes and service levels.

## SETTING GOALS FOR THE MAINTENANCE PARTS MANAGEMENT GROUP

Key performance indicators will be covered in more detail in a later chapter. Critical financial data such as inventory turnover, inventory losses, and inventory reserves should be published (at least as a percent of goal) so that the organization can feel empowered to look at the areas that require focus.

At the highest level: maintenance parts excellence should deliver in seven key areas:

- Parts availability and parts acquisition time
- Distribution quality
- Parts quality (shared with procurement)
- Parts costs; (shared with procurement)
- Inventory turnover (shared with finance, operations, and maintenance)
- Inventory reserves (shared with finance)
- Order systems availability (shared with IT)

## ORGANIZATIONAL CONSIDERATIONS

The actual organizational chart for a simple maintenance parts management operation will support the tree block diagram elements with KPIs aligned (and overlapped) depending on the enterprise size and what the focus is for the enterprise business and maturity dynamics. A simple sample of a maintenance parts organizational chart could look as presented in Figure 4.3.

As part of the maintenance parts management policy and strategy:

- if multiple stock rooms are required to meet the business needs (e.g., country-wide service levels commitments are in place) or
- small satellite inventory caches are to be supported (including inventory items kept in service vehicles),

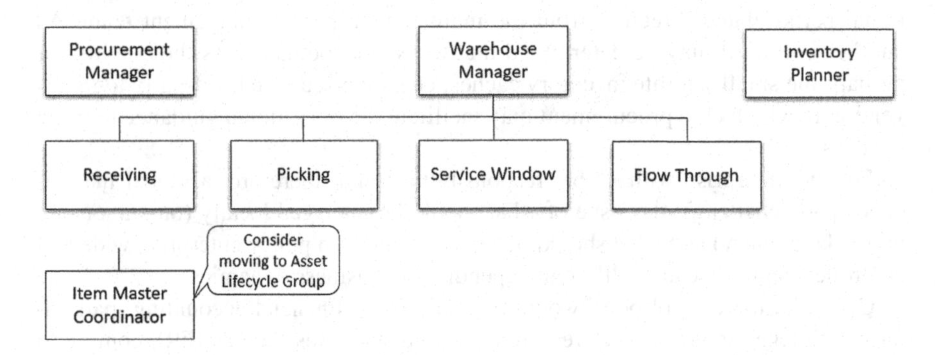


**FIGURE 4.3** Example of a maintenance parts management organizational chart. (Adapted from copyright 101 MPE Org and Metrics ©2020 from Maintenance Parts Management Excellence Training edited by Barry. Reproduced by permission of Asset Acumen Consulting Inc.)

then a reporting hierarchy for this type of infrastructure will need to be set for the enterprise.

For the network of stock rooms scenario – some of the guidance that would need to be provided would include the following:

- How do we keep track of service levels and demand needs across the network?
  - Planned/unplanned maintenance parts orders
  - Service level goals and tracking
  - Inventory tracking and controls
- How do we manage replenishment and related inventory quantities across the network?
  - Stock room on-hand inventory
  - Stock room in-transit inventory
  - Work order reserved inventory
  - Stock room available inventory
  - Automating reorder points and economic order quantities
- How do we manage procurement in the network?
  - Through a central site and redistributed
  - Direct from supplier to each network site
  - Hybrid of above options (Figure 4.4)

## MAINTENANCE RESPONSIBILITIES IN MANAGING INVENTORY INTEGRITY

The maintenance technician may be determined to be responsible for inventories in small satellite inventory caches and service vehicles. At the same time, a local stockkeeper may be needed for a specific network stock room. The local stock room stockkeeper (in the network and not centrally located) may report directly to the local maintenance manager across that country while taking dotted-line operational parts-related direction from the maintenance parts management team. As another option, it may be determined that the stock rooms across the country, or perhaps the small satellite inventory caches, be supported by a third-party logistics vendor, in which case procurement may facilitate the operational guidance to these vendors.

For the enterprise with global responsibilities and, therefore, a global maintenance parts challenge, the issue of what should be managed locally (on-site or in a particular country) and what should be managed through policy enterprise-wide and controlled centrally can be different depending on business dynamics.

Centrally, the control point would typically be in financial accounting, particularly with legal and regulatory requirements and standards such as IFRS coming to completion globally. Other considerations influenced by maintenance policy and parts management policy can be:

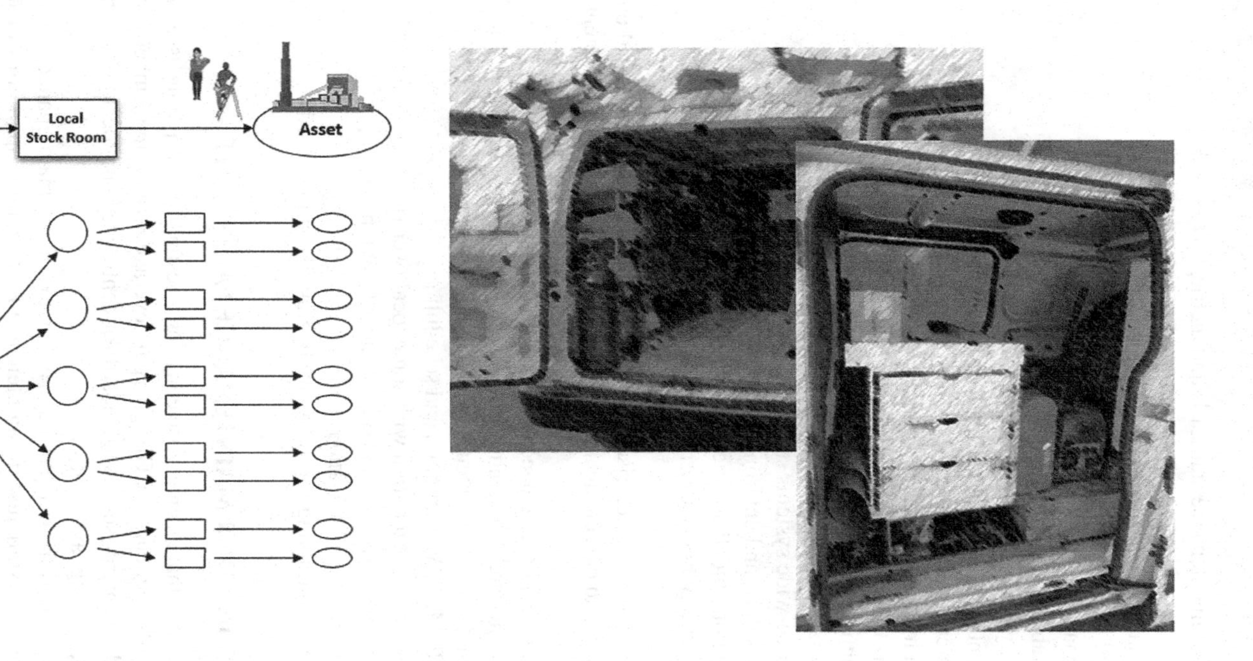


**FIGURE 4.4** Dynamics of a maintenance parts support network. (Copyright Figure 6.5 ©2011, adapted from *Asset Management Excellence, Optimizing Equipment Life-cycle Decisions* edited by Campbell, Jardine, and McGlynn. Reproduced by permission of Taylor & Francis Group, LLC, a division of Informa plc.)

Operations and maintenance standards and metrics:

- Asset management hierarchies and metrics:
  - Asset health and integrity
- Maintenance leading practices:
  - Reliability standards, methods, and disciplines
- Inventory policies and practices
  - Inventory placement and replenishment policies
  - Content management
- Procurement policies and practices:
  - Consolidated spend analysis
  - Supplier relationship management approach
  - Contractor management
- Enterprise-wide systems standards
- Health, safety, and environmental (HSE) standards
- Training on guidelines, metrics, and standards:
  - Knowledge management coordination

Locally, the country or enterprise may allow some flexibility and audited empowerment to support and execute against some of the following business elements:

- Operations and maintenance measured on both team's metrics:
  - Asset hierarchy, operating context, and work order execution
  - Local team development toward leading practices
    - RCM, TPM, planning, and scheduling
- Inventory policy execution with some local over-rides
- Leverage enterprise-wide system standards and metrics
- Execute HSE standards and report
- Provide local training for the above operations and standards

## INVESTING IN YOUR MAINTENANCE PARTS STAFF

It is recommended that an enterprise consider investing in a corporate asset management "center of excellence" (COE) group to leverage local team members to participate and lead an enterprise-wide standard for centralized and local policy standards management, training, and control.

For every operation, but particularly for extensive maintenance parts operations (e.g., 50 plus staff just in maintenance parts central operations or network combined), consideration should be in place for staff resource plans and resource goals with supporting employee reviews and analysis. Ensuring that the maintenance parts operation teams understand their role and drive value will create an environment that encourages employees to act responsibly and innovate. Employee metrics and employee feedback surveys can help create this feeling of being valued.

Maintenance parts management areas of operation work well with a culture that promotes "quality." Providing fundamental training in quality disciplines and methods and how a program may be viewed and approved by management creates an environment where staff can take the initiative and feel empowered.

Employee motivation and empowerment are integral components in achieving maintenance parts excellence, for example, having employees feel they are "making a difference" in their organization and being recognized for contributing to employee morale.

Employees stay where they are when they are:

- Competitively paid
- Given an opportunity to learn (training) and be mentored
- Invited to be involved in real business issues and challenged to solve them
- Empowered and given the tools of empowerment that are supported by management
- Trusted to do the job provided and drive to reach Key metric targets
- Recognized for their contribution (small recognition, awards, promotion)

Rotating staff across and through the maintenance parts operation can also help employee motivation. They can get:

- exposed to various functions within maintenance parts
- get an opportunity to learn additional inventory, procurement, or material management skills
- understand how finance, operations, or maintenance depend on a robust parts operation for success

Rotating encourages learning and innovation and will boost morale. While a new team member is mastering the new tasks, there is a risk that some errors will be made. Mentoring and close supervision will refine new staff training and mitigate this risk. Also, expect to prepare trainees for success and build in quality control checks aligned with your quality culture to minimize errors.

Maintenance parts staff well-being and morale can be heavily influenced by the enterprise culture and the manager directly managing them and the operation.

- An employee's opinion of their company is directly related to the relationship and opinion they have of their direct report supporting manager
- A poor working culture can undermine management every time!

## PROCESS MANAGEMENT DISCIPLINES IN MAINTENANCE PARTS

Every enterprise must have a culture that supports good health and safety policies and practices. However, not every care is taken to ensure an accident will not happen in day-to-day operations. Every employee who works in a material handling role should understand what is expected of them and their environment when it comes to hazardous goods handling (i.e., chemicals, but can also be heavy or sharp objects), fork truck protocol, compressed gas storage, and basic shelving ergonomics, access and storage protocols. For example, bulk racking over a specific height must be

**FIGURE 4.5** Warehouse Health and safety moment considerations. (Adapted from copyright 101 MPE Warehousing ©2020 from Maintenance Parts Management Excellence Training edited by Barry. Reproduced by permission of Asset Acumen Consulting Inc.)

bolted to the concrete floor so that a fork truck cannot move or tip it. A method for a staff member to record a potential area that could cause an accident is essential to a good health and safety culture (Figure 4.5).

A process to track and investigate accidents will contribute to the health and safety of a working environment and provide a way for each employee to understand that they contribute to and should consider themselves accountable for the health and safety of their working environment.

Maintenance parts operations is also where policies, procedures, and processes must be documented in terms of how you do business and updated and shared as they evolve. Make sure that this is in place for your staff. There are many reasons this is important. An up-to-date set of policies and procedures will:

- Act as an operational guide and reference to the maintenance parts management and staff
- Tell other business units how you are operating and what they can expect from your operation
- Act as a foundation for discussion if new situations arise, and a procedural change is being requested
- Act as a basis for new staff training
- Be a good baseline should an internal or external operation audit be called

Note – ISO initiatives often ask operations such as a maintenance parts management operation to document their process flows and have them audited to meet the ISO9001 standards.

Maintenance parts management is dynamic. As new assets arrive, the enterprise grows or shrinks, the business strategy and service strategy evolve, and the maintenance parts management strategy has to respond and provide the needed service levels at an optimal cost. Except for the most basic rules (e.g., parts issues and returns, inventory variance management), every process can and should change as the organization grows and changes.

Ensuring adequate staffing is in place for the demands of the maintenance parts operation is critical. Obtainable and quantifiable goals must be in place to support the business volumes. Staff education should support how they are expected to be performing. If a 24-hour-a-day and seven-day-a-week operation for the warehouse is required to support potential urgent (or emergency) parts requests, understanding how that will be supported across the staff mix is essential. As well, assuming there may be quiet times on off hours, helping the staff to understand how they can be productive with non-urgent requests but still be available and able to meet all urgent service levels.

## CASE STUDY FOR EMPOWERING MAINTENANCE PARTS WAREHOUSE STAFF

A large part warehouse set out to train up to 80 of their staff in process quality and transformation disciplines reinforcing the notion that a process issue needs to be assessed then analyzed before getting management approval to implement.

After two days of training for three sets of six staff (18 staff trained) in the pilot training program, they were asked to pick a key quality metric and find an area within that metric that would help improve their maintenance parts operations business value. The three groups worked competitively (mostly on their own time) and presented their assessment findings, analysis, and recommendations to an executive team. The executive measured how well they followed the quality training program approach and awarded a winning team.

The identified savings opportunities were significant! More significant however was the confidence created to empower the first wave of staff trained to go on to solve other issues within their area of influence.

The pilot program generated a demand for follow-on training for the remaining maintenance parts operational staff. All levels were trained (from analysts to fork truck drivers and parts pickers). As a result, the culture and the staff felt a very high level of empowerment. That location was declared the most effective parts operation globally in that Fortune 50 enterprise within two years.

# 5 Change Enablement in Maintenance Parts

## WHY CHANGE MANAGEMENT IS IMPORTANT

Managing change in a maintenance parts management context is similar to any other business environment. Much depends on what is changing, who may be affected, understanding if the stakeholders recognize their required personal commitments and affected business benefits to the change, the number of people and stakeholder groups that a change will impact, and the history of past changes in the given enterprise environment. These influences, and undoubtedly many more, contribute to a leading change program, and its intentions "fail" if the groups cannot work in concert for the change.

In many organizations, change enablement initiatives have had mixed success. Mixed success has been particularly true where IT systems are in place. The root cause of failure in the change efforts is that 60% of issues that contributed to a failure were "people" related. The remaining reasons are 25% for "process" and only 15% for the "IT" portion.

Conversely, this suggests that the change management effort needed specifically to support asset management personnel would be the same ratio to be successful (60% people, 15% process, and 15% technology). I would include maintenance parts management personnel in an asset management community group with this context (Figure 5.1).

Researching past parts and asset management-related projects has revealed that successful projects had specific traits to help them secure their business and operational success (Table 5.1).

## POTENTIAL REASONS FOR CHANGE MANAGEMENT IN MAINTENANCE PARTS MANAGEMENT

A successful change enablement focus will:

- Target the personal "what's in it for me" by role when developing change communications, training, and change readiness plans for each constituent
- Leverage face-to-face engagement via the sponsor, leaders (change champions), and volunteer "change army" (change agents) members in a company's line organizations
- Engage and train change agents to understand the process and technology impacts of the changes, how people move through the phases of change, and help their colleagues adapt

DOI: 10.1201/9781003344674-5

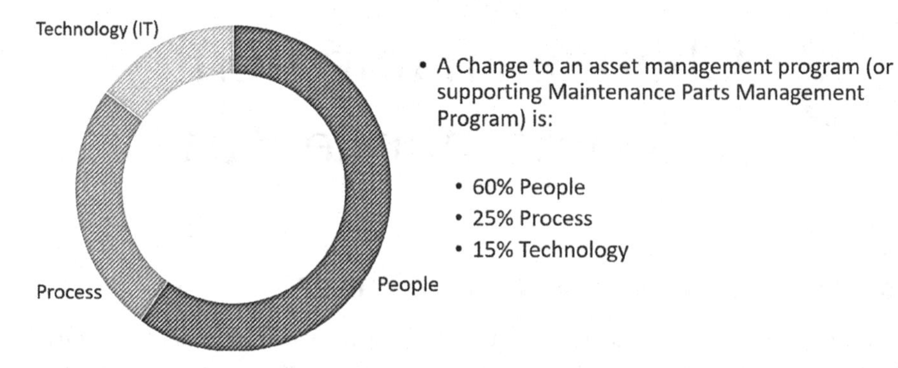


**FIGURE 5.1** Change management effort needed specific to asset management supporting personnel. (Copyright MPE ©2022 from Maintenance Parts Management Excellence Training edited by Barry. Reproduced by permission of Asset Acumen Consulting Inc.)

**TABLE 5.1**
**Change Management "Hind Sight" Can Be 20/20!**

| What Was Evident in the Success of the Change Enablement Program | % |
|---|---|
| Senior management buy-in and leadership | 70 |
| Superuser advocate community | 70 |
| Scalable training and analysis | 70 |
| Communications plan | 70 |
| Train the trainer program (T3) | 60 |
| Aligned communications with program rollout | 60 |
| Stakeholder analysis | 55 |
| What they wished they had more program evidence | |
| Senior management buy-in and leadership | 70 |
| More robust – more frequent communications | 70 |

Adapted from copyright (101 MPE Introduction ©2020) From (Maintenance Parts Management Excellence Training) edited by (Barry). Reproduced by permission of Asset Acumen Consulting Inc.

- Leverage lessons learned from previous maintenance parts programs, EAM, enterprise resource planning (ERP), or other operational implementations/transformations
- Leverage local experience and leading practices in change to contribute to a successfully adopted set of benefits from the planned changes
- Leverage and support training, job aids, and online help during the pre-go-live and post-go-live periods

The well-managed organizational change program ensures that the benefits of the change are realized early after go-live and are sustainable.

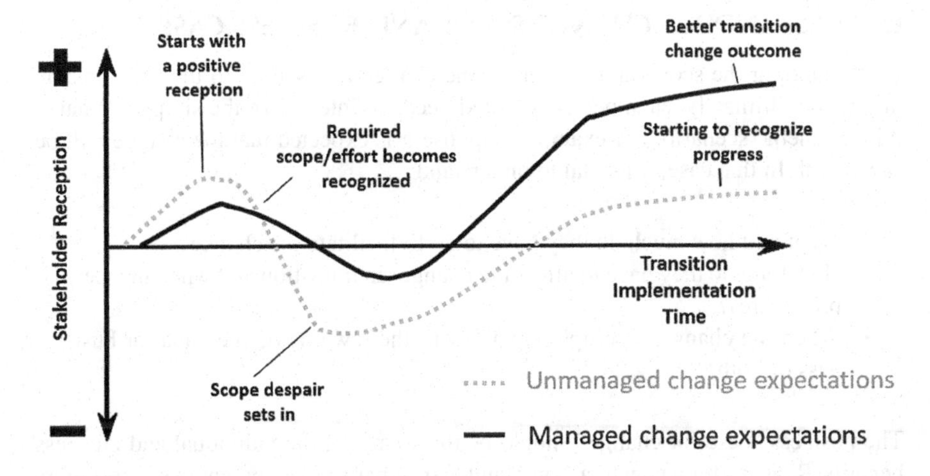


**FIGURE 5.2** Cycle of typical large-scale systems and process implementation. (Adapted from copyright 106 MPE Impact of Change Management in Asset Management ©2020 from Maintenance Parts Management Excellence Training edited by Barry. Reproduced by permission of Asset Acumen Consulting Inc.)

Maintenance parts management may have many changes that go beyond the boundaries of their one business unit and will likely be cross-functional (Figure 5.2).

Examples of these changes could be policy or procedure changes such as enforcing that:

- All maintenance parts Inventory transactions are facilitated through the enterprise parts management system (e.g., no maverick spend or transactions out of the system).
- All parts item numbers, including not stocked items, expected to be needed in an asset lifecycle are understood, documented, and listed in the enterprise parts management system.
- There will be one inventory warehouse in a specific region versus many small satellite stock rooms in a compressed geography.
- All parts must be on the inventory books: parts squirreled away by maintenance technicians will be accounted for and returned to inventory for credit by finance.
- All surplus (unused) parts from work orders will be "identified" and returned to stock.
- All suppliers must be "registered" in the procurement portion of the enterprise parts management system.
- Design engineering will be responsible for establishing an initial spare parts (ISP) list that is "accepted" by maintenance, parts, and finance.
- The enterprise parts management system will have populated all the defined elements and attributes that need to be "set up" in a parts item master for a new part (stocked or not stocked).

## UNDERSTANDING CHANGE SCOPE AND BUSINESS CASE

Understanding the scope and reason for the change is essential to the stakeholders impacted. Ultimately, each person affected needs to internalize the simple "What is it in for them" scenario. For example, suppose it is expected that the change will be successful. In that case, it is vital to understand:

- What and how much do they have to do to facilitate the change?
- How long do they have to affect the change (in transition and when in a new procedure)?
- When the change is complete: what will the rewards be (personal or business or both)?

The business "case for change" should be the source of the individual and business benefits. Reasons for a new focus on maintenance parts management in support of an asset, management community could be to:

- Impact productivity (e.g., people, assets, cycle time)
- Enhance process controls through visibility
- Create business agility opportunities from improved identification of parts requirements across an asset's lifecycle
- Develop improved support and agility of asset refurbishments (turnarounds) or enhancements
- Improve the asset lifecycle costs by better inventory service levels and effective inventory algorithms
- Align cross-functional business units to a single source (system) of the truth for communicating needs, ability to fulfill requirements, and communicate potential shortfalls and risks

## FOUR PILLARS TO SUPPORT CHANGE

Change in the maintenance parts management community can be industry-specific and support multiple asset types (production, facility, infrastructure, IT, and transportation). Change can mean adding new capabilities in analytics supported by new deployment options (on-premise, cloud, or hybrid) or including new IT architectures (performance, scalability, security). In any event, the change ultimately must address the end-user experience.

The user will understand how the new work would be simplified or what risks will be mitigated from the new solution. They will expect that the new solution will: be role-based (e.g., how this will impact them), include new technologies they would understand (e.g., mobile devices, where appropriate), and collaborate and communicate to socialize issues as they arise in the new environment.

Change enablement (also called change management) can be characterized by the analogy of a four-legged stool, where change management is the aggregate of the four legs.

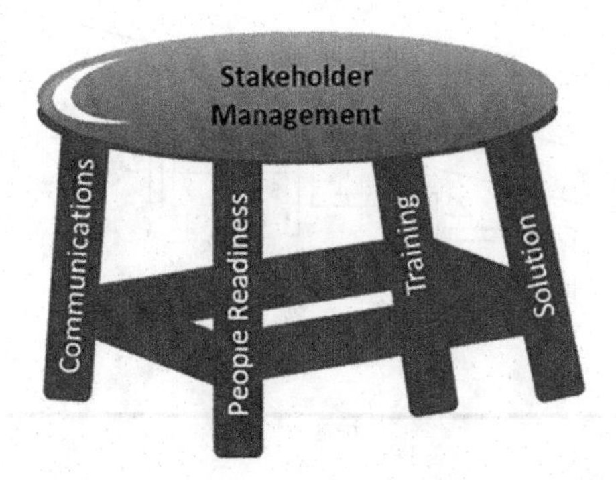


**FIGURE 5.3** Four pillars of a change enablement approach. (Adapted from copyright 101 MPE Introduction ©2020 from Maintenance Parts Management Excellence Training edited by Barry. Reproduced by permission of Asset Acumen Consulting Inc.)

- The first three legs represent the communications plan, the training plan, and the people readiness plan.
- The fourth leg represents the embedded new tool or solution of application functionality that helps direct users through the change – for example, process documentation, workflow, context-sensitive instruction, field-level help, and configuration (Figure 5.3).

All legs support the stakeholders who are affected by the change. The stakeholders will also be assessed and managed by the group, by role, and individually in the stakeholder management plan.

- For a specific maintenance parts management program, there could be a separate focus for change strategy and plan to support stakeholder management, including:
  - The solution
  - Communications
  - Training
  - People readiness

The "case for change" should be understood if the business invests in a substantial change. The identified pain points of what will be "experienced" if no change happens are likely also understood by at least the few who have identified the pain points (issues), opportunities, expected benefits, and threats if no change happens.

Once the initiative has been identified/announced, and management has committed to facilitating a change, it does not mean that the change in the program initiative will be particularly successful. Supporting change means supporting the stakeholders affected. These stakeholders will want to feel that the planned change has a purpose and benefit. The benefits will need to be received as contributing to each person's sense of business community, social connection, and personal support.

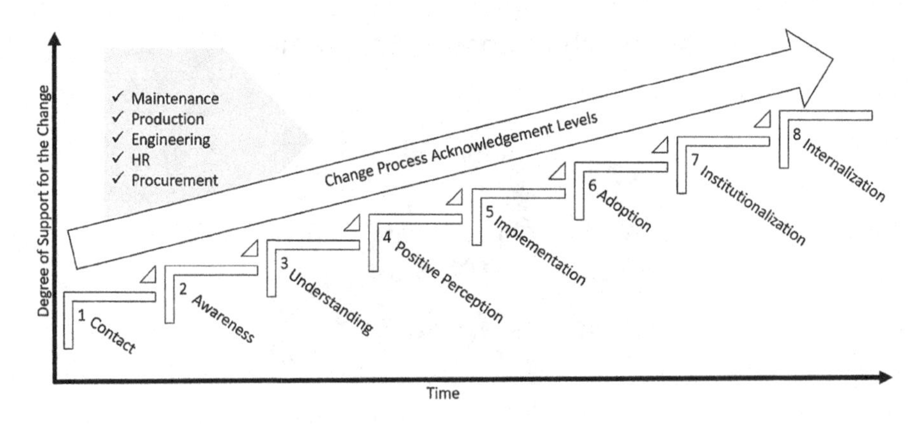


**FIGURE 5.4** Change enablement continuum for impacted stakeholders. (Adapted from copyright 101 MPE Introduction ©2020 from Maintenance Parts Management Excellence Training edited by Barry. Reproduced by permission of Asset Acumen Consulting Inc.)

Changes in maintenance parts management can impact a sub-set of all the stakeholders that touch parts (e.g., store warehouse leadership and material handlers, inventory planner, operations, Maintenance leadership and technicians, design engineering, vendors, finance, health and safety, and legal) (Figure 5.4).

In the change enablement continuum chart, it suggests that over time:

- if you have contacted a stakeholder and declared a change would happen:
  - does this mean they are ready to accept it?
- if you make a stakeholder aware a change is impending:
  - they can say they encountered the change, but does this still mean they are ready to accept it?
- if you meet with a stakeholder and help them understand the impending change:
  - they may confirm that they "understand" the planned change's intent, nature, and impact, but they may still not be ready to embrace it.
- if you meet with the stakeholder, confirm a positive perception of the scheduled change:
  - they may demonstrate that they understand the expected change's intent, nature, and impact and feel optimistic about it, but that does not mean they are ready to implement it.
- if you implement the change with the stakeholder and the enterprise:
  - they may confirm that they participated in the implementation, but that does not ensure understanding if they have adopted it in alignment with the business objectives.
- if you work with the stakeholder to make sure they have adopted the change as designed:
  - that may leave room to confirm if they have fully embraced it to get the expected benefits each time they work on that change.

- if you work with the stakeholder to make sure they have embraced the transition to the degree that it is "institutionalized" into the fabric of their process:
  - that may still leave room for times when the new changeset is missed or omitted.
- if you work with the stakeholder to make sure they have embraced the change to the degree that it is "internalized" and becomes a "habit" and in the fabric of their "values":
  - then that total change enablement has been successful and completed.

## POTENTIAL PROJECT RISKS IN CHANGE

The organizational change management approach should work to address the challenges organizations face over an extended change effort. For example, some of the lessons learned from past projects suggest ensuring that the program sponsor (Executive) *"is seen leading"* this project so that the staff recognizes the program's priority.

Some past program change risks have been identified (Table 5.2):

A successful change management program will have the change leadership fully engaged from the project initiation phase through requirements blueprinting, solution design, building the solution (including testing, integrating), deploying, and

**TABLE 5.2**
**Potential Project Change Management Risks and Mitigations**

| Potential Risks | Mitigation Tactics |
|---|---|
| Executive/ leadership risks | • Confirm strong support, alignment, and commitment from the leadership team<br>• Ensure that the sponsoring executive team is aligned, visible, and engaged.<br>• Promote clear communications and continuity across the leadership team<br>• Build and sustain the change and organizational leadership capacity |
| Organizational risks | • Define and promote the benefits to each stakeholder group<br>• Ensure strong personnel are assigned to ensure project success and the sustainment of the existing business process (Keep the lights on while driving to the new model)<br>• Align leadership and staff to integrate and prioritization of the known related initiatives and projects<br>• Promote a proactive work style to move away from the reactive (crisis-management) approach that may have created the need for change |
| Employee risks | • Consider potential attrition risks in (longer-term) projects and potential replacement attributes/staff<br>• Allow the leadership to own their difficult organizational decisions<br>• Promote that this is not a "fix-all" technology or process transformation<br>• Work to achieve employee buy-in and corporate energy for the duration<br>• Search out and address change fatigue and cynicism (if evident) by those affected |
| Maintenance parts transformation team risks | • Avoid post-project dependency on external consultant resources<br>• Build the required skills and capacity of the expected technology/software transformation team<br>• Manage the long-term team dynamics and effectiveness throughout the project<br>• Encourage effective, empowered decision making |

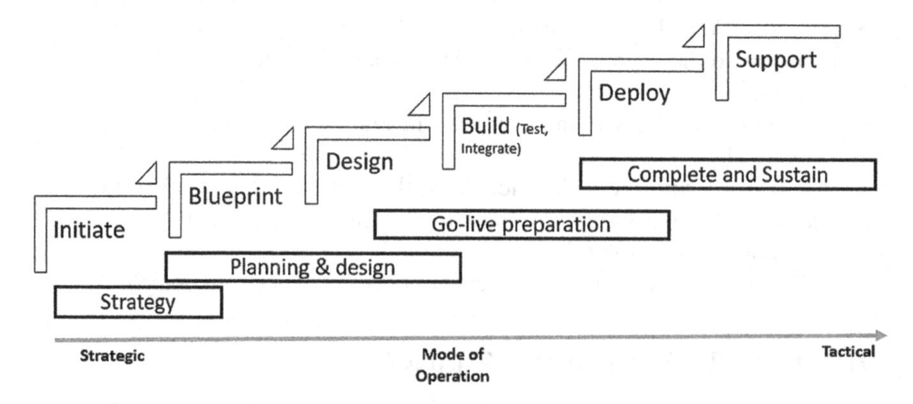


**FIGURE 5.5** Throughout a successful project - the organization change management program is fully integrated into the project lifecycle and activities. (Adapted from copyright 106 MPE Impact of Change Management in Asset Management ©2020 from Maintenance Parts Management Excellence Training edited by Barry. Reproduced by permission of Asset Acumen Consulting Inc.)

sustaining support through the project/program support cycle. Often, the change lead is asked to be accountable for reporting the business benefits, how well the solution has been implemented, and how well it has been adopted and internalized by the key stakeholders impacted. In preparation for this mantel or responsibility, the change lead will be active throughout the project, from early strategizing to planning and designing how the stakeholders will be "addressed" through more tactical activities at the go-live and solution sustainment timelines (Figure 5.5).

The strategizing can include:

- Building vision clarity for the change and the leadership team's commitment to deliver
- Identifying potential organizational solutions to maximize the targeted change benefits
- Communicating the planned change strategy as influenced by early assessments
- Collecting change history, change impact assessments

Planning and design can include:

- Planning the management of change (stakeholder management, communication planning, training requirements and development, team building)
- Designing the proper structure, changes to behavior, and HR processes to integrate the new process and technology

Go-Live preparation can include:

- Managing the delivery of training and communications
- Developing job profiles, security, and updates to HR processes (if applicable)

Sustaining the change through to completion can include:

- Working to confirm new methods are accepted ongoing habits
- Reporting on benefits realized

## PERFORMING STAKEHOLDER ANALYSIS

Each of the planned procedure changes could impact one or many potential stakeholders. Coordinating such a change would be difficult without understanding the detailed scope elements of each change and how each stakeholder will be affected.

Understanding the change history and culture affected is useful when the change leader drives a program with many impacted areas. For example, through interviews or a questionnaire, the change lead could get some insight into whether the enterprise – or at least specific stakeholders are affected – has a cynical expectation of any change or may have change fatigue. Areas to consider in an interview/questionnaire set could include how the organization has performed in past initiatives in the areas of:

- Change vision
- Change leadership
- Building commitment
- Sustaining change
- Configuring change
- Managing change (Figure 5.6)

During the strategy development, a base set of truths need to socialize how the change is defined. An outline that would include the foundational facts about the

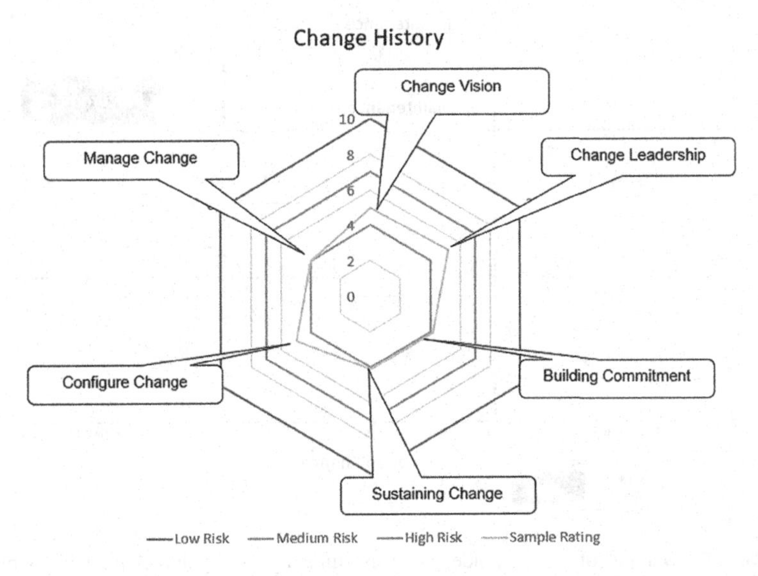


**FIGURE 5.6** Example of a change history summary chart. (Adapted from copyright 106 MPE Impact of Change Management in Asset Management ©2020 from Maintenance Parts Management Excellence Training edited by Barry. Reproduced by permission of Asset Acumen Consulting Inc.)

change allows the change lead/advocates to talk to those potentially impacted and then confirm the impact and how they feel about the planned change. It is helpful to perform this type of feedback testing process multiple times before and after solution deployment to understand what actions are needed to get the stakeholder group to an advocate status at the time of go-live and into sustainment.

Examples of foundational facts should include:

- The purpose for the change or transformation
- The client's vision, values, and beliefs about change
- The overriding imperative for change (the "burning platform")
- Characteristics of the change:
  - who will be affected
  - why they will be involved
  - where the change will happen
  - what the change will look like
  - when the change takes place
  - identification of implementation stakeholders
  - sponsor, agent, and target communities;
  - an approach for identifying and managing resistance to implementation

A mapping of the status collected from each cycle of the impact interviews can be graphed and reported to the program sponsor for considered actions to get the key stakeholders where they should be at Go-Live (Figure 5.7).

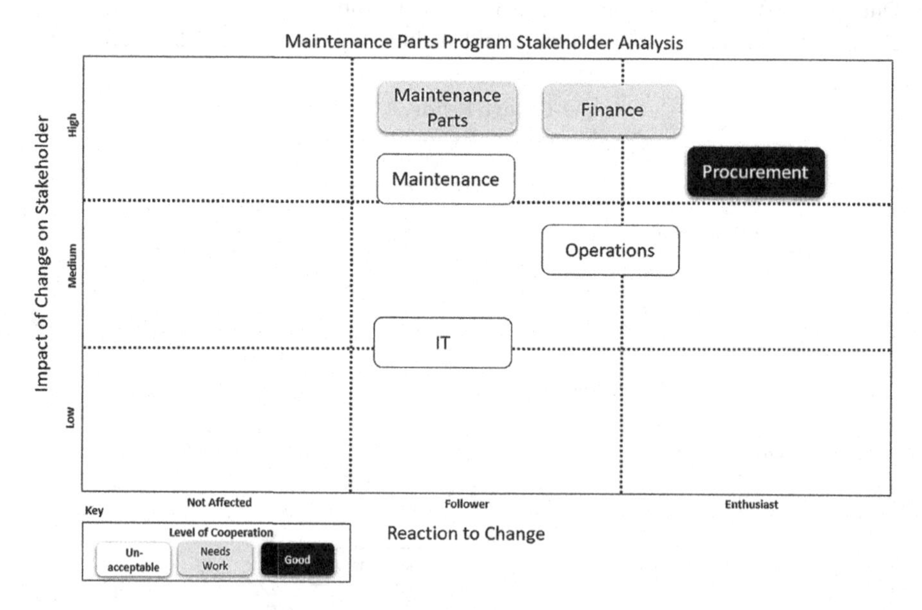


**FIGURE 5.7** Sample of a stakeholder analysis summary at a point in time for a project. (Adapted from copyright 106 MPE Impact of Change Management in Asset Management ©2020 from Maintenance Parts Management Excellence Training edited by Barry. Reproduced by permission of Asset Acumen Consulting Inc.)

## PEOPLE READINESS CHANGE LEAD ROLE/TASK EXAMPLES

Figure 5.8 shows a sample scope of potential deliverables a change lead could facilitate (Table 5.3).

A leading change program is robust and tailored to prepare each stakeholder (employee) for a smooth and safe transition through the planned solution implementation and respects and leverages the enterprise culture.

The change lead will:

Lead by cultivating a change-centric culture from the top to the entire organization.

- The focus should be on:
  - Leadership commitment, visibility, and accountability
  - Active engagement of affected staff
  - Empowerment of change leaders/advocates at all levels

Make change matter to the business and the affected stakeholder by understanding the benefits and embedding the activities of a supporting change management program.

- The focus should be on:
  - An understanding of organizational/personal benefits
  - An appropriate level of visible senior leadership support
  - Adoption of new skills/behaviors
  - Commitment to being a change champion
  - Driving to the planned change benefit realization

Build and create enterprise cultural endurance by further developing and formalizing change expertise and systematically building site/corporate-wide change capabilities.

- The focus should be on:
  - Supporting change through visible leadership and change advocates
  - Training to promote change management methods
  - Building change capabilities and experience
  - Providing a positive stakeholder change experience as a foundation for their next transition

The ongoing (pre-and post-go-live) change program should consist of the following elements:

- Change impacts
  - Impacts are understood for every affected role/shift
  - Change support activities are prescribed and executed
  - Impacts by affected stakeholder – documented in a "journey map" (e.g., a day in the life)

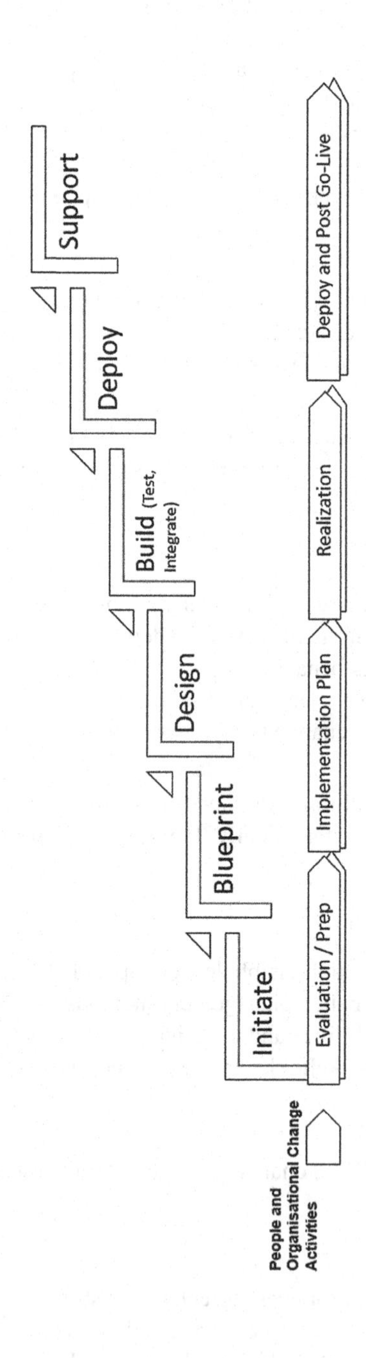


**FIGURE 5.8** Typical change enablement overlap with the formal project phases and the total project lifecycle. (Adapted from copyright 106 MPE Impact of Change Management in Asset Management ©2020 from Maintenance Parts Management Excellence Training edited by Barry. Reproduced by permission of Asset Acumen Consulting Inc.)

**TABLE 5.3**
**Typical Change Enablement Deliverables, by Project Phase, from a Systems Implementation Engagement across the Total Project Lifecycle**

| Evaluation/ Preparation | Implementation Planning | Benefits Realization | Deploy, Go-live, Support |
|---|---|---|---|
| • Confirm vision, mission, goals, & objectives<br>• Confirm management sponsorship/buy-In and assigned teams<br>• Facilitate change history assessments<br>• Facilitate change readiness assessments<br>• Facilitate a stakeholder analysis<br>• Develop & lobby communication strategy and plan<br>• Introduce training strategy/scope | • Assess & understand organizational impacts<br>• Confirm high-level job changes (if any) and performance measures<br>• Develop and lobby change transition plan<br>• Develop relevant communication material<br>• Develop detailed organizational changes (if any) and role/job descriptions, and metrics<br>• Establish training plan<br>• Provide executive sponsor updates | • Set up a transition support network<br>• Develop training material/media<br>• Develop culture change initiatives<br>• Deliver training | • Deliver user support activities<br>• Provide go-live transition support<br>• Ensure continuous improvement<br>• Confirm target benefits realized roadmap<br>• Confirm change complete |

- Change advocate network
    - Create a network of leaders and peers to act as program change advisors and advocates
    - Set up these designates as the single point of contact (SPOC) for outbound communications trials
    - Drive shoulder-to-shoulder support with and through these advocates
- Communications
    - A website of status could be available to the stakeholders and updated regularly
    - Newsletter created for/by leaders and advocates
    - Leverage face-to-face campaigns such as meetings, roadshows, promotions
    - Multiple communications campaigns could be leveraged, including podcasts, emails, management updates, company newsletters, town halls, department updates
    - Build on awareness, clarity, value, and "What is in it for me?"

- Affected stakeholder preparation and training
  - Promote the notion of "just in time" training
  - Prepare specifically for each group impacted by the go-live
  - Develop process handbooks/guides (if applicable)
  - Create "journey map" references or "quick reference guides"
  - Provide on-site and central support as needed

## PROJECT COMMUNICATIONS CONSIDERATIONS

Multiple communications tactics are engaged (beyond announcements and training) in a well-directed change plan. An organization's culture can dictate if a communications session needs to be across various shifts, off-site, or if campaigns engaged need posters, emails, management meetings, newsletters, websites, or one-on-one interviews. The following figure shows an example of a simple communications plan (matrix) (Table 5.4).

Each enterprise may have standards to be maintained when communicating with its employees. Employee agreements may be culturally influenced or more formally in place as influenced by employee agreements.

When considering employee communications, the who, what, where, and how to communicate (e.g., which media to consider) must be thought through.

Communication should include:

- The target audience for the organization's communications with the following:
  - Stakeholders affected/impacted by the maintenance parts management change
  - Stakeholders who can impact the maintenance parts management change or influence changes
- Communication that explains the expected benefits and performance of the planned maintenance parts management change, such as:
  - The implementation of the scheduled shift in maintenance parts management
  - Any changes in resource process work environment or other requirements
  - How these changes affect objectives, risk, and value
  - Changes in the organization's context
- The expected frequency of the organization's communications is influenced by:
  - Stakeholder and other applicable requirements (including legal and regulatory requirements)
  - Maintenance parts management objectives and plans
  - When a significant change occurs
  - After an unexpected incident occurs, to enhance or restore the reputation
  - During development meetings

**TABLE 5.4**
**Sample of a Simple Communications Plan Matrix**

| Initiative | Audience | Medium | Frequency | Design Points (Example) |
|---|---|---|---|---|
| Program Newsletter | Operations, maintenance, finance, maintenance, parts, procurement | Email, podcasts | Bi-Weekly | Branding of project<br>Project updates |
| Poster Campaign | Operations, maintenance, finance, maintenance, parts, procurement | Notice boards | Monthly | Branding of project |
| Extended Team Meetings | Operations, maintenance, finance, maintenance, parts, procurement | Face to face | Monthly | Project design updates |
| Department Meetings | Operations, maintenance, finance, maintenance, parts, procurement | Face to face | As scheduled | What this means to them |
| Executive Meetings | Executives and steering committee | Conf call and Face to face | Monthly | Project updates |
| Executive Letters | Operations, maintenance, finance, maintenance, parts, procurement | email, videos | Quarterly | Why this is important to our organization |
| Road Show | Operations, maintenance, finance, maintenance, parts, procurement | By central location | Before the "new" process start | Branding<br>Readiness<br>Q&A |
| Workshops | Operations, maintenance, finance, maintenance, parts, procurement | Face to face, podcasts, videos | Before the "new" process start | Benefits<br>To Be Process<br>Training |
| User acceptance testing (UAT) | Operations, maintenance, finance, maintenance, parts, procurement | Face to face | Before the "new" process start | Validation of "to be" processes |
| Training | Operations, maintenance, finance, maintenance, parts, procurement | Face to face, T3 | Before the "new" process start | As required, JIT<br>T3 |
| Helpdesk | Operations, maintenance, finance, maintenance, parts, procurement | Phone. email | Post-Transition Support | Sustainment support |

## TRAINING CONSIDERATIONS FOR CHANGE ENABLEMENT

Change enablement lead roles:

- Provide leadership and develop/manage the overall change strategy
- Provide direction on change methodology, tools, and quality, change activities & scope
- Provide change management templates and guidance on how to use them
- Lead and participate in interviews, discussions, and workshops
- Provide change management issue resolutions and recommendations
- Help and provide advice on selecting stakeholders and stakeholder groupings
- Collaborate with project teams for change activities

Training lead roles:

- Build the training strategy and plan and will advise and provide leadership to the development of the training courseware and the training rollout and will assist in train-the-trainers (T3)
- Leverage the solution attributes (or if the software is involved – leverage the IT Integrator to provide SW), training assets, training documentation templates, and tools
- Enterprise trainers will often need to tailor the training materials, related courseware, and data to align with the culture maturity and business requirements. They will often lead the training logistics and training delivery so that the end-user is being taught by one of their own from their own culture

Getting a target audience to understand a training point requires four task-based exercises. The point to be taught needs to be "exhorted, explained, shown examples and experienced." This could be called the 4E approach. It does apply to implementing the education of students, including adults. Formal training techniques recognize the need to "involve" the targeted learner in the learning process. A learner-centered approach can consider activities that will promote the student's involvement in the knowledge transfer. Facilitating collaboration activities with the targeted students can include:

- Group projects
  - Having the target students develop training or data content
  - Staging presentations on the topic
  - Holding competitive activities (including debates, presentations, business cases, etc.)
  - Role-playing
- Gamifying the challenge needed to be solved by posing the gap/issue to the group

Happily, a more formal, holistic, and publicly available method can be leveraged by the Training lead. The ADDIE Methodology is commonly used to describe adult training development and delivery. The ADDIE process method promotes a five-stage notion of "analyze, design, develop, implement, and evaluate."

Within the ADDIE process definition, the need to understand the student audience is essential. It provides the opportunity to tailor the training, communication, and skills transfer uniquely to each stakeholder group's needs.

The Change Teams work with the Training lead and leverage the Corporate Culture to deliver:

- Comprehensive training plans with detailed role-based curricula
- Tools, materials, and job aids
- T3 (Train the Trainer) programs
- Post-training support/plans
- Training evaluation to confirm capabilities and readiness
- Training resources (as required) (Figure 5.9 and Table 5.5)

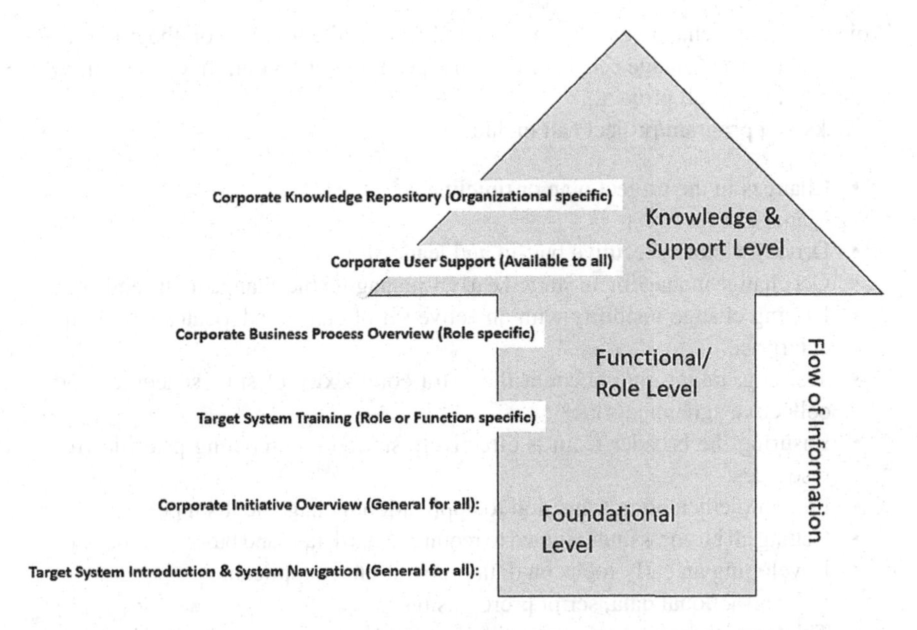


**FIGURE 5.9** Example of a training program and set of target training levels. (Copyright MPE ©2022 from Maintenance Parts Management Excellence Training edited by Barry. Reproduced by permission of Asset Acumen Consulting Inc.)

## TABLE 5.5
## Example of a Leading, Robust and Rigorous Training Program Supporting an Adult Training Approach

| Target Training Level | Training Focus | Training Initiatives/Tasks |
|---|---|---|
| Knowledge & support level | Corporate knowledge repository (Organizational specific) | Presentation materials, FAQs, job aids, EAM/Parts work instructions, standard operating manuals |
| Knowledge & support level | Corporate user support (available to all) | Business and system support contacts, day-one set up, policy and process reminders |
| Functional/ Role level | Corporate business process overview (stakeholder group/role-specific) | Detailed processes, roles and responsibilities, asset mgmt concepts, work management lifecycle concepts, inventory & procurement concepts, business values: configuration workbooks, data quality |
| Functional/ Role level | Target system training (role or function specific | Technology start scenarios/steps, relevant business scenarios, step-by-step work instructions, hands-on system exercises, go-live prep, User support |
| Foundational level | Corporate initiative overview (general for all) | Overall mission/vision statement, scope, supporting policies, high–level processes, change impacts, transition plan, reasons for the change, Impact insights |
| Foundational level | Target system introduction and system navigation (General for all) | Sign In/out, navigation basics, common usage |

Project risks and change enablement-related risks will always be on the mind of the Change lead. The Change role is often agile, even if the program they are working to change is a waterfall project.

Risks to a program/project can include:

- Changes in the project plan or timeline
- Manage scope creep
- Driving visible executive buy-in and leadership
- Get change management started early – managing the change to the end-user
- Driving change visibility with an active set of change advocates across the enterprise
- Training, taking into account the extra complexity of shift schedules and collective agreement rules
- Ensuring the broader team is effectively staffed – competing priorities for resources
- Post-implementation transition to application management support
- Getting all business units aligned to requirements details and business processes
- Developing an early focus on data preparation (if applicable)
  - Transactional data, script prerequisites
- Thinking through and purposefully resource your report development requirements
- What will we do if, after the program goes live, it does not work?
  - Do the stakeholders have contingency training materials to support them?
  - Can we go back?
  - Do we have an escalation path to manage an event?
- Managing through the program "critical success factors"

## EIGHT CRITICAL PROJECT SUCCESS FACTORS FOR A CHANGE LEAD

Eight critical success factors can be helpful as a framework for a change lead to determine the health of the program they are supporting. A simple audit to assess where they believe they are on these listed project status elements can provide insight into a project's success.

The eight critical success factors to confirm are as follows:

1. Is there an accepted and understood compelling need for change?
2. Is the communicated direction clear?
3. Is leadership committed and recognized as leading?
4. Are there targeted, effective communications planned for this transformation?
5. Is the transformation project management well established and being executed?
6. Are there measurable project and transformation goals?
7. Is broad-based stakeholder participation present and being promoted?
8. Is there a tenacious single-program focus?

This list can be beneficial for a change lead to assess the expected "people impact" health of a transformation project.

# 6 Key Performance Metrics for Maintenance Parts Excellence

## THE IMPORTANCE OF KPIS

Key performance indicators (KPIs) or business metrics are critical to evaluating process performance, communicating to management and staff critical areas of progress, and signaling when and where action must be taken to achieve a defined business goal.

KPIs can provide a "score board" for how the organization performs. It will empower staff to work toward crucial focus outcomes when adequately supported. However, too many metrics in a business unit can create confusion and drive suboptimal behavior.

Guidance for KPIs would be to create a handful of key metrics. Four to six key performance indicators are optimal at any "business dimension" level. The KPI selected should suit each business level measured, and it should drive the desired company and employee behavior.

A simple model for parts distribution for an original equipment manufacturer (OEM) could be mapped between providing the desired quality parts at the desired service level and a competitive price.

A manufacturer selling their parts for maintenance may ask themselves:

- What drives my costs (and, therefore, price competitiveness)?
- How do I deliver a quality service?
- What should my KPIs be? (Figure 6.1)

## MAINTENANCE PARTS METRIC COMPLEXITIES

Defining KPIs metrics in the journey to maintenance parts excellence is not a simple task.

First, it is helpful to define the scope of the role of the maintenance parts management operation. As an example:

- Is it a single location supporting a single campus of assets?
- Is it a country-wide network of stock rooms supporting assets in multiple locations?

DOI: 10.1201/9781003344674-6

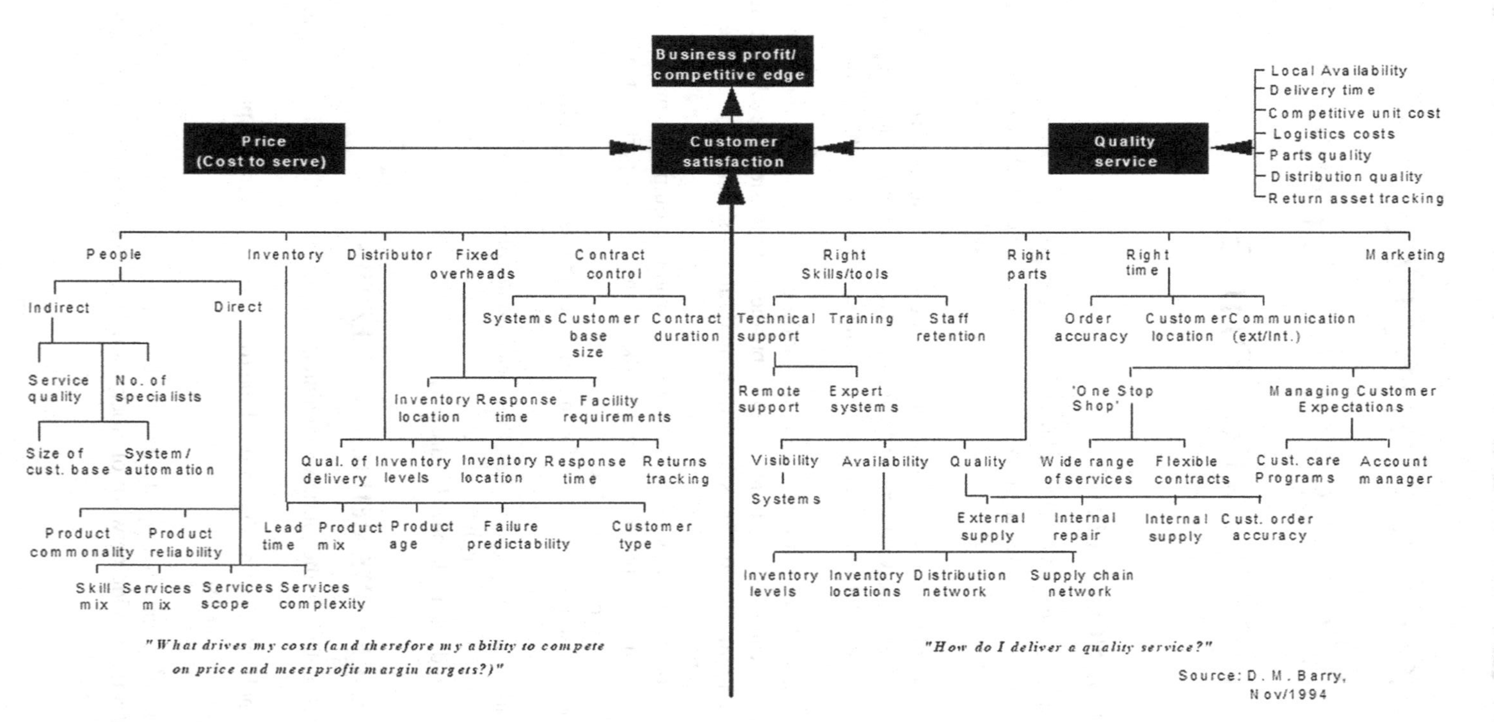


**FIGURE 6.1** Generic model of after-sales logistics service delivery. (Adapted from copyright 102 MPE Org and Metrics ©2020 from Maintenance Parts Management Excellence Training edited by Barry. Reproduced by permission of Asset Acumen Consulting Inc.)

- Is it a central warehouse plus multiple country-wide networks of stock rooms supporting assets in various locations in each country?
- Do you have customers of your parts service operation outside the enterprise? In other words, are you also in the parts sales business?
- Is your parts supplier from a single source, or are your suppliers from multiple entities in multiple countries?
- Do your Maintenance Technicians require access to the maintenance parts 24 hours a day, seven days a week?
- What is your policy for hours of operation for internal and external disbursements?

These are just a few questions that need to be understood and suggest that defining the scope of the maintenance parts business you are setting out to measure is essential in achieving "excellence" for that targeted scope.

A "not-so-simple" scope could include a multinational enterprise where the measures are primarily limited to the in-country activity, as shown in Figure 6.2.

In this model, the maintenance parts operation is responsible for procuring parts from their internal/external suppliers to a central warehouse and supporting multiple (in-country) stock rooms. The parts are disbursed to the maintenance technician through direct orders or a work order process. All new (unused) and used parts are returned either to the local stock room (if a new, unused part) or to a used parts disposition site for considered parts repair when the (in-country) demand warrants a repaired part to be "refurbished" to an "as new" state.

Most maintenance technicians' view of parts support and the level of service they receive can be summed up in one sentence – "Do I get the parts I need when I need them?" The maintenance parts management team needs to recognize this as their primary mission, but the business dynamics will be more complex.

If asked in a workshop what the maintenance technician needs from their parts operation, you are likely to hear comments such as:

- Do you have the part in stock? How many do you have?
- Are you out of stock?
- Can I return the part if not used?
- Is it a quality part?
- Is the part repairable?
- Can you kit parts for planned maintenance?
- Can you support emergency parts needs?
- Can you get parts at a competitive cost?
- What is your pick, pack, delivery cycle time, and quality level?

Across an enterprise, the maintenance technician and his department colleagues are not the only business unit interested in successfully executing the maintenance parts operation. The maintenance parts operation may be assigned to serve fully:

- Maintenance technicians
- Parts repair staff

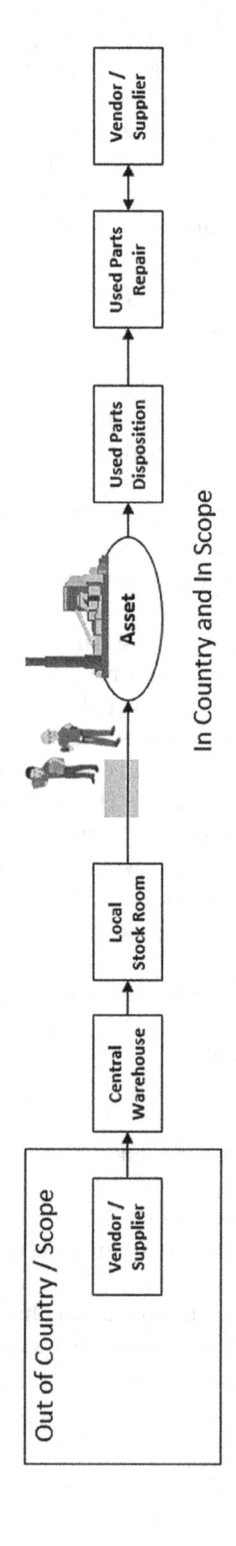


**FIGURE 6.2** Dynamic parts support logistics model. (Adapted from copyright 102 MPE Org and Metrics ©2020 from Maintenance Parts Management Excellence Training edited by Barry. Reproduced by permission of Asset Acumen Consulting Inc.)

- Maintenance inventory staff
- Procurement
- Design engineers
- Accounting/finance
- HR
- Audit
- Health/safety/environment

Local stock rooms working at a significant distance from the central warehouse can depend on the central warehouse and network support. They often do not have full empowerment to stock everything the local maintenance technician requests. The local maintenance parts stockkeeper may be looking for locally influenced indicators of success to support what they believe is their mission (Figure 6.3).

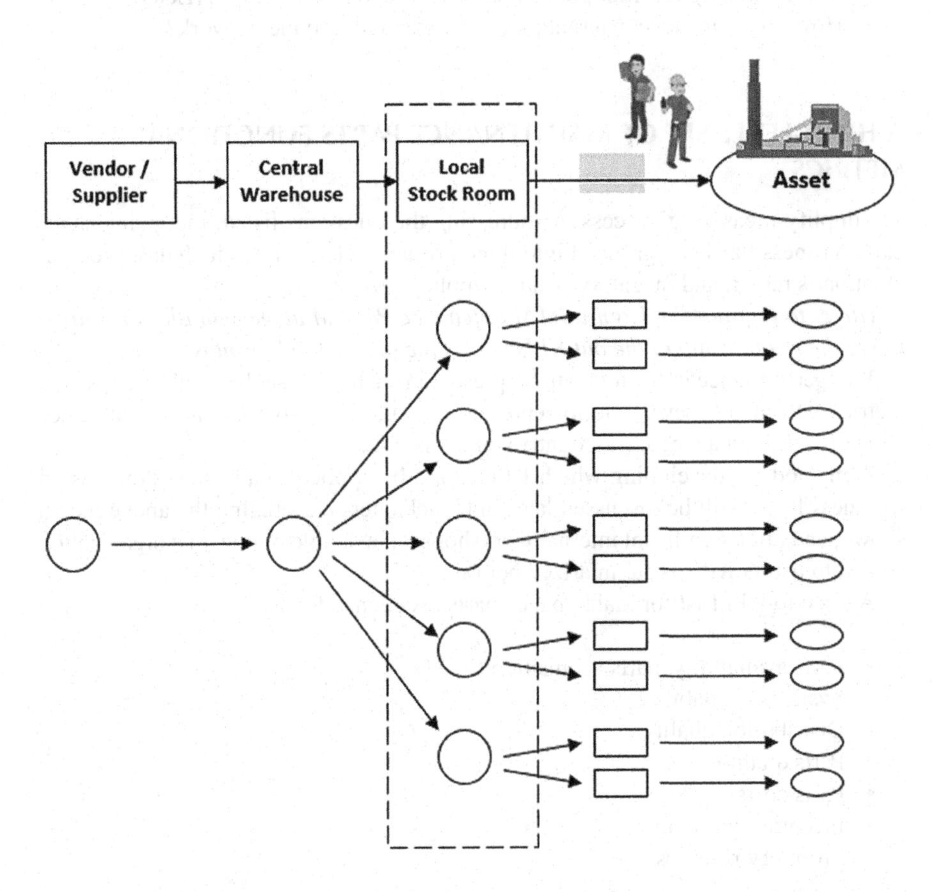


**FIGURE 6.3** Dynamic parts support network model. (Adapted from copyright Figure 6.5 ©2011 from *Asset Management Excellence, Optimizing Equipment Life-cycle Decisions* edited by Campbell, Jardine, McGlynn. Reproduced by permission of Taylor & Francis Group, LLC, a division of Informa plc.)

The local stockkeeper is likely focused on the local needs of each technician. This role will want to understand:

- How do we keep track of local maintenance technician forecasted needs, such as:
  - Planned/unplanned maintenance parts support requests
  - Local maintenance parts service levels
  - Local and regional inventory tracking controls
- How do we manage replenishment across the parts distribution network:
  - Managing quantities by location such as:
  - Stock room on-hand inventory
  - Stock room in-transit inventory
  - Work order reserved inventory
  - Stock room available inventory
  - Automating re-order points and economic order quantities (EOQs)
- How do we manage external parts procurement into the network?

## A HIGH-LEVEL SET OF MAINTENANCE PARTS FUNCTIONAL METRICS

To simplify measuring success in addressing the questions listed, the maintenance parts business has been grouped into three primary sub-groups – logistics, procurement/parts repair, and inventory planning/policy.

*(Refer to Figure 4.2 – High-level Maintenance Parts Management Block Diagram to see the relationship of the out KPIs versus the primary sub-groups).*

Key performance indicators will be present in each of these three sub-groups, and metrics will also be available to represent the success of the overall maintenance parts operation in a single (in-country) business unit.

A method for developing which KPIs could be applied to a business dimension (business level) will be discussed later in this chapter – leveraging the above model shown, and after significant internal workshops, an example of how one organization agreed to list its KPIs is summarized below.

Agreed to KPI List for maintenance parts excellence focus:

- Parts availability, parts acquisition time
- Systems availability
- Distribution quality
- Parts quality
- Parts costs
- Inventory turnover
- Inventory reserves

One or all three sub-groups in the maintenance parts operation (logistics, procurement/parts repair, and inventory planning/policy) can support these metrics. Each of the three sub-groups would have its own KPIs to drive its focus efficiencies.

For example, the "logistics" sub-group key metrics, including the central warehouse and the network of local stock rooms, could focus on distribution quality, measured as the right parts delivered with the correct quantity to the right place within the right place committed time. Such a metric could be received, as feedback, from a maintenance technician process (phone, SMS message, process survey, etc.). Parts quality would also be a metric for the sub-group procurement and parts repair and the sub-group logistics, considering that:

- a Part could be shipped from the supplier "damaged"
- a technician could return a part to stock "as good," but the part could become damaged, or
- a material handler could damage it as part of their day-to-day processes

Expanding the maintenance parts operational KPI grouping as a process makes the potentially high-level supporting KPIs relatively straightforward regarding where they may fit.

For the overall maintenance parts operation, the business metrics may be, as previously stated, parts availability, parts acquisition time, systems availability, distribution quality, parts quality, parts costs, inventory turnover, and inventory reserves (Figure 6.4).

For the Logistics sub-group, the high-level business metrics may be:

- Distribution quality: Right part, right quantity, to the committed place, in the committed time
- Inventory accuracy
- Inventory variances
- Dock to stock cycle time, receiving accuracy

For the procurement and parts repair sub-group, the high-level business metrics may be:

- Procurement:
  - Supplier relationships
  - Parts cost
  - Service level agreements met (SLAs) and vendor performance
- Used parts disposition:
  - Percentage of parts identified as repairable
  - Percentage of repairables successfully returned
  - Percentage of maintenance parts to be fulfilled through a used part repair process
- Parts repair:
  - Yields achieved from parts repair attempts
  - Out of box failure of repaired parts, second Returns within 90 days of disbursement
  - Average cost/repair
  - Credit value back to finance for used parts repaired

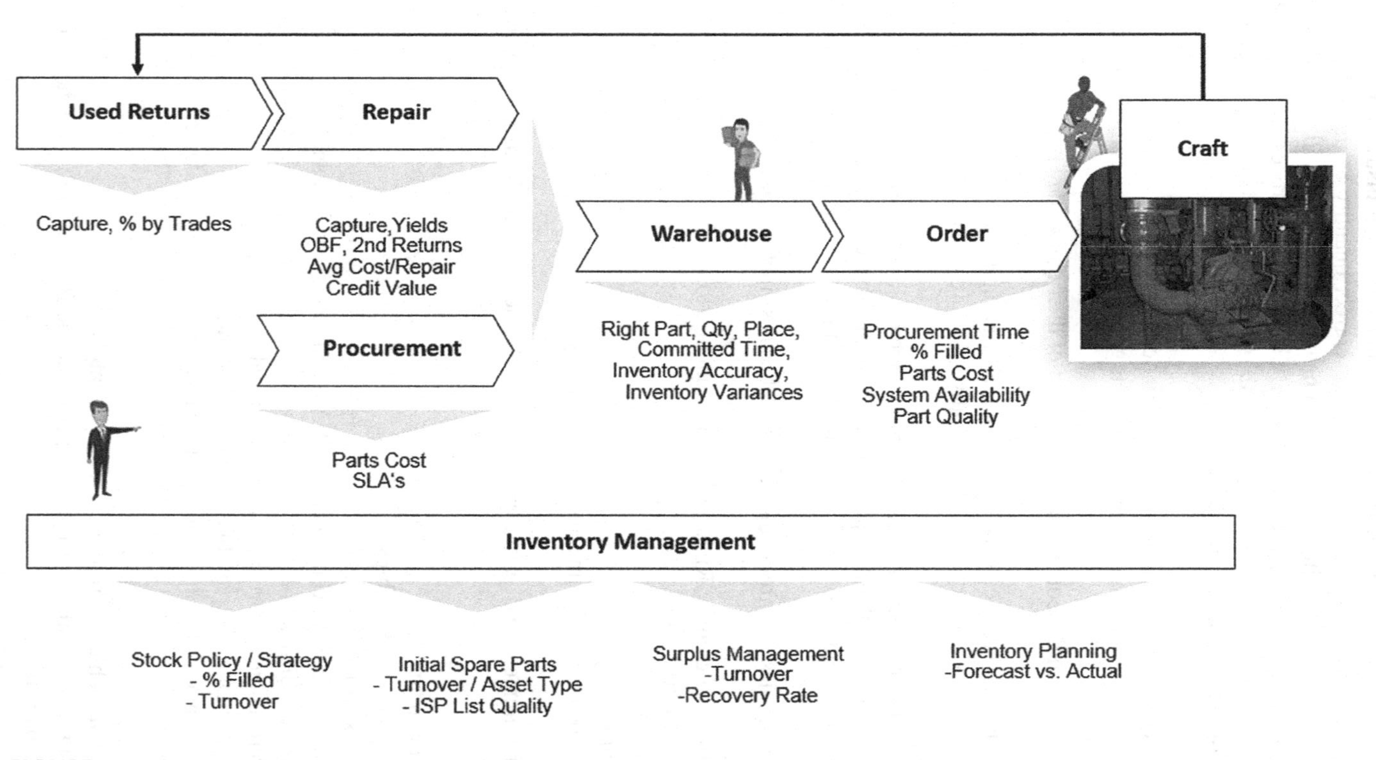


**FIGURE 6.4** Generic maintenance parts flow model (high-level KPIs). (Adapted from copyright Figure 6.4 ©2011 from *Asset Management Excellence, Optimizing Equipment Life-cycle Decisions* edited by Campbell, Jardine, McGlynn. Reproduced by permission of Taylor & Francis Group, LLC, a division of Informa plc.)

For the inventory planning and policy management sub-group, the high-level business metrics may be:

- Stock policy/strategy:
  - Parts availability
  - Inventory turnover
- Initial spare parts – also a metric for the design engineer and maintenance management:
  - Inventory turnover by asset category and type
  - Initial spare parts list quality (also known as recommended spare parts RSP) – also a metric for the design engineer and maintenance management
  - Critical parts list effectiveness
- Inventory planning:
  - Inventory forecast versus actual
  - Stocking policy support (stocked versus not stocked)
- Surplus management:
  - Inventory turnover
  - Surplus recovery rate (cycle time)

When setting out to test your current metrics or starting fresh to pursue maintenance parts excellence, many company stakeholders and environmental and economic influences will need to be considered. The following will need to be considered when developing a new KPI focus:

- Will a particular KPI define a primary value for the business unit (or sub-business unit) it supports?
- Is the performance indicator a primary metric (KPI) or a contributing metric (PI)?
- Will a lagging or leading (an early indication of failure) metric best serve the business and behavior?
- What business units (business dimension level) and subunits need metrics?
- What are the primary values that business dimension provides and the current pain points in that set of processes?
- Should my maintenance parts KPIs only measure the parts business and not the maintenance business?
- Do I have a client value (client facing) metric?
- Can I promote balanced behavior in my primary metrics (balanced scorecard)?

## A BALANCED SCORECARD FOR MAINTENANCE PARTS MANAGEMENT

Confirming that parts operation's business unit metrics are balanced and not one key metric at all costs may be appropriate (Figure 6.5).

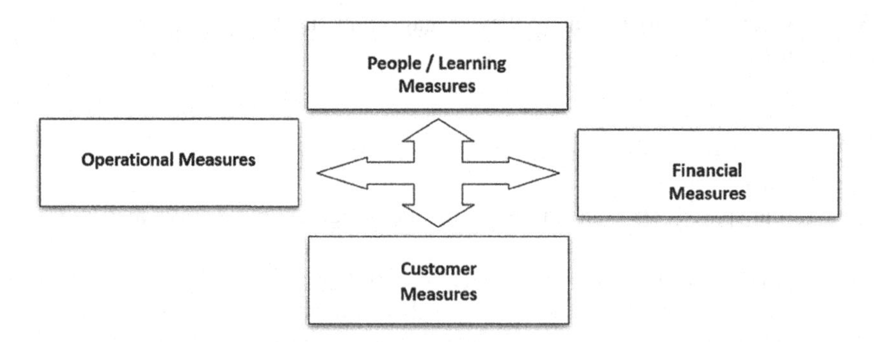


**FIGURE 6.5** Common balanced scorecard elements for a Maintenance Parts Excellence-focused business unit. (Adapted from copyright 102 MPE Org and Metrics ©2020 from Maintenance Parts Management Excellence Training edited by Barry. Reproduced by permission of Asset Acumen Consulting Inc.)

A balanced scorecard approach strives to obtain a balance among these four essential aspects of the business:

- Operational measures
- People/learning measures
- Financial measures
- Customer measures

Many combinations of a balanced scorecard exist, but all will typically contain these high-level elements.

A maintenance parts business unit exists to create or support value. In the case of a maintenance parts operation, its primary customer is maintenance, but it also serves operations, design engineering, and finance, to name a few. "maintenance" may be an appropriate primary choice for many maintenance parts operations to simplify who their primary customer is.

It is also essential to recognize that business priorities will change from time to time depending on the business priorities.

Finance will have specific expectations from the maintenance parts management group, inventory and parts management security stewardship, inventory value management, service levels, and inventory turnover. Typically Finance also works with the maintenance parts organization to forecast inventory provisions for when scrap and other losses need to be predicted in their inventory accounting. At a high level, however, the Finance requirements in a KPI are typically the same as the highest executive requirements for the enterprise, given they both will be looking for asset/inventory capital valuations, profits and losses.

People/learning measures can be established, leveraging enterprise management and HR staff assistance. A few considerations in this space could be:

- Employee sick leave frequency
- Accident rate
- Hours of training per employee per year

## BUSINESS DIMENSION LEVELS AND VALUE DRIVER INFLUENCES ON BUSINESS UNIT KPIS

For maintenance parts excellence targeted operational metrics, the specific business unit level or business dimension needs to be called out. For example, is the KPI for the material handler and the logistics sub-group? Or Is the KPI for the middle management or the director in charge of all maintenance parts management or for the executives that may have maintenance parts management reporting to them? These levels can represent a business dimension even within the same business unit.

The measurements should be specific, performing operational-level elements, and backed up with metrics derived from transactional or surveyed data. The higher up in the business unit, the more subjective the metric may become as a lower business dimension contributes to the higher business dimension's consolidated metrics (Table 6.1).

The strategic level business dimension measures the company's strategies to achieve Return on Asset, business growth, gain market success, and improve competitiveness throughout their supply chain or value stream.

Here it would help if you were looking for a business integrity focus, including:

- Corporate alignment to annual report
- Society/sustainment responsibilities
- Good governance

The operational level manages business as usual (BAU) business operations/processes throughout the value stream. Here it would help if you were looking for a capital and asset integrity focus, including:

- Return on investment/cost-effectiveness
- Asset lifecycle management, parts contributions to overall equipment effectiveness
- Health, safety, environment (HSE) compliance
- Operational alignment

The functional business dimension should support processes that manage functions and support Operational requirements that contribute to the organization's desired

**TABLE 6.1**
**Concept Table Introducing Business Dimensions in a Business Unit**

| Level | Business Dimension | Stakeholder Level |
|---|---|---|
| Subjective metrics | Strategic | Executive |
| Primary process metrics | Operational | Middle management |
| Objective metrics | Functional/process | Supervisor/staff operations |
| Asset-specific dynamics | Asset Specific | Asset functional view |

performance and value, including financial, HSE, operations, and Strategy. Here it would help if you were looking for a Process Integrity Focus, including:

- Operational support, maintenance parts management execution
- Supporting functions (design engineering, maintenance, operations, finance, vendor)
- HSE elements, resource scheduling

The asset-specific business dimension should support asset-specific functional metrics that likely could align with a reliability-centered maintenance (RCM) analysis. Asset-specific (operating context) OEE metrics could be applied here along with asset lifecycle costing metrics and HSE metrics

## KPI DEVELOPMENT PROCESS

Once the business unit dimensions are understood, we can follow a KPI development process to establish KPI options for each dimension, then identify designated value drivers key to the success of Maintenance Parts Excellence.

With the business dimension and value driver matrix built (see Table 6.3 with KPI development business driver matrix), the KPI development team would define/identify and qualify business pain points by matrix section (see Figure 6.6 showing a KPI development flow).

With the business dimensions identified, a workshop can facilitate identifying why each dimension exists. The simple question can be "What values drive the behavior of this business dimension?" Then simply, a list will emerge to be prioritized.

A high-level example list of "value drivers" for maintenance parts excellence could be:

- Process quality
- Productivity
- Costs
- Health, safety, and environment protection

Mapping business dimensions with the new list of value drivers would look like Table 6.2. Some content may become reasonably obvious to populate once the business dimension stakeholders are aligned with value drivers.

To take one value driver (inventory cost-effectiveness) across the three business dimensions could look like Table 6.3:

An example of a partially populated matrix could have alignments with strategy, and distribution quality may be:

- Internal/external customer satisfaction

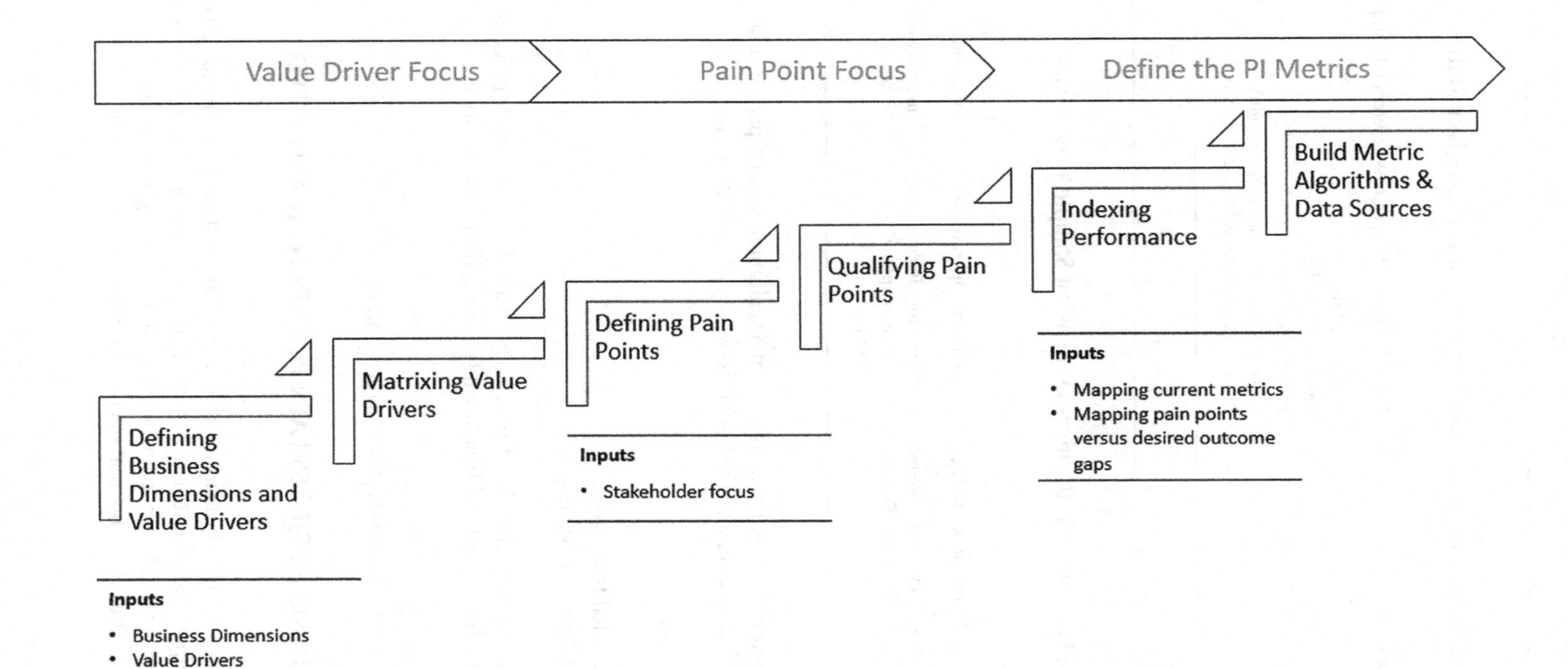


**FIGURE 6.6** KPI development process slow. (Adapted from copyright 102 MPE Org and Metrics ©2020 from Maintenance Parts Management Excellence Training edited by Barry. Reproduced by permission of Asset Acumen Consulting Inc.)

**TABLE 6.2**
**Example of Parts Management KPI Development Business Driver Matrix Options**

| Business Dimensions | Parts Business Drivers Cross-reference Examples |
|---|---|
| Strategy | Parts availability levels |
| Operational Primary | Distribution quality |
| Functional/Sub-processes | Parts cost/quality |
| | Process and systems quality/availability |
| | Inventory effectiveness (by asset, type, or location) |

**TABLE 6.3**
**Inventory Value Effectiveness Business Dimension Sample Versus Value Driver Matrix**

| Value Driver | Dimension: Strategy | Dimension: Operational | Dimension: Functional |
|---|---|---|---|
| Driver: Inventory value effectiveness | • Inventory Turnover<br>• Inventory Reserves | • Inventory Integrity<br>• Inventory policy surplus | • Inventory controls violations |

An example of a partially populated matrix that has alignments with parts availability levels (a value driver) and operational (a business dimension) may be:

- Parts availability
- Inventory integrity
- Percentage filled from stock
- Parts referral responsiveness

An example of a partially populated matrix that has alignments with process and systems quality (value driver) and functional (business dimension) may be:

- Ease of ordering
- Data integrity in the maintenance parts systems

## UNDERSTANDING BUSINESS PAIN POINTS THAT MAY NEED MEASURING

The potential lists of areas to measure would describe current "issues and requirements" placed upon the maintenance parts team to manage. These "issues and requirements" can be declared "pain points." Brainstorming pain points can help determine the essential elements that need to be addressed to achieve maintenance parts excellence.

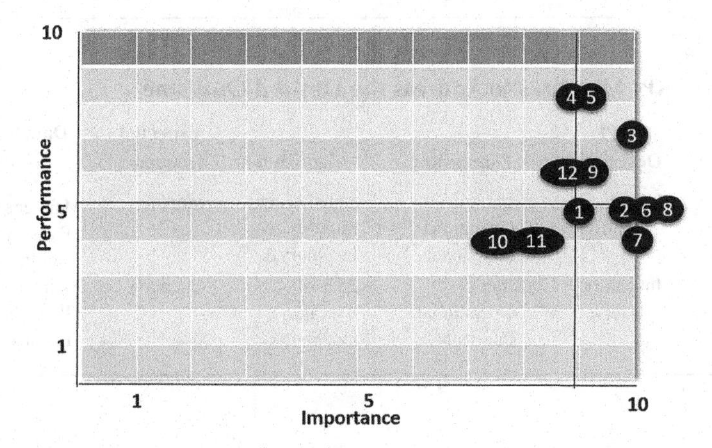


**FIGURE 6.7** Sample importance – Performance grid for pain points. (Copyright MPE ©2022 from Maintenance Parts Management Excellence Training edited by Barry. Reproduced by permission of Asset Acumen Consulting Inc.)

Example pain points may be:

- Poor inventory policy due to static reorder quantities enforcement
- Item master data quality and integrity issues
- Vendor reliability and related service levels and lead times
- New defective parts discovered by the maintenance technician
- Limited planned work leading to parts expediting

Often there could be 40–100 pain points. With these issues identified, the next step is to categorize them. To which business dimension does each pain point align? To which value driver does a pain point align? Note: a pain point could align to multiple business dimensions or value drivers.

Once listed, determine each pain point's impact on the business, perhaps on a scale of 1–10. Then determine how well the current maintenance parts management team performs what is needed to mitigate or minimize the consequence of the failure.

Mapping an Importance to performance grid against the multiple pain points will help prioritize which issues should be addressed early and which could be addressed in a later initiative focus (Figure 6.7).

In the short term, the prioritized focus should be on the most important pain points (perhaps the top 50%) but currently not performing well (bottom 50%). So, the "lower-right" pain points on the shown Importance to Performance grid should be confirmed as short-term priorities. What would be left is approximately 25% of the listed pain points to be addressed by the KPI development team.

## CONVERTING PAIN POINTS TO PERFORMANCE INDICATORS

The pain points should be confirmed as the prioritized ones to focus on and rebrand the pain point as a "positive" statement, not a negative comment. So, for example, a pain point of "poor parts availability" could be rebranded as "parts availability."

**TABLE 6.4**
**What Each KPI Measures to Address the Desired Outcome**

| Desired Outcome | PI Description | Dimension | Algorithm | Expected Frequency | Data Source (System) |
|---|---|---|---|---|---|
| 1.0 Turns | Inventory Turnover | Strategy, Operational, Functional | Annual Usage/ Monthly Avg Inventory $ | Monthly | Maintenance Parts System and Finance |
| <25% of Average Inventory | Inventory Reserves | Strategy, Operational, Functional | <25% of Average Inventory | Quarterly | Maintenance Parts System and Finance |

**TABLE 6.5**
**What Are the Short and Long-term Goals and Timelines for Each KPI**

| Performance Indicator Characteristic | Metric | Long-term Performance Target | Near-term Performance Target (Next 6–12 Months) | Near-term Performance Target (Next Year-end) |
|---|---|---|---|---|
| Inventory Turnover | Annual Usage/ Monthly Avg Inventory $ | 1.0 Turns/ year | 0.7 Turns/year | 0.8 Turns/year |
| Inventory Reserves | <25% of Average Inventory | <25% of Average Inventory | <25% of Average Inventory | <25% of Average Inventory |

You can now focus on the Performance Indicator, its description, algorithm, expected frequency, and data sources with identified priorities (Table 6.4).

The Performance Indicator (PI) agrees and sets short and long-term targeted goals (timelines) for each metric (Table 6.5).

## MANAGING SHORT- AND LONG-TERM GOALS

It is essential to recognize that the enterprise may change its overall business priorities from time to time. It is common to see an enterprise ask the maintenance parts operation to move their dominant KPI value driver focus from parts availability service levels to distribution quality, and later to parts costs and/or inventory cost-effectiveness, to systems quality, and then back to parts availability service levels over a few months or years (Figure 6.8)

For sizeable multinational maintenance parts operations supporting infrastructure across multiple countries, performance will be a high focus, given that even a slight deviation from a target service metric or inventory level could have consequences that could cost the company millions of dollars.

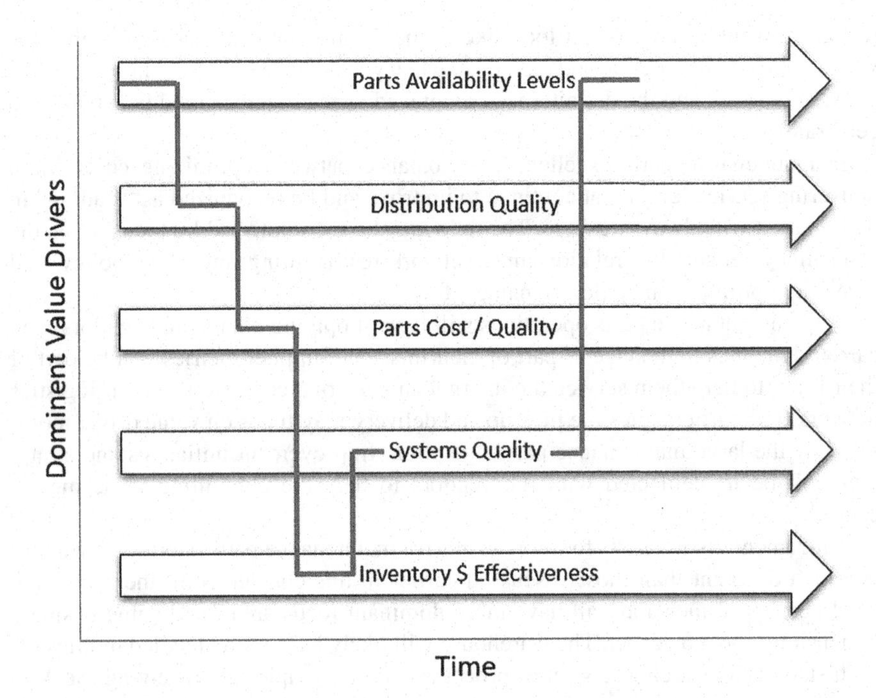


**FIGURE 6.8** The dominant value driver focus may change from time to time in a given enterprise. (Copyright MPE ©2022 from Maintenance Parts Management Excellence Training edited by Barry. Reproduced by permission of Asset Acumen Consulting Inc.)

Successful and more extensive maintenance parts operations will have their value drivers published. They can leverage "total quality management" techniques (TQM) to maintain a sense of empowerment for management and employees across their enterprise. They will declare "short and long-term" goals for their value drivers and measure and reward their management and staff against the achievement of the dominant value driver.

Published short-term goals could include a commitment to:

- A factor of ten times reduction in distribution defects
- A focus on distribution fulfillment cycle time
- Create value driver advocates with support training in quality, transformation, and change enablement techniques across each business dimension level
- Create a cross-functional team (including internal/external customers) to understand better what services are needed as their markets evolve
- Recognize and celebrate success when key targets are met or exceeded

Published long-term goals may be an extension of the short-term goals. They will likely also include a vision for their enterprise's future maintenance parts excellence definition and achievement. Published long-term goals help management and staff recognize what success might look like in the future. Defined success is helpful if you strive for that direction and recognize when you achieve it. If the long-term goals are well defined, and the definition is recognized as met, there should be a reason to celebrate!

In maintenance parts excellence, the balance between optimizing costs while delivering service levels that continue to improve and be recognized as "leading" in their market will always be there. The ratio may be accomplished by focusing on the asset lifecycle, supplier relationships, network restructuring, inventory policy, and inventory optimization tactics, to name a few.

Large maintenance parts operations will have supplying plants (and vendors) and carrier dynamics to manage as part of their mission. Supplier metrics will be part of their focus to help them service the internal and external customers in their logistics network. For carriers, on-time pick up and delivery may focus on value driver monitored by the large maintenance parts operation. Improvement initiatives and strategies will be implemented with the vendors to help the commitments be met or exceeded.

Fundamentally the KPIs for more extensive maintenance parts operations will not be much different than those discussed earlier in this chapter. Still, the functional level business dimension will have more dominant focus areas, and some business dimensions may go deeper. The warehouse will likely have more detailed metrics for each step of the warehouse's set of processes. Two examples are receiving (dock to stock cycle time and receipt accuracy) and parts picking (will also have picking cycle time and pick, pack, and ship accuracy).

Customer satisfaction will be even more critical to the reputation of the large maintenance parts operation looking to achieve excellence. A process will be in place to collect distribution and service quality feedback. A customer satisfaction management flow would be in place to address any complaints or shortfalls to support customer data collection (Figure 6.9).

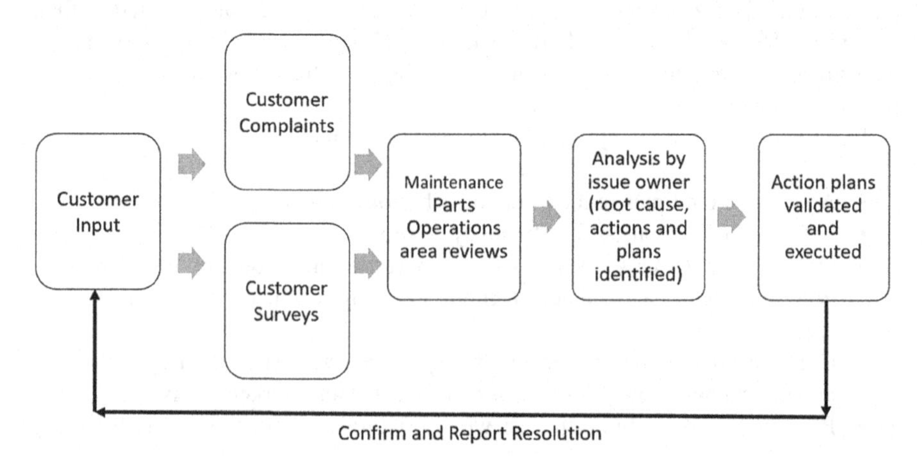


**FIGURE 6.9** Customer Satisfaction Management flow for Maintenance Parts Excellence. (Copyright MPE ©2022 from Maintenance Parts Management Excellence Training edited by Barry. Reproduced by permission of Asset Acumen Consulting Inc.)

## EXAMPLES OF SOME OF THE MAINTENANCE PARTS METRICS

The following tables show some examples of KPIs and PIs that have been used in some leading organizations. The critical issue is that the data needs to be available and trusted for these metrics to influence the stakeholders' behavior.

### KEY PERFORMANCE INDICATORS (KPIs) (TABLE 6.6)

**TABLE 6.6**
**Algorithm Examples of Maintenance Parts KPIs**

| Metric | Equation | Primary Owner(s) |
|---|---|---|
| Parts Availability | $\dfrac{\textit{Total Parts lines fulfilled}}{\textit{Total Parts lines ordered}}$ | Inventory Planner |
| Parts Acquisition Time | $\dfrac{\textit{Asset impact down time waiting for Parts}}{\textit{Total Parts orders}}$ | Maintenance/Inventory Planner |
| Distribution Quality | $\dfrac{\textit{Reported miss in fulfillment quality}}{\textit{Total lines disbursed}}$ | Stock room |
| Systems Availability | $\dfrac{\textit{Total time the Parts systems is available}}{\textit{Total hours in a time period}\left(\textit{month}\right)}$ | IT |
| Parts Quality | $\dfrac{\textit{Total Parts pieces reported as defective}}{\textit{Total Parts pieces disbursed}}$ | Stock room |
| Parts Costs | $\dfrac{\textit{Specific Parts cost from stock}}{\textit{Specific Parts cost from local retail}}$ | Procurement |
| Inventory Turnover | $\dfrac{\textit{Annual Usage}}{\textit{Monthly Average Inventory value}}$ | Finance |
| Inventory Reserves | $\dfrac{\textit{Monthly Average Inventry value}}{\textit{Percent of Acceptable Risk}}$ | Finance |

### STOCK ROOM METRICS (PIs) (TABLE 6.7)

**TABLE 6.7**
**Algorithm Examples of Some High-level Stock Room PIs**

| Metric | Equation | Primary Owner(s) |
|---|---|---|
| Right part, quantity, place within the committed time | $\dfrac{\textit{Reported miss in fulfillment quality}}{\textit{Total lines disbursed}}$ | Stock room |
| Inventory accuracy | $\dfrac{\textit{Location and quantity variances in a stock room}}{\textit{System declared location and item quantity}}$ | Stock room |
| Inventory value variance | $\dfrac{\textit{Value variances to System inventory value}}{\textit{Total value of a stockroom's inventory}}$ | Stock room/Finance |
| Dock to Stock | $\dfrac{\textit{Actual Dock to Stock cycle time}}{\textit{Dock to Stock target cycle time}}$ | Stock room/Receiving |

## Source and Refurbishment (PIs) (Table 6.8)

**TABLE 6.8**
**Algorithm Examples of Some High-level Source and Refurbishment PIs**

| Metric | Equation | Primary Owner(s) |
|---|---|---|
| Part Cost | $\frac{\textit{Specific Parts cost from stock}}{\textit{Specific Parts cost from local retail}}$ | Procurement |
| Vendor Service Level Agreements (SLAs) | $\frac{\textit{SLA Variance}}{\textit{SLA target}}$ | Procurement |
| UPR Parts Returned | $\frac{\textit{Pieces actually received at UPR station}}{\textit{Pieces expected Returned}}$ | Stock room/Finance/ Inventory Planner |
| Percent of UPR parts declared (Capture) | $\frac{\textit{Items formally in UPR designation}}{\textit{Number of Items with UPR potential}}$ | Maintenance/ Procurement |
| Parts Refurbishment Yields | $\frac{\textit{Pieces successfully refurbished as new}}{\textit{Pieces entering refurbishment process}}$ | Procurement |
| Second Returns from refurbished parts | $\frac{\textit{Refurbished pieces returned as defective with 90 days}}{\textit{Pieces successfully refurbished as new}}$ | Procurement |
| The average cost of parts refurbished | $\frac{\textit{Total Refurbishment cost by item}}{\textit{Pieces successfully refurbished}}$ | Procurement |
| Repair value credit to parts expense ratio | $\frac{\textit{Average cost of repair}}{\textit{Cost of a new purchased part}(\textit{item})}$ | Procurement/Finance/ Maintenance |

## Inventory Planning (PIs) (Table 6.9)

**TABLE 6.9**
**Algorithm Examples of Some High-level Inventory Planning PIs**

| Metric | Equation | Primary Owner(s) |
|---|---|---|
| Parts Availability | $\frac{\textit{Total Parts lines fulfilled}}{\textit{Total Parts lines ordered}}$ | Inventory Planner |
| Inventory Turnover | $\frac{\textit{Annual Usage}}{\textit{Monthly Average Inventory value}}$ | Finance/Inventory Planner |
| Initial Spare Parts (ISP)/ (RSP) Policy cycle time | $\frac{\textit{Missed ISPs by Commisionning}}{\textit{ISP in place for Commissioning}}$ | Design Engineering/ Inventory Planner |
| ISP Quality | $\frac{\textit{ISP pieces accepted into Inventory Policy}}{\textit{Number of Items potentially used in asset lifecycle}}$ | Design Engineering/ Finance/Inventory Planner |
| Inventory Planning Forecasting | $\frac{\textit{Inventory value forecast}}{\textit{Actual inventory value}}$ | Inventory Planner/ Finance |
| Second Returns from refurbished parts | $\frac{\textit{Refurbished pieces returned as defective with 90 days}}{\textit{Pieces successfully refurbished as new}}$ | Procurement |
| The average cost of parts refurbished | $\frac{\textit{Total Refurbishment cost by item}}{\textit{Pieces successfully refurbished}}$ | Procurement |
| Repair value credit to parts expense ratio | $\frac{\textit{Average cost of repair}}{\textit{Cost of a new purchased part}(\textit{item})}$ | Procurement/Finance/ Maintenance |

# 7 Maintenance Parts Warehousing and Logistics

## INTRODUCTION TO THE MAINTENANCE PARTS WAREHOUSE FLOW

As suggested in Chapter 3, maintenance parts management and a traditional manufacturing organization's supply chain will look quite different. Maintenance parts management exists to serve and support the operational asset's return on investment. Therefore, the maintenance parts must be close to the asset when called upon to mitigate the risk attributed to its functional failure.

This chapter will discuss the role of the maintenance parts warehouse – how parts would typically flow, some of the mechanical processes and organizational elements to consider, and a more defined set of metrics to consider for success.

A warehouse should be designed to pick, pack, and ship a part with the most efficient cycle time at the highest level. If parts are issued to the maintenance technician in an over-the-counter scenario, the most active parts should be stocked relatively close to the counter. However, maintenance parts warehouses may have multiple ways to disburse parts from its stewardship. Moving working (functional) parts out of a warehouse may include orders for:

- Emergency/urgent orders (unplanned maintenance)
- Planned maintenance or overhauls
- Supplying satellite stock rooms or maintenance technician vehicle stock
- Shipping surplus parts back to a supplier or a new owner
- Shipping new defective parts back to the supplier
- Scrapping parts (Figure 7.1)

With the intention of a maintenance parts operation to serve the installed operational Asset base, the warehouse parts flow should align with this critical timeline expectation.

DOI: 10.1201/9781003344674-7

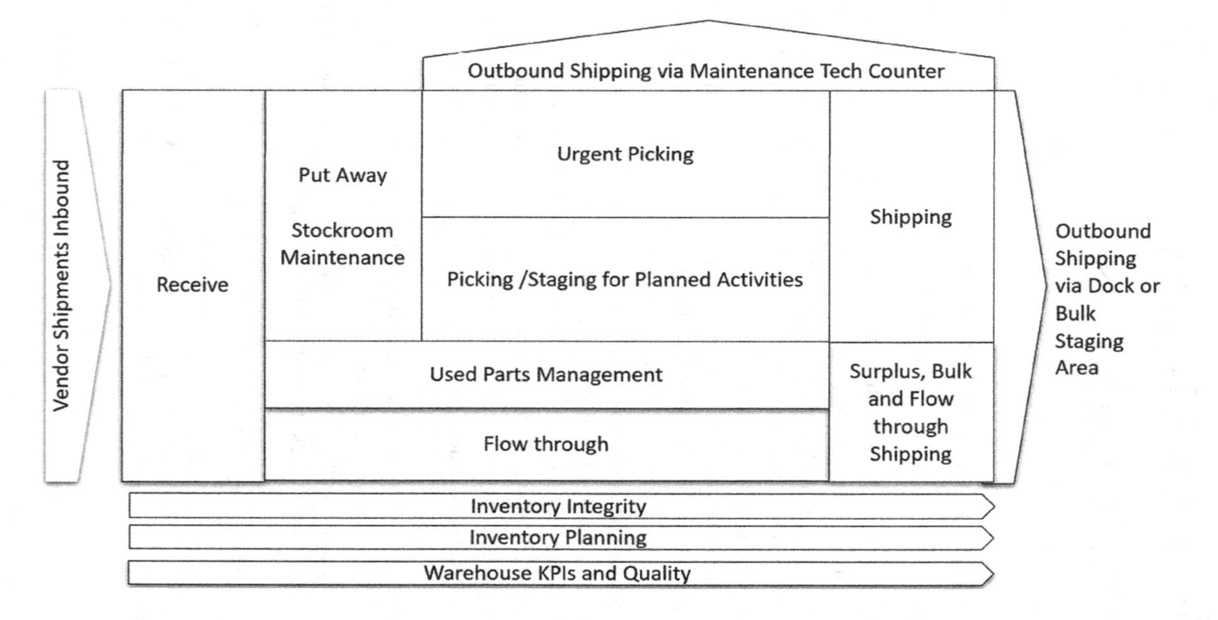


**FIGURE 7.1** High-level maintenance parts warehouse flow model. (Adapted from copyright 103 MPE Warehousing ©2020 from Maintenance Parts Management Excellence Training edited by Barry. Reproduced by permission of Asset Acumen Consulting Inc.)

## WAREHOUSE SETUP CONSIDERATIONS

Many factors influence an optimal layout for a maintenance parts warehouse. All of them are interrelated. The optimal layout factors can include:

- Service level functional requirements (why the warehouse is in place and who it serves)
- Material flow (how should parts flow through the warehouse to provide the required service levels)
- Parts physical stocking dynamics (what are the dynamics of the current parts stocked and expected future activities)
- Equipment & facility constraints (what physical challenges do we need to work with to leverage existing facility resources to provide the warehouse service levels)
- Security and environmental requirements (what are the specific security and environmental requirements for our location and operating schedule) (Figure 7.2)

A systematic approach should be established to collect these requirement factors to ensure the warehouse can deliver the desired service levels efficiently and effectively. An optimal warehouse setup will:

- Ensure sufficient space is allocated for expected inbound and outbound receiving, shipping, and related staging
- Minimize potential congestion in aisles and multiple touches by a material handler to facilitate a transaction

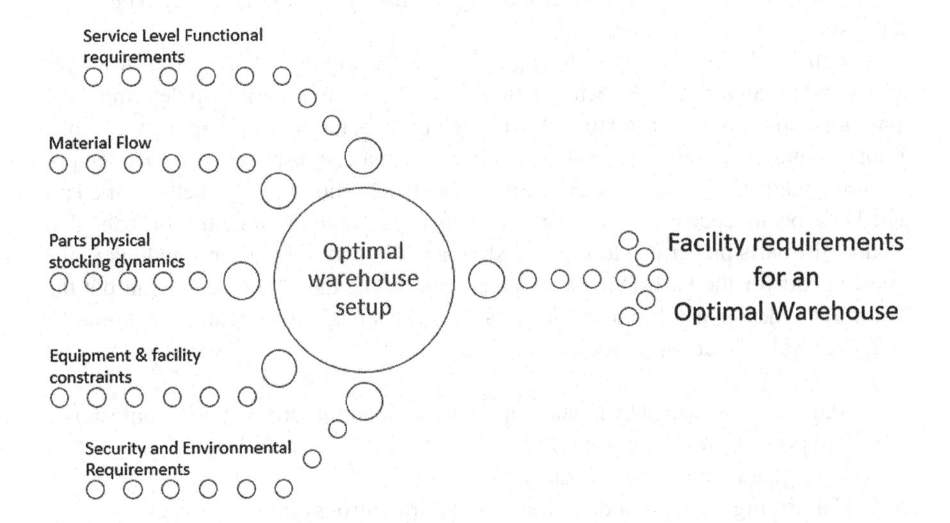


**FIGURE 7.2** A high-level Maintenance Parts warehouse influence model. (Adapted from copyright 103 MPE Warehousing ©2020 from Maintenance Parts Management Excellence Training edited by Barry. Reproduced by permission of Asset Acumen Consulting Inc.)

- Create efficient material flows between functional areas within the warehouse operations to minimize time-consuming bottlenecks or long distances traveled
- Contribute to parts location controls to ensure enough space exists to support the expected stock quantities
- Contribute to minimal damage of materials being put away into stock
- Maintain a disciplined approach to assigning parts that align with the characteristic dimension of each Part and the expected stocking quantity (e.g., tiny Parts may most effectively fit in a high-density stock drawer, while large amounts of medium-sized parts may better fit in a bulk rack that can be handpicked ergonomically for safety and speed)
- Include the use of conveyor systems if you expect volume demands and "pick" types (e.g., end-of-day replenishment to satellite storerooms) to warrant it
- Provide practical staging areas for filling kitted orders while aligning with daily work order schedules

Starting with service levels, a clear definition of what is expected will help the warehouse align to deliver on these expectations. For example, potential parts order demand types in a maintenance parts warehouse may be demonstrated as some of the activities listed in Table 7.1.

The warehouse layout should account for the apparent needs for things such as:

- shipping and receiving areas (such as a dock and the expected aisle space) to comply with safety regulations
- physical shelving, storage equipment

It also needs to account for the work and staging space to fulfill the expected demands at the service levels committed.

A high-level Material Flow diagram is shown in Figure 7.1. Material flow processes and volume rates between functional warehouse areas will help determine the space allocations and their relationship to each process to provide at least a workable, if not optimal, layout. For example, parts received may be inspected and then temporarily staged in a cart designated to be put away in a section of gray shelves. The Part will have been received in the parts system to show as a "quantity-on-hand." If needed, the parts picker can look to the shelf and bin and, if not seen, can look in the received cart for the Part. During a quieter time in the day, the receiver can put the Part into the designated bin to complete a "dock to stock" process and cycle time.

The complete receiving process may include:

- Staging in receiving to validate shipment and quantities to shipping manifests)
- Parts inspection (if appropriate)
- Parts quarantined (if applicable)
- Confirming the receipt quantities into their control system
- Labeling each Part with a stocking identification number
- Staging the received Parts into a cart for putting away to stock later
- Finally put away to stock

**TABLE 7.1**
**Partial List of Warehouse Transactions and Expected Service Levels**

| Stakeholder Served | Expected Service Level | Warehouse Dynamic |
|---|---|---|
| Maintenance technician | 15 minutes- pick, pack, and ship for urgent orders | Highly active parts stocked close to the counter |
| Maintenance technician | Pick up at the warehouse counter | |
| Maintenance technician | Orders due "next day" are kitted and staged | |
| Maintenance technician | New parts returns processed within 24 hours of receipt | Requires processing area for new parts returns |
| Maintenance technician | Used parts returns processed within 48 hours of receipt | Requires staging area for used parts processing and storage |
| Remote maintenance technician | 15 minutes- pick, pack and ship for urgent orders | Courier service available – as needed – close to the counter |
| Satellite stock room | 15 minutes- pick, pack and ship for urgent orders | |
| Satellite stock room | The next day, orders were picked, packed, and shipped by the end of the day for overnight delivery | Overnight courier may be used – shipped via the dock |
| Satellite stock room | Replenishment orders are picked, packed, and shipped bi-weekly by the end of the day for overnight delivery | Overnight courier may be used – shipped via the dock |
| Satellite maintenance vehicle | Overnight replenishment or parts used the previous day | Requires a second or third shift in place to facilitate order picking and packing or vehicle put away |
| Inventory returns to vendor | Picked on an as-required basis new defective or surplus returns) | It may need to be staged for consolidated shipping (e.g., monthly) |
| Surplus scrapping | Surplus parts scrapping perhaps two times a year | Requires a staging area to hold picked "scrapped" parts until approved by Finance to ship for disposition |

This set of tasks is the first opportunity to correct the "in-stock" quantities and adequately locate them in the right place. The receiver may also identify parts that were ordered directly to a maintenance technician's order (work order) and pass it through to the Maintenance Technician (with the help of the picker if appropriate). Used Parts received would be placed on pallets based on the predetermined disposition of a specific part or "item" number.

The picking process may include:

- Accepting the order for the maintenance technician (either over the counter or preferably through a mutual system)
- Printing the pick list, picking the part (s)
- Packing the part(s)
- Shipping the order through a "third party" courier, local van, or the part counter

Optionally, the parts order could be staged for a specific work order if designated when ordered.

The preparation of a simple material flow diagram (as shown in Figure 7.1) illustrates and quantifies the material flows between functional areas within the warehouse. It will assist in developing (options) alternative block layouts within the warehouse structure and plan. Aligning the warehouse design with expected volumes will help confirm the staffing plan needed to deliver a well-executed warehouse operation.

The Maintenance Part's physical stocking requirements need to be understood and captured as a component of the system item master data to support planning and execution insights for proper handling and storage of each Part. For example, the stocking characteristic can include requirements such as:

- Temperature, humidity, or prolonged sunlight exposure constraints
- Special shipping requirements (fragile)
- High-value designations that required additional tracking or security
- Serialized tracking or simple traceability process requirements
- Weight and cube dimensions
- Workplace Hazardous Material Information System designation (WHMIS) requirements
- Shelf-life controls
- Substitutions (or same as designations)
- "Where used" (primary equipment designations), to name just a few (Figure 7.3)

If available and reliable, this data can help the warehouse planner analyze where a part should best be stocked when combined with recent disbursement history (e.g., the last year or six months).

**FIGURE 7.3** Environment can have a significant impact on the value of the Part stocked. (Copyright MPE ©2022 from Maintenance Parts Management Excellence Training edited by Barry. Reproduced by permission of Asset Acumen Consulting Inc.)

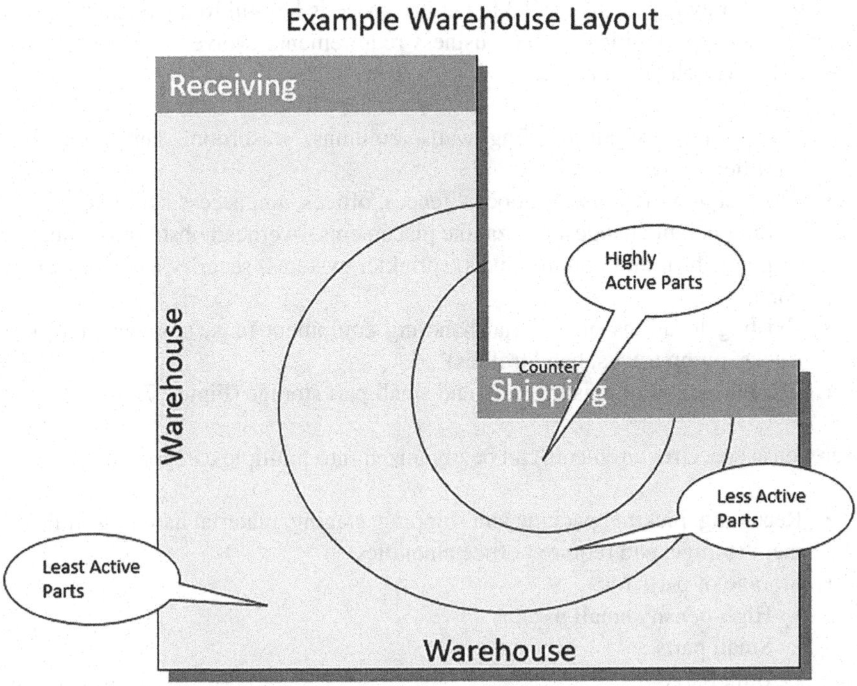

Example Warehouse Layout

Receiving
Highly Active Parts
Counter
Shipping
Warehouse
Less Active Parts
Least Active Parts
Warehouse

**FIGURE 7.4** High level of an existing parts warehouse, the outbound counter, and radar format for placement of parts by activity. (Adapted from copyright 103 MPE Warehousing ©2020 from Maintenance Parts Management Excellence Training edited by Barry. Reproduced by permission of Asset Acumen Consulting Inc.)

Performing a "Bin Trip Analysis" of disbursements from the past year can reinforce where parts should be stocked in relation to the most likely outbound location of the warehouse. For example, in Figure 7.4, a large "L"-shaped warehouse facility located its maintenance technician counter in the inside center of the "L." With the parts counter located centrally, all the parts clustered around this most likely outbound warehouse location. The warehouse was set up with a radar format for parts placement by historical activity. An analysis of "bin trip activity" can be used to support and complement the warehouse mission to deliver time-sensitive order fulfillment. The most active parts recently ordered are stocked closer to the shipping counter/dock. This ensures:

- Faster picking cycle time
- Higher service levels
- Less work for the stock personnel when responding to immediate needs

The parts analysis of "bin trip activity" can be completed through a simple parts disbursement history Pareto list.

A scaled drawing of the available facility space is helpful to appreciate the constraints (and opportunities) as the business requirements evolve. Therefore, a base facility drawing should include:

- The location of all building walls, columns, washroom, and eyewash facilities
- The location of doorways, docks, fences, offices, and access controls
- Ceiling height changes, mezzanine placements, overhead obstructions such as pipes, heat and cooling ducts, sprinkler systems, security cameras, and zones
- Existing locations of material handling equipment (e.g., storage for fork trucks, pump trucks, hand trollies)
- The placement of existing bulk and small part storage (Figure 7.5)

Warehouse space requirements can be organized into multiple categories:

- Receiving, picking, packing and shipping, staging, material handling equipment storage, and required office amenities
- Storage of parts for:
    - High-density small parts
    - Small parts
    - Bulk parts
    - High-density bulk parts
    - Uniquely shaped parts (e.g., pipes)

Gray Shelf Small Parts Storage

Bulk Rack Large Parts Storage

High Density Small Parts Storage

High Density Large Parts Storage

**FIGURE 7.5** Basic warehouse storage options. (Copyright MPE ©2022 from Maintenance Parts Management Excellence Training edited by Barry. Reproduced by permission of Asset Acumen Consulting Inc.)

- Utilizing the higher unused facility space available with solutions like:
  - Mezzanine to leverage available warehouse heights for slow-moving parts
  - Carousels for leveraging warehouse heights and automation assistance in the put-away and picking processes
- Considerations of environment for:
  - The actual environment (e.g., temperature, humidity, access to sunlight, airflow) (Figure 7.6)

Typically, a maintenance parts warehouse's most significant required space is not for storage of maintenance parts. Workable aisle space is needed for each section of shelving used. Manual pick "gray" shelving is typically used for smaller parts storage and ideally has 1 meter (3 feet) in each shelf facing aisle. High-density cabinet drawer shelving for tiny parts would work best with 1.5 meters of aisle space. These cabinets require the least space to pick a part with the drawer fully open and move the cabinet with a pump truck, if necessary, later. Bulk racking needs 3 meters of space (or more) across each shelf access point to allow the style/type of fork truck used in your warehouse to safely and comfortably place or remove pallets. High-density racking will need to factor in the design of the racking (loading from the same side or opposite sides), and space for staging pallets are not required but are in the way of the pallet you are looking to pick (Figure 7.7).

Acceptable aisle widths and overall storage heights significantly impact the required warehouse space. Space considerations must be allocated to the receiving, shipping, and staging areas if parts orders are "kitted" for planned maintenance or turnaround activities. Considerations would also need to complement the warehouse storage and their aisles. Kitting materials to support planned maintenance or turnaround activities

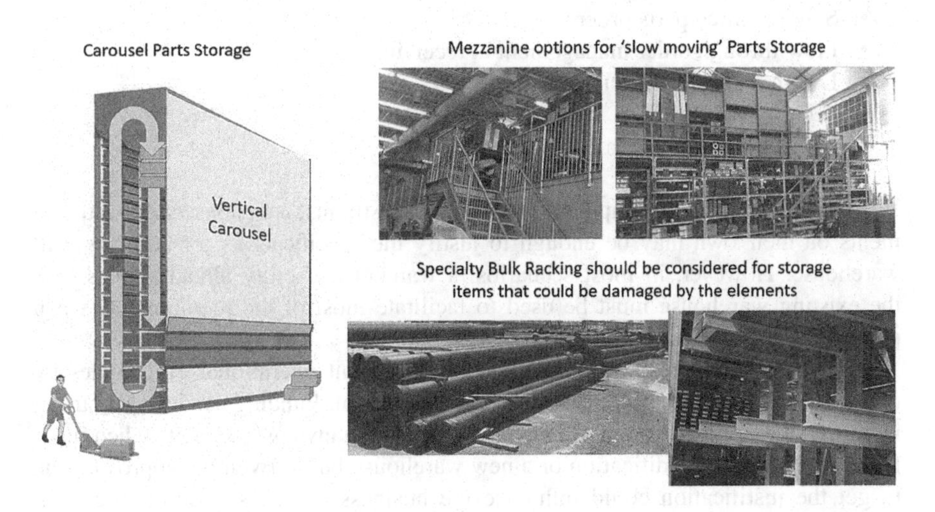


**FIGURE 7.6** Additional warehouse storage options. (Copyright MPE ©2022 from Maintenance Parts Management Excellence Training edited by Barry. Reproduced by permission of Asset Acumen Consulting Inc.)

**Aisle spacing for gray shelves should be 1 meter**

**Fork trucks should be able to safely maneuver**

**FIGURE 7.7** Model for considering aisle space in warehouse requirements sizing. (Copyright MPE ©2022 from Maintenance Parts Management Excellence Training edited by Barry. Reproduced by permission of Asset Acumen Consulting Inc.)

can significantly reduce the time a maintenance technician or crew needs to wait for parts. Trash requirements should also be allocated for space, material handler efficiencies, and enterprise recycling commitments.

## BUILDING WAREHOUSE SPACE AND EQUIPMENT REQUIREMENTS

When working through your warehouse flow and space requirements, make sure you address how you will meet the needs for:

- Bin trip analysis (as described earlier)
- Picking slips
- Staging kitted parts orders
- Flow-through order management at receiving
- Health and safety disciplines
- Security (24/7)
- Environment management (heat, humidity)

When working through equipment and facility constraints, warehouse space requirements on their own may be enough to justify the specifications for a newly built warehouse. However, in most scenarios, a warehouse facility already exists, and the existing warehouse must be used to facilitate most of the requirements when practical.

Many warehouse operations can create significant operational efficiencies by optimizing their current facility capital and equipment. Funding can be a constraint in getting the optimal warehouse design. There are many cost elements when justifying a warehouse modification or a new warehouse build. Even the approval time to get the justification could influence the business case. As mentioned earlier, drawing the requirements will help confirm the starting point. The existing warehouse layout and known Parts dimensions will help to influence and affect the expected space requirements for the planned changes. A review of the current design in hand will demonstrate how it may or may not meet the future parts stocking requirement options. These options can be worked through to see how requirements can be achieved with the existing facility or how you can modify the existing facility to maximize the service levels while waiting for the new extension or build (Figure 7.8).

An approach to estimating the stocking requirements of an existing warehouse can be to look at how you handle the current mix of stocking requirements and then forecast the expected differences (Figure 7.9).

The equipment considerations should include documenting the requirements for:

Receiving Equipment (Table 7.2):

Flow-through, parts put away, new and used parts processing equipment (Table 7.3):

New and used parts returns processing equipment (Table 7.4):

Parts picking packing and shipping equipment (Table 7.5):

A processing insight that can help move parts return receipts faster from the dock or the counter is to use colored labels filled out by the maintenance technician to

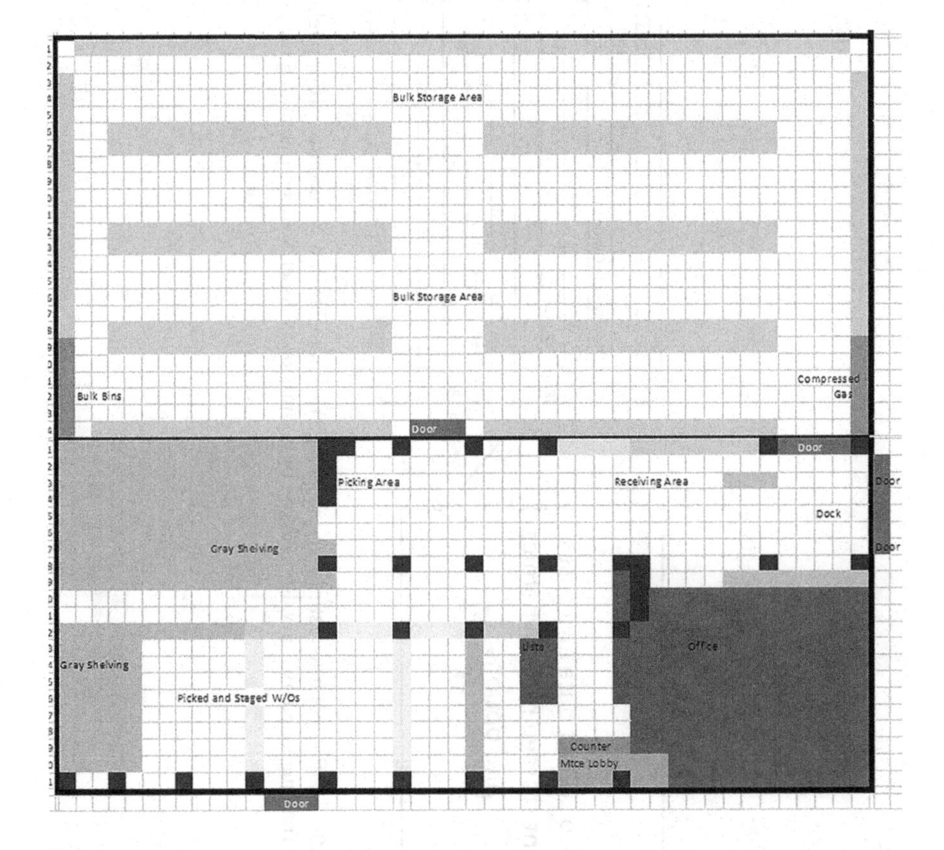


**FIGURE 7.8** Example of a simple mapping of a current parts warehouse facility. (Adapted from copyright 103 MPE Warehousing ©2020 from Maintenance Parts Management Excellence Training edited by Barry. Reproduced by permission of Asset Acumen Consulting Inc.)

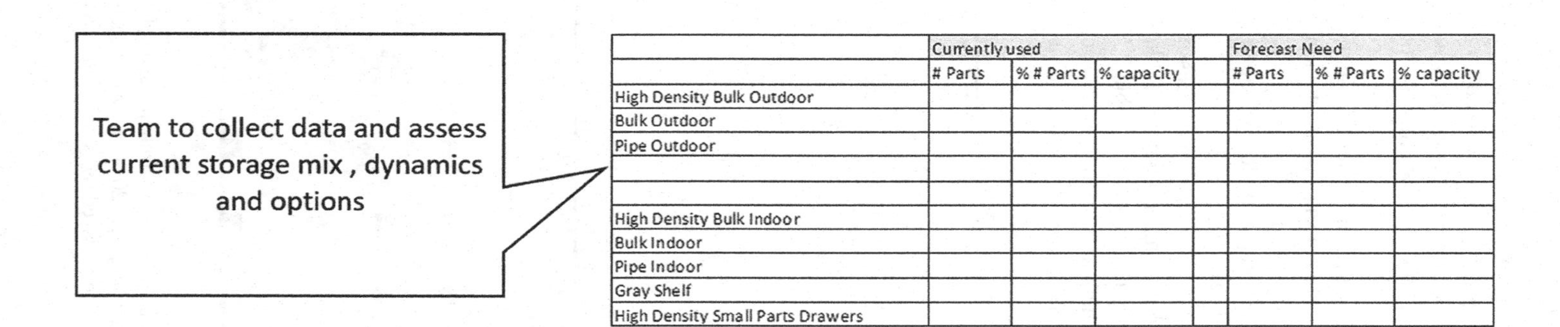


| | Currently used | | | | Forecast Need | | |
|---|---|---|---|---|---|---|---|
| | # Parts | % # Parts | % capacity | | # Parts | % # Parts | % capacity |
| High Density Bulk Outdoor | | | | | | | |
| Bulk Outdoor | | | | | | | |
| Pipe Outdoor | | | | | | | |
| | | | | | | | |
| | | | | | | | |
| High Density Bulk Indoor | | | | | | | |
| Bulk Indoor | | | | | | | |
| Pipe Indoor | | | | | | | |
| Gray Shelf | | | | | | | |
| High Density Small Parts Drawers | | | | | | | |

**FIGURE 7.9** Starting point for estimating the storage requirements of a warehouse. (Copyright MPE ©2022 from Maintenance Parts Management Excellence Training edited by Barry. Reproduced by permission of Asset Acumen Consulting Inc.)

**TABLE 7.2**
**Equipment Considerations for the Receiving Area**

| Location | Equipment Consideration |
|---|---|
| Docks | • Scissor lifts to trucks<br>• Inside lights for trucks to load/unload via fork truck or pump truck<br>• Wheel chalks |
| Receiving staging area to validate bulk receipts | • Bulk racking for receipt staging<br>• Access to high-density bulk storage for fast-moving chemicals<br>• Access to "put away" staging area |
| Fork trucks | • Fork truck stations<br>• Eyewash for batteries at fork trucks station |
| Environment | • Good airflow, not air-conditioned (separated from an environmentally controlled warehouse)<br>• Water sprinklers and lighting suitable for a warehouse |
| Other | • Computer to add receipts, phone, desk, WIFI capable<br>• A place for drivers to sit, a washroom with appropriate security for drivers |

Note – Understanding the receiving cycles (daily/weekly) is needed to estimate dock size and workflow dynamics (for example, if they receive only in the morning and ship in the afternoon).

**TABLE 7.3**
**Equipment Considerations for the Parts Put Away and Flow-Through Areas**

| Location | Equipment Consideration |
|---|---|
| Staging area<br>Access to Fork trucks as needed | • From receiving<br>• Leverage staging shelves (on wheels) for smaller parts<br>• Direct to stock for larger and pallet loads of Parts<br>• Work area to receive/validate parts identification, quantity, quality<br>• Make inventory adjustments as appropriate<br>• Work with the parts picking team to confirm the work orders parts they are expected to support<br>• Notify the work order owner that parts have arrived and are ready for pick-up/shipping<br>• Leverage MSDS sheet storage<br>• Environment<br>• Good airflow, humidity-controlled / air-conditioned – part of an environmentally controlled warehouse |
| Other | • Computer to add receipts, phone, desk, WIFI capable |

Note – Understanding receiving cycles (daily/weekly) is needed to understand the best "flow-through."

**TABLE 7.4**
**Equipment Considerations for the New and Used Parts Processing Areas**

| Location | Equipment Consideration |
| --- | --- |
| Staging area<br>Access to Fork trucks as needed | • From parts receiving<br>• Leverage staging shelves (on wheels) for smaller parts<br>• Direct to stock for larger and pallet loads of Parts<br>• Work area to receive/validate parts identification, quantity, quality<br>• Make inventory adjustments as appropriate<br>• Work with the parts picking team to confirm the work orders parts they are expected to support<br>• Notify the work order owner that parts have arrived and are ready for pick-up/shipping<br>• Leverage MSDS sheet storage<br>• Environment<br>• Good airflow, humidity-controlled / air-conditioned – part of an environmentally controlled warehouse |
| Others | • Computer to add receipts, phone, desk, WIFI capable |

**TABLE 7.5**
**Equipment Considerations for the Pick, Packing, and Shipping Areas**

| Location | Equipment Consideration |
| --- | --- |
| Counter for urgent orders to be disbursed to the maintenance craft | • Depending on the package size and courier, urgent shipping can be out the same docks as receiving or over the counter |
| Order area | • Bulk staging area for fork trucks<br>• Shelving for staging/consolidating parts that were picked as an urgent order<br>• Need an area to package and – if required – weigh potential shipments<br>• Work area to validate parts identification, quantity, quality, and pack and create shipping labels<br>• Make inventory adjustments as appropriate<br>• MSDS sheet storage<br>• Prepare shipping packing slips |
| Environment | • Good airflow, humidity-controlled/air-conditioned – part of an environmentally controlled warehouse |
| Others | • Computer to add receipts, phone, desk, WIFI capable |

Note – Urgent orders are expected to be processed in 15 minutes and shipped or picked up within the same "2–4-hour" period from when ordered.

identify how it should flow and be quickly identified and processed. In the example shown in Figure 7.10, an orange label tells the receiver that this is a used part, and a gray label tells the receiver this is a part that should be scrapped. Finally, a green label informs the receiver that the part returned is considered new and should be returned to stock for full credit.

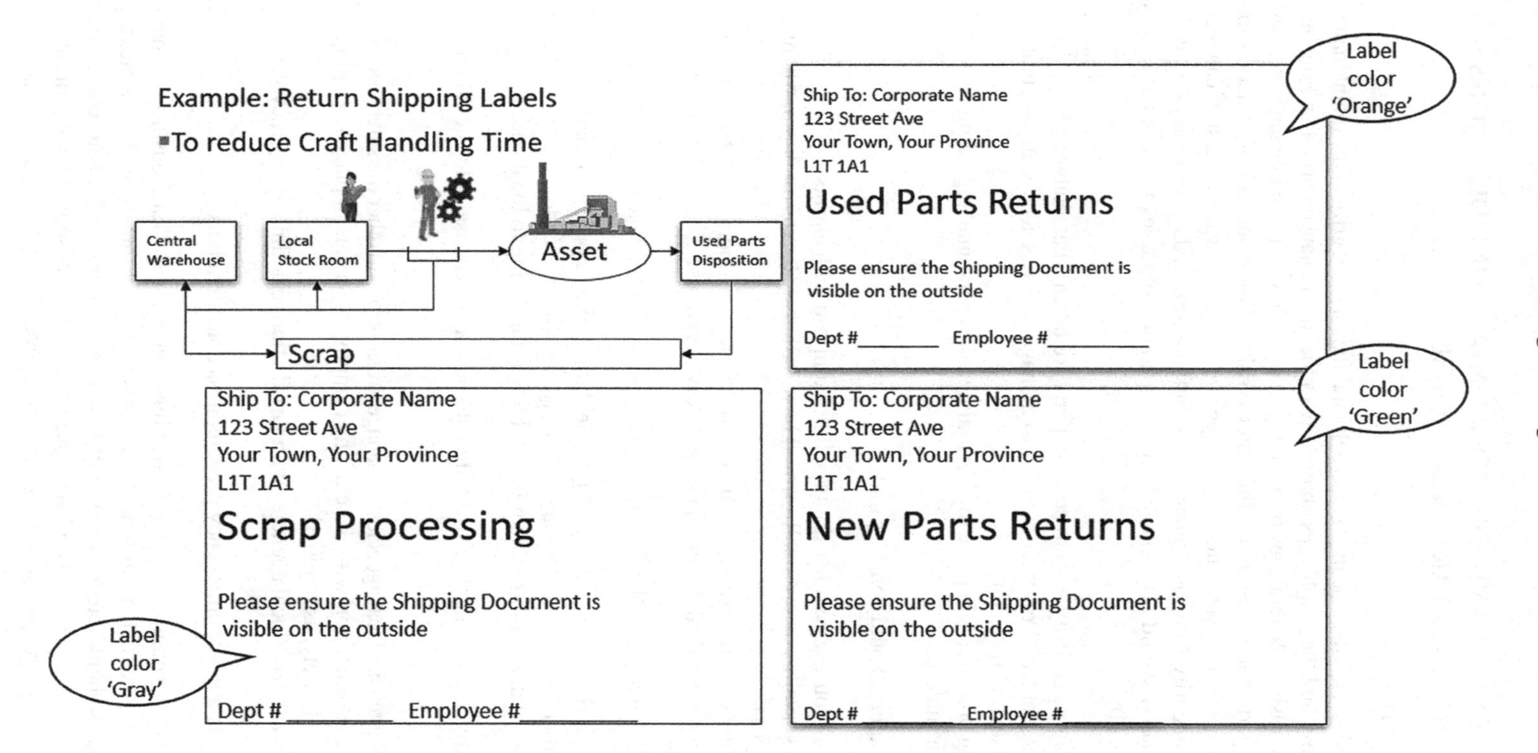


**FIGURE 7.10** Example return shipping labels that expedited parts returns sortation and processing at the Receiving dock. (Adapted from copyright 103 MPE Warehousing ©2020 from Maintenance Parts Management Excellence Training edited by Barry. Reproduced by permission of Asset Acumen Consulting Inc.)

## SECURITY, SAFETY, AND ENVIRONMENTAL CONSIDERATIONS IN THE MAINTENANCE PARTS WAREHOUSE

### Security and Access

A security culture needs to be in place to help maintain the integrity of the maintenance parts inventory and the employees charged with the maintenance parts' handling and storage stewardship. Warehouses provide high service levels to maintenance personnel and the parts controlled under their stewardship. Warehouse security should be managed to secure the type of item stored appropriately. The warehouse staff can then manage the integrity deterioration or risk of theft 24 hours a day, seven days a week.

Consideration should be in place to have separation of duties for off-hour access to warehouses by:

- Leveraging security staff to confirm parts are taken (if available)
- Leveraging RFID or other controls to manage who has access and accurate disbursement recording
- Having procedures in place to alert maintenance personnel regarding receipt of their materials

Warehouse safety should consider that:

- All warehouse access fire doors will be equipped with alarms
- Video surveillance optionally would be installed at all exterior doors, where appropriate
- A security system (e.g., security card readers at doors) will be used to control access to all external doors of the warehouse

Fire prevention strategies should include:

- Water sprinkler systems through the bulk pallet systems, otherwise as needed above gray shelves and offices as according to code
- Fire extinguishers throughout the building (confirming to local code)

To track, support, and manage life Safety Issues, a warehouse delegate should be assigned to:

- Coordinate with local agencies to manage fire control and other life safety issues
- Ensure coordination with the existing campus systems is in place regarding personnel safety, as required
- Coordinate with local agencies with jurisdiction regarding emergency vehicle access
- Facilitate a list of emergency preparedness considerations

Safety should be paramount in every working environment. Warehousing will have its own set of potential hazards and concerns. An environment with tall racking and fork trucks should create sensitivity to everyone's surroundings to ensure that movements are executed safely. Compressed gas containers should be appropriately stored, and the culture should be in place to reinforce "safety first" in every action.

**FIGURE 7.11** Example warehouse safety hazards – Compressed gas cylinders not chained and bulk racks, not floor bolted. (Adapted from copyright 103 MPE Warehousing ©2020 from Maintenance Parts Management Excellence Training edited by Barry. Reproduced by permission of Asset Acumen Consulting Inc.)

- Should bulk racks be bolted to the floor when fork trucks can be present?
- Should gas tanks be secured from falling and protected from warehouse equipment?
- What are the chemical safety and WHMIS considerations for material handling and shipping? (Figure 7.11)

## MAINTENANCE PARTS WAREHOUSE TASKS

A maintenance parts warehouse set of tasks will follow the material flow previously shown.

Parts need to be received (Table 7.6):

**TABLE 7.6**
**Example of Typical Parts Warehouse Tasks**

| Typical Task Frequency | Task Description |
|---|---|
| Daily | • Fill all urgent orders when received<br>• Refer urgent orders that cannot be fulfilled<br>• Fill kitting/staged orders as required<br>• Look for previous day's errors<br>• Look for stock exceptions<br>• Look for non-filled orders (must refer, substitute, or cancel all part numbers (items) with a non "Available" fill code)<br>• All stock location changes must be filed with the daily report<br>• All stock receipts (replenishment packing slips) must be filed with the daily report<br>All inventory adjustments approved by management |
| Weekly | • Stock room integrity check:<br>• Verify count<br>• Stock location variance:<br>• Verify location<br>• Action "stock actions" – (down-level parts, scrap actions, part modifications, etc.)<br>• Return surplus inventories to the central site |
| Monthly | • Stratified Sampling inventory audits (or annual inventory audits) |

## ORGANIZATIONAL REQUIREMENTS

The maintenance parts warehouse organizational requirements must be understood and aligned with the maintenance parts warehouse flow model. The flow will differ, depending on the size of the warehouse, the number of parts stocked, and the volume of parts transactions (receipts, pick, and ships) that happen on any given day or week. Organizational requirements could be:

- Shift coverage required:
  - 24 hours a day and seven days a week
  - Prime business hours – Monday to Friday
- Dock activity volume:
  - multiple daily receipts or
  - receipts two days a week
- Receiving activity volume:
  - 100 a day or
  - 15–20 shipments a week

- Stock – Put away:
  - 250 lines a day or
  - 250 lines/month
- Stock picking and prepared for pick-up:
  - 15,000 items/month Perhaps 6,000 lines/month, 300 lines/day or
  - 1,500 items/month Perhaps 600 lines/month, 30 lines/day
- Counter service orders:
  - 15,000 items/month Perhaps 6,000 lines/month, 300 lines/day or
  - 1,500 items/month Perhaps 600 lines/month, 30 lines/day
- Shipping (including exports):
  - Multiple shipments a day or
  - 1–2 shipments a week
- New parts returns:
  - 150 new parts return lines/month, or
  - 15 new parts return lines/month
- Used parts management (if applicable):
  - 10–200 months or
  - 1–2 months
- Scrap:
  - Monthly or
  - Twice a year
- Inventory cycle counts:
  - Avg 900 counts/day or
  - Avg 90 counts/day
- KPIs:
  - Reported daily/weekly or
  - Reported monthly/quarterly

Depending on the size and the disbursement volume of the warehouse, it may be managed by 1–10 employees or an entire staff of more than a headcount of 100. However, the primary roles are the same regardless of the size and volume.

## INVENTORY INTEGRITY MANAGEMENT IN THE MAINTENANCE PARTS WAREHOUSE

Traditional organizations annually "count everything in a weekend to achieve inventory control service levels." As a result, they then adjust stock balances and values. Unfortunately, this method is inefficient, often inaccurate, and usually under strict time and resource constraints.

Exception inventory validation counts happen when the stockkeeper notices that the "on-shelf" quantity does not match what the system suggests should be there.

Many inventory "audit count and control" methods could be applied in a maintenance parts Warehouse. Two approaches to consider are:

- The complete count of inventory once a year, leveraging:
    - Stock identity validated against the description, and changes noted, reducing duplicates, increasing parts recognition efficiency, and helping standardize the catalog
    - Confirmed stock locations
    - Inventory accuracy levels (dollar value and quantity variances) can act as a performance measure
- Perpetual stock count or cycle counting is a better alternative. You set up numerous counts, typically less than 200 items, to be counted regularly. Then, all inventory (at least A and B class) is counted at least once over a year. The benefits of cycle counting are:
    - More accurate counts, as the number of items is relatively small and variances can be easily researched
    - Stock identity validated against the description and changes noted, reducing duplicates, increasing parts recognition efficiency, and helping standardize the catalog
    - Confirmed stock locations
    - Inventory accuracy levels (dollar value and quantity variances) can act as a performance measure

Many finance organizations have accepted the perpetual stock count process as a stratified sampling approach that counts the dollar value of the parts multiple times in a year but likely not every stocked item in a given year. From a financial control perspective, this is considered better control. Still, it does not help to find all missing parts and, therefore, may contribute to an urgent "out-of-stock" situation later in the life of the warehouse.

## LEVERAGING THE 5S DISCIPLINE IN MAINTENANCE PARTS WAREHOUSING

The 5S discipline came from Japan's Total Quality Management technique and is modified to fit the English language. The key concept behind 5S is to arrange items or activities in a manner that promotes workflow, such as bin trip analysis, tools kept at the point of use, work set in an order such that material handlers should not have to bend to access materials repetitively, and workflow paths to be modified to improve efficiency. The benefits of such an approach are in having:

- A tidy workspace
- Assigning every practical tool a location
- Eliminating the opportunity to waste time looking for things

This approach in the warehouse will empower the staff and help boost morale. As a team in the warehouse to determine:

- *what* should be kept
- *where* it should be retained
- *how* it should be stored (Table 7.7)

**TABLE 7.7**
**5S Description Table**

| Original Term | English Term | Description |
|---|---|---|
| Seiri | Sorting | Review all the tools, materials, etc., in the plant and work area and keep only essential items. Everything else is stored or discarded. |
| Seiton | Straighten or Set in Order | Focus on efficiency. When translated, this means to "Straighten or Set in Order." *It is not focused on sorting, cleaning, or sweeping (the following "S").* Instead, it promotes arranging the tools, equipment, and parts to promote workflow. Tools and equipment should be kept close to where they will be used (e.g., to optimize the workflow path), and the process should be set in an order that maximizes efficiency. A bin trip analysis could be an example of this activity. |
| Seisō | Sweeping or Shining: Systematic Cleaning | Promotes keeping areas neat and clean, particularly at the end of a shift. For each work area, everything is restored to its place. This step makes it easy to know what goes where and enables confidence in where everything should be. Tidiness and cleanliness should be part of the daily work culture |
| Seiketsu | Standardizing | Standardized work practices or operating in a consistent and standardized fashion.<br>This original term could also be thought to mean sanitizing or cleaning. Everyone knows what their specific responsibilities are to keep above 3Ss. |
| Shitsuke | Sustaining | Standards should be met and reviewed. Creating a culture that supports the first 4S steps should set the organization for success. Now you need to sustain it and measure it to sustain it. People do what you "inspect" more than what you "expect," the new way to operate. Reviewing and confirming that a new culture standard is met is helpful to ensure success. |

Note that a sixth phase, "Safety," is often added to the 5S discussions. However, purists may argue that adding safety is unnecessary since following "5S" will result in a safe work environment.

## AUTOMATION AND TECHNOLOGY IN MAINTENANCE PARTS WAREHOUSING

A new set of warehouse requirements should be designed based on current and future needs.

Some insights to help speed up parts orders and handling can include:

- Mobile devices, where the orders are placed through a mobile device by the maintenance technician and printed at the parts warehouse
- Mobile devices used by the material handler to support transactions such as:
  - Physical counts
  - Issues and returns processing
  - Inspection
  - Transfers and receiving

Barcoding can work well if the parts and shelves are already labeled. However, labeling costs (if not in place) need to be justified against manual parts verification process costs. The order picker confirms the location, item number, and parts description against each picking list as part of their picking discipline. Often the picking accuracy is not materially better when a bar code reader is being used versus being compared to a disciplined manually parts picking process.

The benefits of warehouse material storage and handling automation may not outweigh the risks and costs. Automation likely does not provide a sizable financial benefit if the warehouse (or stock room) is relatively small and has low volumes. However, the labor and space-savings benefit of an Automated Storage and Retrieval System (ASRS) can be significant for extensive warehouse operations for bulk storage and smaller parts storage such as a carousel.

ASRS solutions in a maintenance parts warehouse can risk:

- The ability to pick quickly
- Backup if power or support IT system is not available
- IT automation requirements

The capital costs associated with automated equipment and the inherent reduction in operational flexibility have generally resulted in "carousels" relegation to large and complex distribution centers. These systems must be fully cost-justified, employing operating and throughput service improvements before considering them over the conventional manual pick methodology. Carousels will create savings and offer risks if an urgent part request and the support system is unavailable due to missing support systems or power.

# 8 Sourcing and Refurbishing Parts

## INTRODUCTION TO PROCUREMENT

Procurement can consist of many support activities in an interconnected enterprise. Procurement will have relevant attributes, buyer motivations, vendor attributes, and processes specific to supporting good procurement controls.

Today, each connected employee can leverage the enterprise's suite of work tools for procurement, whether at the office, working from home, or in the field with a mobile device. Technology can create simplified and automated purchase transactions with the oversight of the enterprise procurement team and transaction approvals.

Specifically, to support maintenance parts management and related maintenance service outsourcing, procurement will have a direct role in supplier selection, supplier relationship management, and transaction facilitation.

Once the maintenance parts stocking or replenishment requirements are understood, strategic commodity sourcing, supplier relationship management, and transaction execution are standard business responsibilities for a leading practice procurement organization (Figure 8.1).

From an organizational perspective, procurement could report to finance or operations or be part of the maintenance parts management organization.

Within the maintenance parts management group, the inventory planner typically helps understand what commodities are needed to be acquired and when for inventory stores. maintenance or operations may identify their needs and create the purchase request if a part, commodity, or service is needed to be applied directly (e.g., to a maintenance work order) without going through the maintenance parts management set of processes.

Procurement will help refine these requirements. For example, strategic sourcing will require a better understanding of the needed commodities. This set of processes will analyze past and planned spend in these areas. Procurement will assess the market dynamics for these commodities, coordinate the request for information (RFI) and request for proposal (RFP) activities, and develop a suitable commodities strategy for the organization. Supplier relationship management would empower procurement to engage and manage selected suppliers, contract specifications, pricing, compliance, and assess supplier performance to establish service-level agreements. Leveraging the internet, most companies and personnel can execute transaction management via an automated interface between the two companies. They leverage business to business (B2B) transactions and allow procurement to focus more on the first

DOI: 10.1201/9781003344674-8

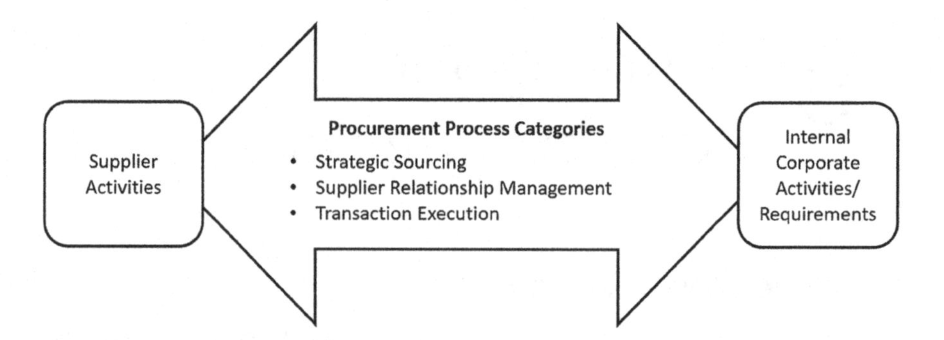


**FIGURE 8.1** High-level procurement process responsibility model. (Adapted from copyright 104 MPE Inventory Management ©2020 from Maintenance Parts Management Excellence Training edited by Barry. Reproduced by permission of Asset Acumen Consulting Inc.)

two procurement activities (strategic sourcing and supplier relationship management). Being connected through the internet helps an enterprise manage parts catalogs, purchase orders, order requisitions, order confirmations, and receipt confirmations. The procurement group can have less of a hands-on role in the transaction, and the maintenance parts management folks can complete the transaction cycle with receipt validation.

## THE PROCUREMENT STEPS OF A PARTS REQUEST

A high-level parts order flow may look as shown in the following figure. A simple order for materials may bring the purchase request (PR) through a process to confirm that the material is not already in stock and available or even in surplus. The original requester must ensure that the requested materials are set up as an "item" in their enterprise system database. The system should be able to display items that are already defined with its related attribute data, such as:

- Specifications
- Alignment to the equipment maintained
- Preferred vendors identified with inspection requirement documents (for the receiving group to perform)
- Product, supplier, and vendor's performance service-level agreements and evaluation criteria documented
- Order approval processes
- Invoice approval processes (Figure 8.2)

*A simple purchase request for materials to support a maintenance task requires a cross-functional group to work their process promptly for the order to be executed well.*

If a process step falters or a member of the process team fails to execute well, the original requestor suffers the results of inferior supplier support, poor vendor

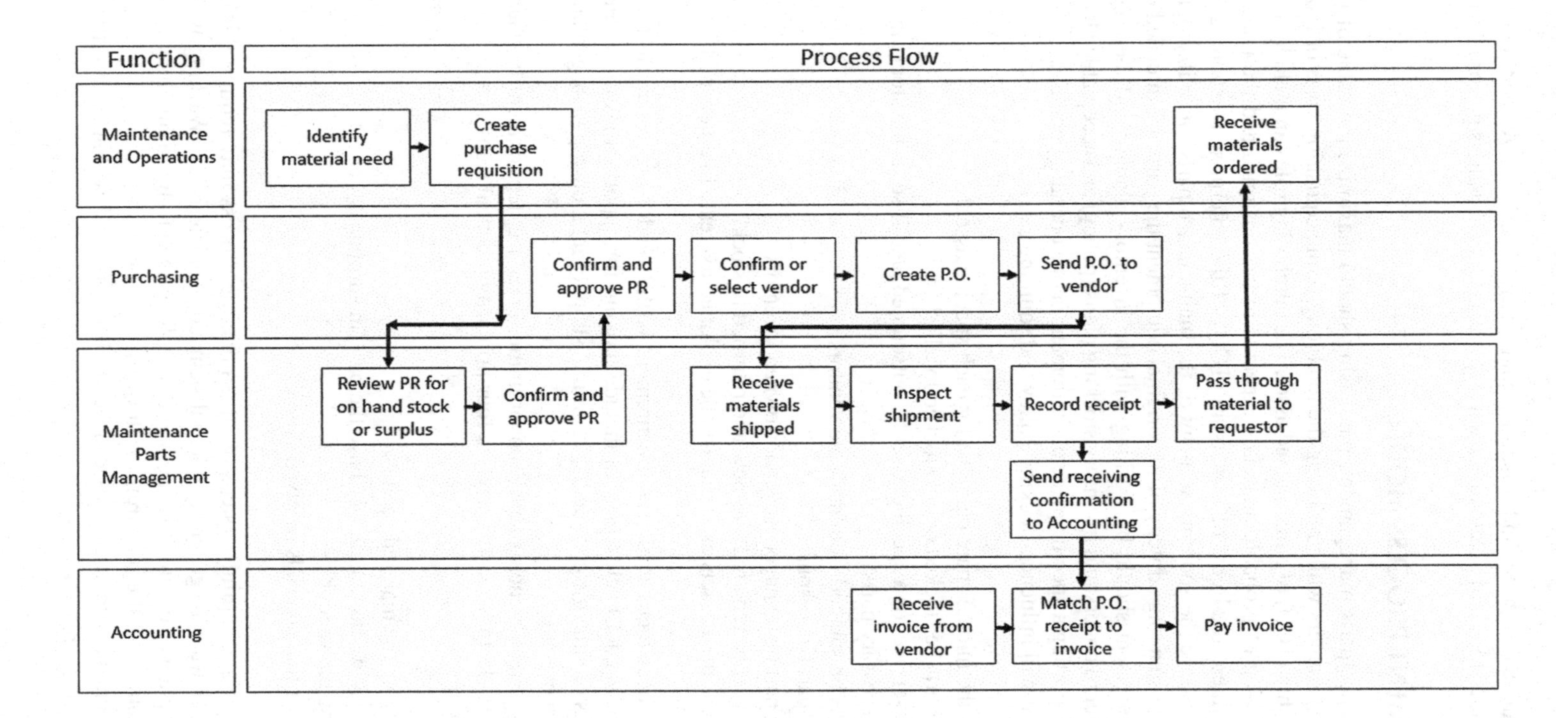


**FIGURE 8.2** High-level procurement support process diagram for a technician purchase request (PR). (Adapted from copyright 104 MPE Inventory Management ©2020 from Maintenance Parts Management Excellence Training edited by Barry. Reproduced by permission of Asset Acumen Consulting Inc.)

perceptions, and a drawn-out and frustrating purchasing experience. A good process that is well understood and executed at each step creates good business performance.

## PROCUREMENT PROCESS MODEL

Like maintenance parts management, purchasing should maximize the value of its service to its customers while reducing the total long-term acquisition costs or total cost of ownership (TCO) of a maintained asset. The purchasing role can include supply chain thinking of probable materials expected to be consumed and the related consumption rates. Total cost of ownership (TCO) of the enterprise assets (equipment) will influence the sourcing strategy of the maintenance parts. Still, it should also control the sourcing strategy of the original equipment the parts are intended to support. Procurement should be working with the inventory planner and the asset design engineer to understand inventory stocking plans (policy) and acceptable alternatives for supporting a specific piece of equipment in one operating context.

Insights that will influence stocking policy and sourcing can be:

- What is the initiate spare parts list for each specific asset?
- Should parts be stocked or not stocked locally?
- Can the service for a specific asset be outsourced, and if so, is the stocking of the necessary parts?
- Can parts be sourced from multiple vendors?
- Can the part be co-sourced?
- Can the part be managed as vendor-owned inventory?
- Can the part be repaired – as new and returned to stock?
- Do surplus used assets exist that can be exploited to create like-new parts?

One of the critical focus areas for procurement is the vendor process cycle time (service level). Vendor cycle time can include looking for opportunities to execute across all areas of strategic commodity sourcing, supplier relationship management, and transaction execution.

Process mapping may help identify opportunities to reduce steps or identify where the step bottlenecks may exist. There are four primarily practical steps in a procurement transaction:

- Recognizing the material need
- Confirming we do not already have the part (in surplus)
- Notifying the vendor
- Paying for the materials received

Looking for ways to minimize steps or streamline the process may require thinking about better ways to generate material demand characteristics. Working with the inventory planner on stocking policy and replenishment algorithms may help optimize and automate the procurement flow.

The procurement responsibilities can be further expanded within each critical scope category (Figure 8.3).

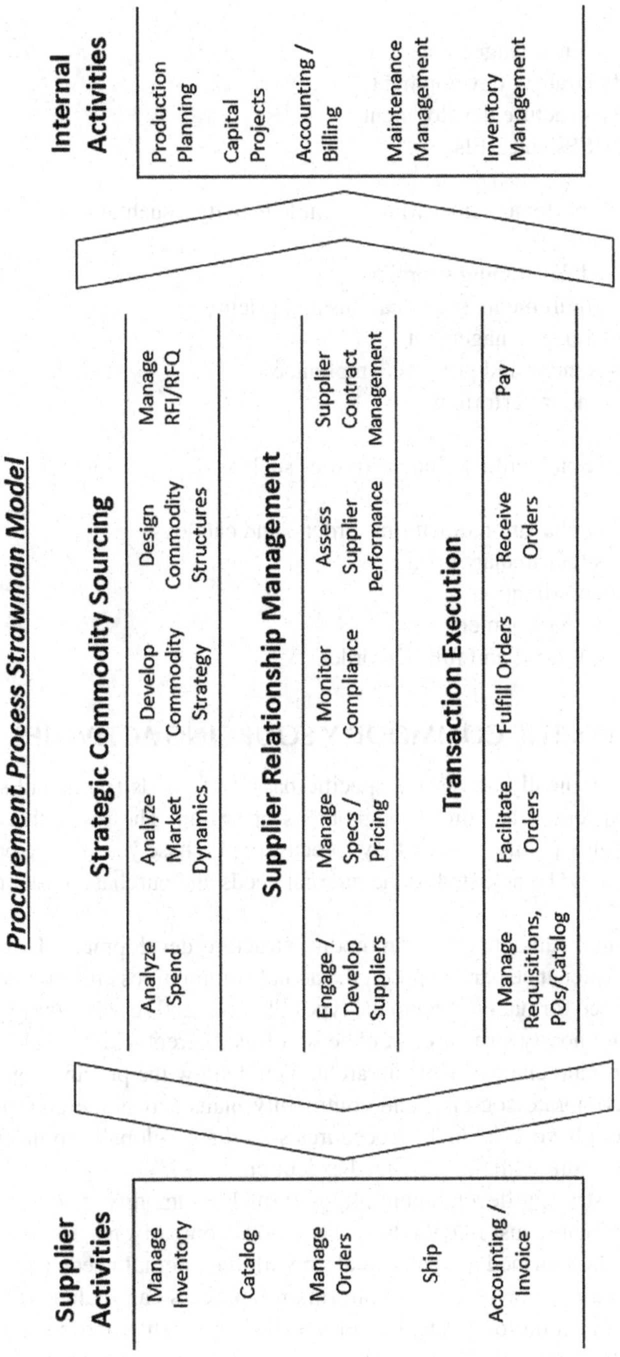

**FIGURE 8.3** Procurement process strawman model. (Adapted from copyright 104 MPE Inventory Management ©2020 from Maintenance Parts Management Excellence Training edited by Barry. Reproduced by permission of Asset Acumen Consulting Inc.)

Strategic Commodity Sourcing would include activities such as:

- Spend analysis
- Market dynamic engagement
- Commodity strategy development
- Commodity structure development
- Managing RFPs and RFIs

Supply Relationship Management would include activities such as:

- Engaging and developing suppliers
- Managing requirements specifications and pricing
- Supplier contract management
- Monitoring vendor and process compliance
- Assessing vendor performance

Transaction Execution would include activities such as:

- Managing purchasing requisitions, orders, and catalogs
- Facilitating order management
- Fulfillment coordination
- Order receipt management
- Paying for approved and fulfilled orders

## EXPANDING ON THE COMMODITY SOURCING ACTIVITIES

A spend analysis typically assesses a specific basket of goods (parts and services) the enterprise expects to acquire throughout its operation. The research compares vendors' pricing and quality to what they forecast to purchase yearly. The basket set group of goods would be adjusted as the internal needs and purchasing requirements evolve.

There are formal standards for commodity structure development. However, in the context of maintenance parts support, the actual commodities created may follow a formal or localized format tailored to the types of goods and services they primarily purchase. The commodity structure could be set up as a parent–child relationship for crucial assets for maintenance. This hierarchy would allow the procurement organization to share corporate sourcing and commodity plans across the enterprise. All departments would have a defined "procedures set" shared globally to unify to one consolidated procedure with decentralized execution.

A commodity strategy development aligns formal commodity definitions to the enterprise requirements, just like the basket of goods comparisons. The commodity strategy should be confirmed annually to align with the internal needs and purchasing requirements as they evolve. The confirmation process can include developing and documenting negotiation strategies for various commodity and contract types. Commodity strategies should be supported by, and communicated to, the customer and aligned with other procurement strategies (specifically the supplier strategies).

Commodity strategies need to define when contract benefits should be confirmed, and these contract types should be clear, concise, and intuitive. Upon negotiating contracts, the buyer needs to know the commodity requirements and align with contract types such as consignment, cost plus, and blanket.

As the business evolves, new assets are used to deliver enterprise value. As the vendors to support these assets change, a market may emerge to support new and existing assets. Procurement is in place to understand how the supplier market evolves. Procurement engages in opportunities that will bring:

- Better investments (assets) and products to the enterprise
- Better-valued services to the enterprise
- Competitive solutions to support the existing assets in the enterprise

Procurement may see an opportunity to establish a direct new product buy procedure. New items that have never been acquired before, whether inventory or non-inventory items, should have clear, concise instructions for executing a purchase; regarding supplier selection strategies when the product set or environment changes. This process is known as dynamic market analysis.

Of course, procurement has a role in supporting RFPs and RFIs. Their support can consist of rules for engagement, acting as the mediator between the business unit looking for a new product, part, or solution and the courting vendors. Procurement would typically place themselves in a position to negotiate formal purchases fairly for all parties.

At times, the use of "supplier self-service" capability to process RFQs and RFPs could be leveraged, if applicable. For example, this self-service automation could eliminate buyers having to copy, collate, package, and send RFQ packages to bidders. Instead, a formal supplier selection process would be deployed across the enterprise.

A "well-defined" supplier selection process would be utilized and communicated to internal and external sources resulting in clear expectations by all parties. This selection process would benefit the enterprise by aligning corporate requirements with supply risk management. It also helps to eliminate potential conflicts of interest in the enterprise. Procurement will guide when an existing supplier should fill the requirement instead of a new one. Procurement will lead a strategy designed to capitalize on the supplier selection procedures. The strategic processes would be developed further into the mechanics of "when to use" an existing or new vendor/supplier.

## EXPANDING ON SUPPLIER RELATIONSHIP MANAGEMENT ACTIVITIES

Procurement would create new suppliers (vendors) where none exist. Procurement will engage and develop potential vendors to do business and services with their enterprise in a mutually beneficial way. Typically, an enterprise will consolidate parts purchased from a consolidated group of vendors to maximize their spend leverage with the given supplier and minimize the price and cost of each acquisition transaction. A centralized procurement can leverage and formalize existing supplier development processes across all key suppliers. They can also consolidate, standardize, document, and communicate the techniques used for supplier development across operations and suppliers. This process unites the departments together and aligns suppliers accordingly.

Contract management is part of supplier relationship management and manages procurement contracts initiated and implemented through the supplier development and commodity development policy. Therefore, contract management strategies and tactics must support the procurement mission and business strategy and achieve the desired results.

Managing requirements specifications and pricing: Procurement would facilitate the grading and vendor/solution selection once the vendor receives a proposal. This selection process can be an extension of an RFP process. Still, in many cases, this is an extension of an open Purchase Order or existing working agreement between the enterprise and a specific vendor.

Procurement would act as the holder of all supplier contracts and manage when an agreement must be renewed or re-established through a formal RFP process. Contract management can include developing standardized terms and conditions with local flexibility that can be used across the enterprise. For example, it may consist of a legal procurement requirement that has but is not limited to free on board (FOB), workers' compensation board (WCB), Insurance, etc. In addition, all areas of purchase order legalities needed to be standard across the enterprise operations (may differ by country) and understood and utilized across the business in a harmonized process to eliminate potential risks.

Vendors will have service levels that they would typically commit to in any purchasing agreement. In addition, the suppliers would be tracked for vendor compliance, particularly on critical metrics (i.e., 95% on-time delivery, price variances, and quality shipment variance behaviors).

Procurement can track and publish vendor performance with vendors selected and coached and performance measured. Typically, these reports would be shared with the vendor and the internal enterprise staff so that opportunities for improvement can be.

## EXPANDING ON TRANSACTION EXECUTION ACTIVITIES

For purchasing transaction execution, the ideal scenario is to set up as much of the purchasing activity as an automated transaction as is practical and maintain essential purchasing controls.

Managing purchasing requisitions, orders, and catalogs may mean:

- Having orders automatically set up as requisitions in the maintenance management system or parts management system
- Collecting the purchasing approvals and automatically sending the orders to Predetermined vendors as set up in their procurement system
- Closing the process with a receipt and invoice "2" or "3-way" match

Item lists or catalogs would be integrated to select the correct item and the desired price, vendor, and timeline. The accounts payable process ensures that suppliers quickly, accurately, and cost-effectively meet the requirements of processing invoices. Typically, processes and systems are set up to eliminate duplication of effort, speed up the process, and allow visibility and control of all aspects of the payable process.

If the process is in place, then procurement has also automated the order management facilitation, order fulfillment, order receipt management (with the help of the receiver in maintenance parts), and the invoice's paying with the services from accounts payable. However, procurement would manually facilitate these four steps if not in place.

## LEADING PRACTICES IN PROCUREMENT

Evidence of leading practices in strategic commodity souring can include:

- Working with design engineering to collect the required maintenance parts data elements as a component of the vendor selection process
- Establishing commodity strategies based on business needs/strategies
- Analyzing spending for common commodities as well as alternative sources for OEM-centric components
- Outsourcing elements of procurement that may be effectively satisfied locally
- Quality control is managed with a professional supplier process control (SPC) to make an exit, inspection, and quality control redundant.

Leading supplier relationship management can include supplier collaboration disciplines such as:

- Managing obsolescence with the vendor
- Workshops with suppliers to determine solutions
- Collaboration on regulation control
- Partnering in capital expenditure or construction projects
- Coordinating access to electronic catalogs, electronic procurement, and electronic data interchange
- Sharing forecasts
- Vendor co-development of new products or components
- Collaborative process improvement
- Co-development of the supply chain by coordinating distribution methodology and services levels to meet the needs of the enterprise
- Working with suppliers to bundle goods and services in one consolidated contract

Vendor/supplier feedback would be solicited regularly to ensure good communications and a fair exchange of opportunities. Supplier feedback could be included as:

- Part of the regular formal review process and meetings
- Inclusion in an annual supplier conference
- Supplier surveys

Evidence of leading practices in sourcing transaction management can include:

- Actual requirements are communicated electronically daily to the carrier
- Import/export process for actual items received and hence modification of duty
- Regular review of legislation/tariffs to maximize the benefit
- Parts demand is synchronized supply chain demand management
- Automated and integrated supplier warranty claim process with maintenance/operation's contract system to manage validity and pricing immediately
- Integration with parts availability, fulfillment, and maintenance defects tracking
- All methods of claim submission
- A mutually agreed formula for payment exists
- Carrier partnerships are negotiated

## THE ROLE OF TECHNOLOGY CAN CONTRIBUTE TO SUPPORTING PROCUREMENT

The types of technology in procurement can include:

- E-procurement (or EDI)
- Electronic catalogs
- Procurement software module on its own or in a packaged enterprise resource planning systems (ERP) or enterprise asset management (EAM) solution. There is a mix of organizations that are using the procurement module to:
  - Electronically file transfer purchase orders (POs) and receive consolidated electronic invoices, or allow suppliers to "self-bill"
  - Organizations that still E-mail or fax POs and (no electronic capability for POs or invoices)
- Procurement within a new business system implementation (again, a software module on its own or a packaged ERP solution or EAM solution)
- Software for supplier relationship management, request for proposal management, contract management, and expense management
- Providing suppliers with direct access to inventory system for vendor-managed inventory
- Collaborative procurement portal
- Use of supplier websites to punch out to and raise purchase orders
- Data warehousing
- Managing open or automated contracts
- Managing advanced shipping notices
- Automatic transaction steps such as "assumed received"

To support supplier management and development, an enterprise will typically have:

- A procurement system to record and report supplier performance measures
- A Supply plan on a public portal or website for each supplier to view

- Weekly supplier dashboard reporting
- A spend analysis reporting capability/tool that the supplier or the enterprise could use

A procurement organization may be fully automated with electronic catalogs, PR and PO creation and approval, electronic PO transmission, and electronic invoicing and payment.

The automated procurement discipline may include:

- Electronically collected RFP/RFI responses
- PRs and POs raised through an in-house procurement application and either transmitted via electronic file, e-mail, auto fax, or manual fax
- Electronic quotes
- Purchasing through supplier websites to punch out to and raise purchase orders
- Purchasing through a third-party e-procurement service
- Receipt of electronic catalogs for upload into procurement application or organization intranet site (note – only the critical supplier catalogs are usually received, rather than all suppliers)
- Electronic invoicing and auto payment upon price matching (primarily for crucial suppliers)
- Electronic invoicing and manual payment, again for key suppliers only
- Reverse auctions
- "Punching out" capability from their procurement application to a vendor application

## KEY PERFORMANCE METRICS FOR PROCUREMENT

Key performance metrics for procurement will evolve depending on where the enterprise is in its overall process maturity and focus on a specific transformation period.

One enterprise looking to create more automation while improving its procurement processes began a year-over-year focus on:

- The sourcing (staff) expertise in place
- Cost savings
- Supplier quality metrics met
- Purchasing escapes – maverick buying
- Business controls acceptance and audit results
- Client satisfaction
- Percentage of business supported through electronic catalogs
- Number of suppliers able to support electronic orders and payment systems
- PO fulfillment cycle time

The metric performance considerations listed in Table 8.1 shows a partial list of potential metrics that could be used depending on where the organization is in its pain points and maturity.

**TABLE 8.1**
**Examples of Procurement Performance Metrics**

| Performance Metric Considerations | Performance Metric Considerations |
|---|---|
| • $ spend, $ value extracted | • Out of stocks internally |
| • Additional services obtained | • Process improvement |
| • Automated transactions (%) | • Rebates, warranty management |
| • Category Management | • Safety |
| • Communication/relationship success with supplier & internal unit | • Savings versus target |
| • Contract performance – asset lifecycle management, innovation, value add | • Service levels |
| • Contract reviews with strategic suppliers | • Shareholder value |
| • Contracted data provided against asset lifecycle | • Sourcing – good process, visibility, reporting, well-managed transition, value realization |
| • Cost variance to the standard | • Supplier development, differentiation/ rationalization |
| • Customer satisfaction | • Supply rate |
| • Delivery quality | • Transaction costs, Unit costs |
| | • Vendor-managed inventory value |

## IMPROVING PROCUREMENT PROCESSES

To achieve purchasing excellence as part of maintenance parts excellence, the procurement staff's training focus will need to be well established. Much of the required coursework and certifications can be facilitated cost-effectively by using the ADDIE methods discussed in a previous chapter and taking full advantage of online learning tools. Online training tools are particularly effective when the enterprise is global, and training materials must be standardized and served to a worldwide procurement population.

When considering what areas of procurement to focus on when doing an internal assessment, the approach discussed in Chapter 2 can be applied here. An example list of developed initiatives for one small maintenance network is shown in Figure 8.4. These initiatives could be prioritized using a benefit versus ease of effort grid or an importance/performance grid to help the team prioritize what they should plan the work on first (Table 8.2).

All the listed initiatives could be considered or at least tied to leading practices in procurement (Figure 8.4).

Some large enterprise organizations outsource their procurement processes to third-party service companies. As a result, these companies would have demonstrated their ability to leverage their automation and disciplines, and spend volumes to generate savings for the enterprise. Additionally, some third-party logistics organizations provide more than just procurement services. Their services portfolio may include the entire logistics management process continuum, including:

- Logistics strategy
- Planning
- Tools and applications
- Procurement

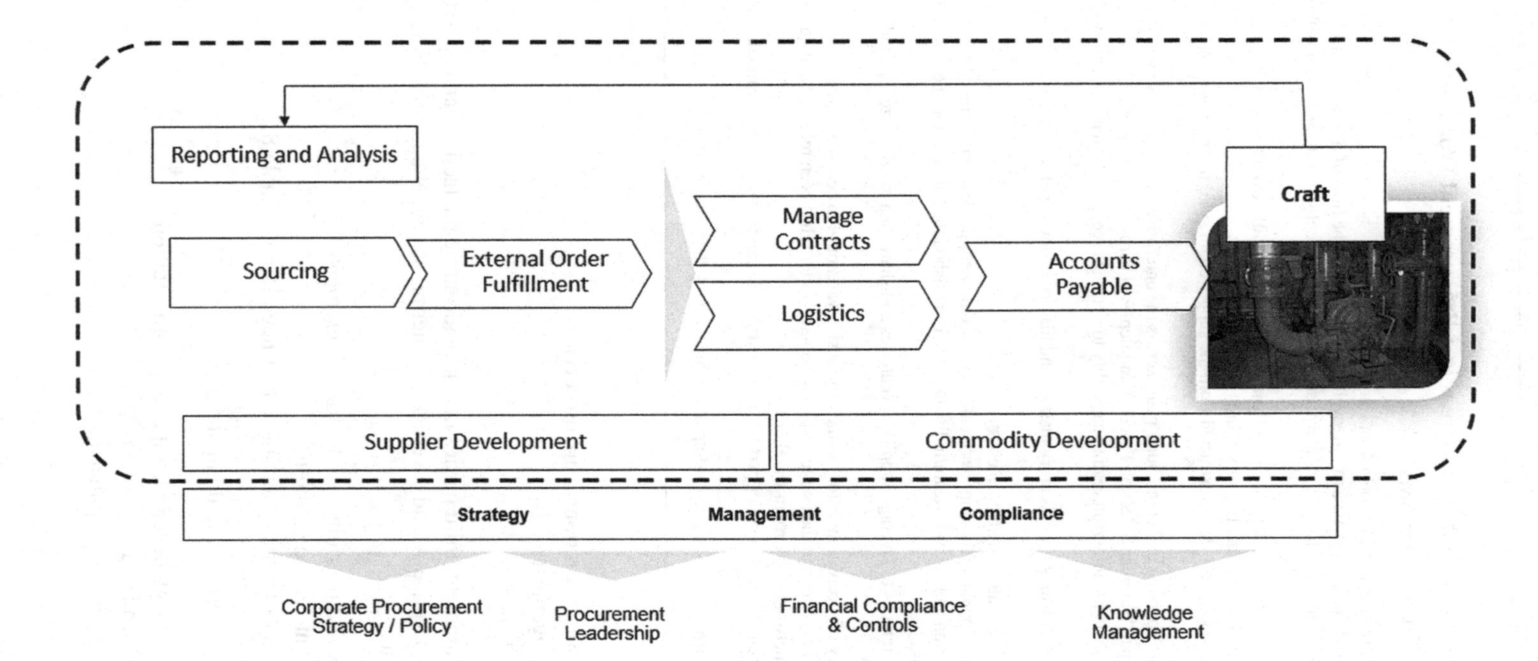


**FIGURE 8.4** Scope of a typical Procurement group. (Adapted from copyright 104 MPE Inventory Management ©2020 from Maintenance Parts Management Excellence Training edited by Barry. Reproduced by permission of Asset Acumen Consulting Inc.)

**TABLE 8.2**
**Example of Proposed First Focus Areas in a Procurement Workshop**

List of Potential Procurement Initiatives

- Clearly define procurement mission and business strategy
- Create a process to promote total asset lifecycle management at the time of new asset build
- Centralize all procurement across the enterprise to one management structure and set of defined processes
- Leverage global contracts to best leverage their global spend or with regional spend
- Establish a direct new product buy procedure
- Ensure key metrics (KPIs) can be extracted from the supply chain system (or Knowledge database)
- Share Corporate sourcing and commodity plans across the enterprise
- Develop a formal "Supplier Selection Process" and deploy it across the enterprise
- Create supply chain planning/forecasting capability to optimize required on-hand inventory balances
- Create the capability to query inventory masters and history to develop information views of the spare parts assets
- Link procurement strategy with business strategy
- Promote the development and empowerment of the procurement process and its people
- Ensure recognition and rewards (team/individual) are established that support the success of the procurement process
- Look for opportunities to leverage supplier inventories and share data to minimize on-hand inventory levels
- Create visibility of received inventory at the central site warehouse to become "in-transit" inventory on the system and then go to "on-hand balance" after it has been received and quantity verified at the network stock room warehouses
- Develop an essential set of supplier performance metrics that can establish supplier monitoring and ranking
- Ensure regular access to an understanding of the supply market environment and changes

- Parts logistics operational support services
- Asset management
- Operational support

Specific to procurement, third parties may provide strategic, tactical, and operational services. The enterprise would leverage its internal strengths and look to the third party to augment:

- A strategy and operating model for procurement
- Procurement-related solutions
- A more extensive spend mix in vendor negotiations (Table 8.3)

Strategic sourcing services could include:

- Category strategy development and implementation
- Category spend and Market analysis
- Stakeholder relationship management

**TABLE 8.3**
**A 3rd Party Procurement Services Model Example**

| Strategic | Tactical | Operational |
|---|---|---|
| • Recommend strategies & savings methodology<br>• Recommend suppliers<br>• Recommend terms & conditions<br>• Recommend policies & procedures | • Spend & market analysis<br>• Execute sourcing plans & strategy<br>• Negotiate & implement Contracts<br>• Supplier relationship management<br>• RFx development and execution<br>• Buying channel assignment savings & compliance tracking | • Inventory planning & forecasting (if not performed by the enterprise inventory planner)<br>• Parts logistics operational support services |

- RFx development and execution
- Contract negotiation and performance
- Buying channel assignments

Commodity management support could include services such as:

- Delivery of spend performance targets
- Execution of sourcing plans and strategy
- Supplier relationship management
- High-value project buys
- Tactical buys based on a clip level
- Savings tracking and compliance monitoring

Significant benefits are achieved by partnering with a reliable third-party procurement company. However, the benefits and costs need to be well understood and mapped out before handing over what may be a core value set in your current enterprise operation.

Some Leading Procurement Practices to consider

## USED PARTS RETURNS AND PARTS REFURBISHMENT

Today, parts procurement has multiple "source categories." For example, it can be sourced from the original equipment manufacturer/supplier or a parts repair/refurbishment source. Managing the used parts returns (UPR) and refurbishing used parts is an emerging and viable part of a maintenance parts procurement strategy for many asset types. The discipline and control of the UPR inbound parts processes are required to make parts repair and refurbishment work.

When the field maintenance strategy is to replace a component rather than performing a field repair, the component (also identified as a spare part) would be flagged as "repairable" and returned via a UPR process and dispositioned as "used part" in the staged used defective inventory that would feed a parts repair process.

The parts repair source is often a mix of vendor and in-house services that repair the defective part and then return it to a central warehouse stock as if it was as good as a new part. The rebuilt (same as new) part would be distributed into the network from the central warehouse as demand for "as new parts" is generated. In these cases, the maintenance parts supply chain is a "two-way" supply chain. As new parts are used in the field, many extracted and defective parts are returned to a central depot to be repaired or refurbished.

In many industries, such as aircraft maintenance, the field service is the physical swap of a field replacement unit (FRU), and the used part is a rotating part (or "rotatable") that will be returned to the local depot/bench for repair, refurbishment, or overhaul, through a UPR process for proper disposition.

A "reverse logistics" process is established to ensure that the used defective part is recognized when ordered (as repairable or used returnable) to support the management of used parts. The repairable part is then tracked (when used) to be returned to the Used Part Return (Disposition) center and staged for repair. From a used parts return disposition center, a returned part can be directed for:

- Repair
- Scrap
- Warranty claims to the OEM/supply vendor (not shown in Figure 8.5)
- Precious metals (other value) reclamation (not shown in Figure 8.5)

The opportunity to fix "as new" a designated "repairable part" back to inventory is initiated when the inventory forecast and replenishment demand call for more specific parts to be repaired and returned to stock. This benefit is that repair costs are typically significantly less than purchasing a new part. In addition, the lower-cost repaired part allows lower-cost quality parts into the inventory network as a "like new" part (Figure 8.5).

A parts assembly component repair can occur in the field but is more likely to be swapped out and returned to the central repair shop using the used parts returns disposition center.

As parts are used/replaced in the field, many are returned to be repaired or refurbished. Used parts returns happen in multiple ways. First, there are situations where complete products or components must be carried or sent back to a local maintenance center for repair if they cannot be repaired in the field. In most cases, the maintenance technician would exchange a defective (bad) part with a spare part and then carry or send (via courier) the defective part back to the used part return center. In this case, the used and defective part must have been identified as returnable in advance when the designated spare part had been ordered and shipped (or carried by the maintenance technician) to the original place of the asset repair.

Often, it is not known which new parts are needed in advance, and accordingly, the maintenance technician orders more spare parts than required. In this case, they must return the leftover (unused) spare parts and the defective parts to the appropriate locations (maintenance parts warehouse). For this type of repair and return set of processes, the necessary logistics controls must be in place to ensure that the part

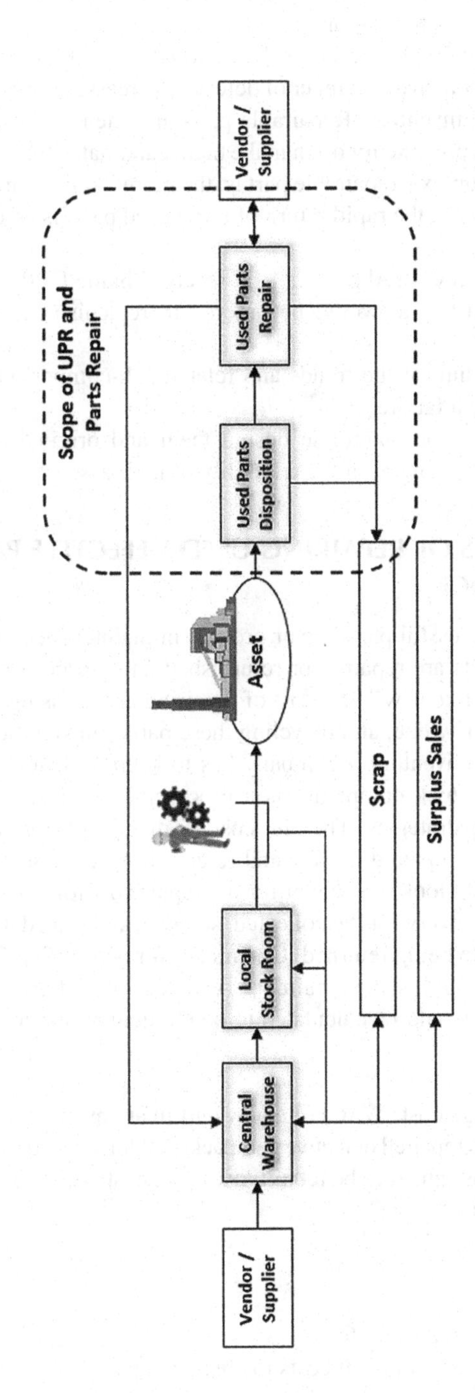


**FIGURE 8.5** Scope the used parts return and parts repair discussion. (Adapted from copyright 104 MPE Inventory Management ©2020 from Maintenance Parts Management Excellence Training edited by Barry. Reproduced by permission of Asset Acumen Consulting Inc.)

arrives at the right destination after arriving at the dock (see Chapter 7 – Figure 7.15 on warehousing and logistics for shipping label insights).

Items designated as repairable parts must be returned as used (defective) parts as rapidly as practical to minimize the number of defective carcasses not being available for repair and returned to inventory. Repairable parts are often expensive and complex. Their high price and complexity often make them candidates for refurbishment/restoration. The more complex a repairable part is, the more likely a part sub-component could fail. In such cases, the rapid return of these used parts is essential to:

- Minimize the number of used parts not yet received in the UPR center
- Reduce the new parts that would be ordered to replenish the new parts inventories
- A better understanding of the trends and related sub-components that are most likely to cause a failure
- Provide feedback to the enterprise design team and original equipment manufacturer (OEM) vendor on the reliability of their assets

## FINANCIAL BENEFITS OF REPAIRING USED DEFECTIVE PARTS AND RETURNING TO STOCK

A unique feature of a successful parts repair process in maintenance parts management is that defective parts are repaired or refurbished and placed back into active inventory when deemed "like new." The cost of repairing and re-using a part can be less than 25%–50% of a new one, and recycling these parts can significantly reduce the amount of purchased inventory a company has to keep on hand. Having parts repaired and returned to stock complements a procurement strategy and complements inventory planning strategies. The unit cost of a part will decrease the more often parts can be reliably repaired as new and returns to stock at a fraction of the cost of new. Finance would look to account for the repair cost for used rebuilt parts and consider some cost recovery for the collected, stocked, and staged used defective parts that incur costs while being returned. One model for accounting for a repaired part back into inventory would be the repaired part value is placed back in inventory. The item inventory value would be calculated to be the item repair cost + used part cost when:

- The weighted average cost (WAC) of a new part in an inventory
- The average cost of a repaired defective part packaged for stocked (repair cost)
- Cost of a used carcass about to be rebuilt: used part cost= 25% of WAC plus some logistics costs

Note:

- The repair cost would account for:
  - The apportioned development costs to create a repair cost
  - The sub-components and labor required to repair the part

    - The packaging costs to prepare the part for being stocked into inventory
    - The infrastructure costs for the parts repair center
- Used part costs would account for:
    - 25% of a WAC part stocked at the time of repair
    - Apportioned logistics fees and overhead for getting the part returned and staged

To repair a part, one must confirm it is a viably refurbishable item. The costs for support tools and equipment for troubleshooting, repairing, and testing the repairable part must be factored into the decision. The development of a repair process and test process needs to be understood. Support tools and equipment can include special-purpose tooling, computerized diagnostic equipment, and customized programming. Also, the cost of supporting technical documentation required to execute repair activities must be factored in. This documentation may be available from the original equipment manufacturer and negotiated at the time of the initial equipment purchase. Once the repair and testing processes are understood, the search for available sub-components and the supply chain for the sub-components must be understood and aligned. In simple terms, a repairable part is a part and labor proposition. If the repairable part's refurbishment cost is above 50% of a delivered new part, it may be better to plan on continuing to buy new part and not set up the infrastructure to return and refurbish repairable parts.

Repair costs are crucial, but asset lifecycle support can be an even more significant influence. If it is expected that the OEM/vendor cannot support your critical assets for the planned duration that you require the support, then a parts repair process may become essential to your enterprise's success. Sizeable single-piece engineered components like shafts, impellers, rotors, and machine casings are costly. Repair or reverse engineering and remanufacturing these items may prove more cost-effective than a discard or replacement tactic.

Another influence on whether to do parts repair or not is considering the cost and reputation of the enterprise to be environmentally sustainable. For example, in some areas, being seen refurbishing volumes of used parts versus filling up landfill sites may influence the organization to accept a higher cost for repair tolerance than others without such a recycling agenda.

If a decision is made to repair a part, the most "cost-effective" approach should be adopted. The cost of using "in-house" staff and processes versus outsourcing the repair to the OEM or a third party should be examined. Outsourcing will also need to include the cost of additional two-way shipping and parts tracking, with each part set serialized to assist with transportation and repair quality tracking. Ensuring the "in-house" staff is qualified and experienced in such repairs is also key to the success of specific tactics selected.

The best time to decide to repair a part is at the asset design stage in the asset lifecycle. Much of the information is available from the asset OEM. The related costs can supplement the best maintenance strategy, including an initial spare parts stocking and repair strategy. A year into a specific asset lifecycle may bring a new approach, but the original supporting data may be challenging to acquire.

Asset and parts repair costs can be provided from the OEM and related repair facilities the supplier recommends to compare to your own "in-house" estimates. Failure modes, related failure rates, and spare replacement rates data from the OEM will be necessary to predict the repair frequency for a total asset lifecycle cost calculation. Repairable parts may break out with the help of the OEM vendor. The data will, of course, not likely be precisely what is experienced when implemented, but the OEM data can be an essential starting place. They can be an opportunity for lessons learned for the monitoring asset types that OEM bids on in the future. In the absence of OEM data, the previous history for like products or Technology may help the enterprise determine if there is a good business case for "in-house" parts repair or outsourcing the parts repair.

Some enterprise organizations elect to repair their parts. This practice should be questioned with many specialized repair vendors available to provide timely, cost-effective quality repairs. Is a parts repair process considered a core competency in that enterprise? Distance from viable repair vendors, perceived quality, and strategically supporting internal staff may influence this decision.

## KEY PERFORMANCE METRICS FOR UPR AND PARTS REPAIR

Used parts returns

- Used parts return management:
  - Flag rotatable/returnable parts
  - Local disposal of "Low-cost" parts
- Warranty credit management:
- Collection for staging "pull" inventory into parts repair processes
- Parts disposition:
  - Scrap credit
  - Disposal of "low-cost" parts

Parts repair

- Repair capture rate:
  - Of all parts rotatable/repairable, do we have a repair process?
- Repair yields:
  - What percentage is successfully repaired of all parts that go through this process?
- Out-of-box Failures:
  - How many are deemed "out-of-box" failures of the repaired parts sent to stock?
- Second return rate:
  - Of the repaired parts sent to stock, how many are deemed failed again within 90 days or one year of repair date?
- Cost of the components of parts/services required for repair:
  - Is the repair cost attractive to be considered a rotatable/repairable part (e.g., 25% cost of new)?

- The total value of repair:
  - What are the offset savings of procuring repaired parts versus a new part?

Maintenance parts repair considerations to account for:

- Leverage a pull from the carcass inventory process based on new parts inventory dynamic needs, not a Kanban repair process when repairing parts. Repairing available carcasses can create an inflated repaired parts inventory and inflate the enterprise's financial books. Critically required repairs may go unfulfilled, while automated replenishment buys more parts from the OEM. Again, the enterprise inventory values go up when less expensive quality parts could have been provided from the carcasses staged in UPR.
- Modifications to refurbished parts should not be instituted without agreement from all stakeholders (e.g., design engineers, maintenance, and operations management).
- A tracking system will be in place for rotatable assets. Create a tracking system to track the flow and quality of all non-rotatable parts in the repair cycle. The tracking system will allow critical parts metrics to be managed, such as out-of-box failure and second return rates.
- Ensure that all new repair processes are proven through an accepted testing method or a closed operating environment. In addition, all repaired parts must pass a mutually agreed testing process and be serialized to track the repair trail and quality.
- Like an asset repair process, review parts repair procedures for completeness and technical accuracy before starting the refurbishment. However, do not accept vague or poorly documented procedures that create too much room for interpretation.
- Use a tag process to ask the maintenance technician to provide the part failure description/insights to help the repair cycle time process, if practical.
- Provide specific targets for cycle and delivery times for each step of the used parts returns and repair process (e.g., used parts returns shipped within 48 hours of use, received within 48 hours of arriving at the dock to be processed for staging, held for monthly consolidation of warranty processing, parts repaired to meet the forecasted needs in new part inventory).
- Suppliers of repaired parts, including "in-house" repair suppliers, are measured for parts repair cycle time commitments met, parts quality, and parts repair costs. Penalties are set for significant parts repair commitments missed, whether from internal or external suppliers to the new part inventory.

# 9 Inventory Planning and Policy Management

## INTRODUCTION TO INVENTORY PLANNING

All maintenance parts value chain elements must be well managed to meet business asset demands across the enterprise asset lifecycle.

The maintenance parts inventory planning role is critical to the success of an asset strategy. The asset strategy supports the enterprise operations supply chain value chain, which, in turn, contributes to the enterprise's value to its customers and shareholders (Figure 9.1).

Maintenance parts inventory stocking levels are declared through an inventory policy. Ideally, stocking levels are determined before a new asset installation, and the stock represents an expected investment that generates the need to experience carrying costs. Inventory carrying costs include the cost of the space required for the parts warehouse and related stock rooms and the cost of the inventory (capital on the books of the enterprise).

What stock needs to be formally planned and supported can be a complex question. Setting an inventory plan (policy) requires multiple business unit communication inputs and many business unit commitments to perform in concert and mutually agree (Figure 9.2).

In a traditional manufacturer's supply chain, we would see:

- Product lifecycle considerations and a supply chain network design (strategic, long-term decisions and actions)
- Supply chain planning (tactical, mid-term decisions and actions)
- A recognized cycle to optimize the supply chain planning
- Supply chain execution (operational, short-term decisions, and actions)

Often, there is, at best, only a distinction between two categories – supply chain planning and supply chain execution.

This chapter will work with the maintenance parts network parameters provided, focus on maintenance parts inventory planning and execution, and address asset parts lifecycle activities and maintenance parts inventory optimization in later sections of this book (Table 9.1).

DOI: 10.1201/9781003344674-9

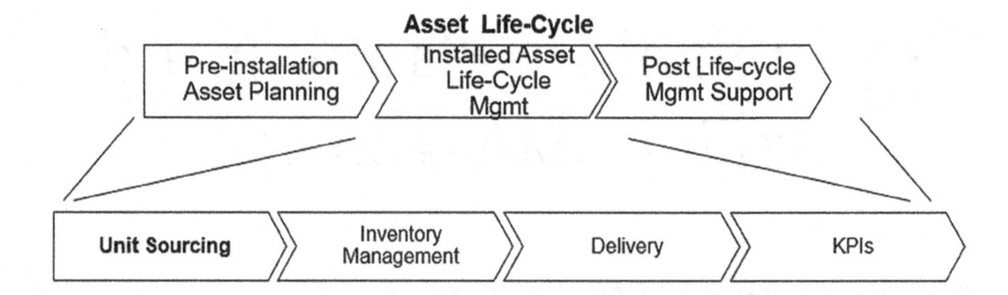


**FIGURE 9.1** The maintenance parts inventory planner's role is critical to the success of the asset lifecycle. (Adapted from copyright Figure 6.3 ©2011 from *Asset Management Excellence, Optimizing Equipment Life-cycle Decisions* edited by Campbell, Jardine, McGlynn. Reproduced by permission of Taylor and Francis Group, LLC, a division of Informa plc.)

## MAINTENANCE PARTS INVENTORY PLANNING

Maintenance parts inventory planning is the process of:

- determining target inventory levels for each part at each stocking location (inventory planning)
- selecting the appropriate replenishment policy for each part at each site (replenishment planning)

There will be trade-offs regarding where the part support should support a specific asset and its operating context. Often it will make more sense to service a need from a second level (echelon), rather than the closest. Expensive items, or slow-moving non-critical parts, will be held at a central warehouse location to reduce safety stock carrying costs. Parts planning must solve the trade-off between an agreed-to-customer service level and the value of inventory required to support it. Algorithms can be used to calculate the perceived "optimal" parts inventory control policy for a specific enterprise. This approach will help maintain the total maintenance parts supply chain cost while meeting all item, asset, and enterprise target market requirements.

## INVENTORY POLICY DEVELOPMENT

The inventory planner would facilitate developing an asset and site-specific inventory policy for ISP/RSP, highly active and moving, medium moving, and surplus inventory. The planner would be expected to forecast, facilitate, and execute the inventory policy based on the mutually agreed to ISP/RSP levels provided by the design engineering team, maintenance, operations, and finance and established disciplines for active inventory to the policy success.

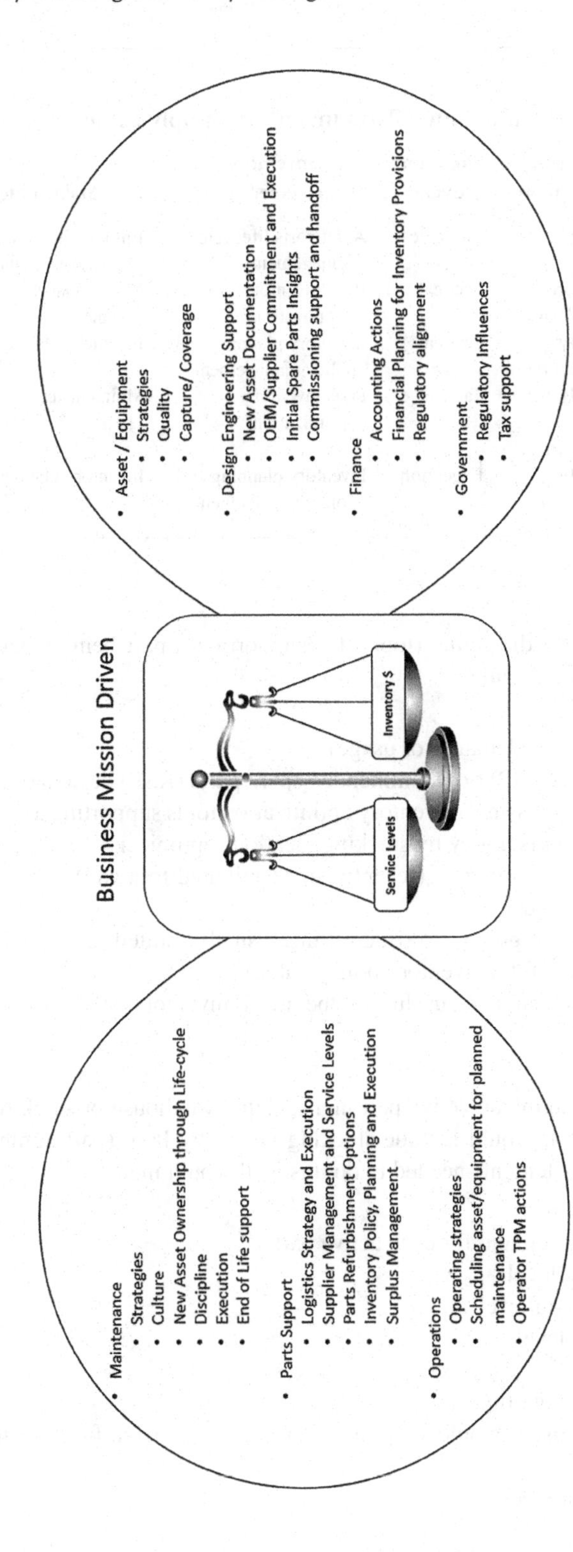


**FIGURE 9.2** The maintenance parts inventory plan requires many stakeholders to input and execute to achieve success. (Adapted from copyright 104 MPE Inventory Management ©2020 from Maintenance Parts Management Excellence Training edited by Barry. Reproduced by permission of Asset Acumen Consulting Inc.)

**TABLE 9.1**
**Planning Cycles in a Maintenance Parts Inventory Supply Chain**

| Time Horizon | Spare Parts Support Activity | Program Cycle | Inventory Management Cycle | Stakeholders |
|---|---|---|---|---|
| Years | Asset parts lifecycle | Strategic | Asset/parts lifecycle management | Engineering, maintenance, Inventory, finance |
| Years | Parts support network design | Tactical | Parts inventory optimization | Maintenance, inventory, finance |
| Months | Parts supply chain planning | Tactical | Inventory planning and policy management | Inventory planner, finance |
| Months | Parts supply chain optimization | Tactical | Parts inventory optimization | Maintenance, inventory, finance |
| Weeks/Days | Parts supply chain execution | Execution | Inventory planning and policy management | Inventory planner, finance |

The inventory policy discipline (tools) for supporting a part being stocked in a specific location can be through:

- Local disbursement demand (or usage)
- Initial spare parts (ISP) or recommended spare parts (RSP) designations
- Some "all-encompassing" Inventory optimization tools supporting all assets and supporting parts policy by stocking location (optimizer)
- A historical transaction date (recently stock initiated to a site) or recently used (date protected)
- A local stock room policy override setting (usually limited to a small percentage of the site total inventory policy value)
- Declaring parts that are surplus to the local inventory stocking policy (Table 9.2)

The dynamics of managing an active part in a specific warehouse or stock room, as described in Figure 9.4, would include stocking policy by Item (part number) and Stock room. The data elements needed to do this well would include:

- Weekly average demand for each active part
- Maximum stock level
- Minimum stocking level
- Safety stocking level
- Reorder point
- Economic ordering point
- How the part is used (would keep the supporting asset from functioning or is superfluous)
- Impact if "out of stock"

**TABLE 9.2**
**Maintenance Parts Inventory Policy Setting Disciplines Available to an Inventory Planner**

| Inventory Policy Discipline | Description | How Determined | Discipline Owner |
|---|---|---|---|
| Active parts management | Parts level support to parts most active by stock room location | Minimum, maximum economic order quantity, safety stock elements for parts with two-line disbursements within a rolling three-month period | Inventory planner |
| Initial spare parts (ISP)/(RSP) | Insurance parts settings | provided by the Design Engineering group (pre-asset go-live) and Maintenance (post-asset commissioning) | Design engineering/ maintenance |
| Optimizer | All-inclusive Min/ Max, EOQ setting based on 40–50 dynamic variables | Continent-wide usage, Asset Reliability requirements, cost of the part, picking, transportation of replenishment, and expediting | Community of design engineering/ maintenance, operations, finance, and inventory planner |
| Date protection | Parts already in stock by location maintain a support quantity of 1 | Replenished to 1 if used based on the date last disbursed or date created in site less than the target date set by location (e.g., one year, two years) | Inventory planner |
| Local override | Local support override setting (usually 1) for local stock room risk needs | Allowed to a maximum of a percentage of the total stock room value (e.g., 2% for stock rooms) | Local stock room, Local maintenance management, and inventory planner |
| Surplus | Parts in stock in a specific stock room that are not in policy | Parts on hand – by a site that is beyond the policy quantity or not in policy (as described above) | Inventory planner |

- Impact if "not stocked"
- Is the part an active substitute for another policy-supported part in the inventory
- Cost of acquiring and handling the part restocking
- Cost of the part/item per unit
- Part/Item quantities in the parts procurement cycle:
  - Quantity on hand
  - Quantity on order
  - Quantity in transit
  - Quantity reserved
  - Quantity available (Figure 9.3)

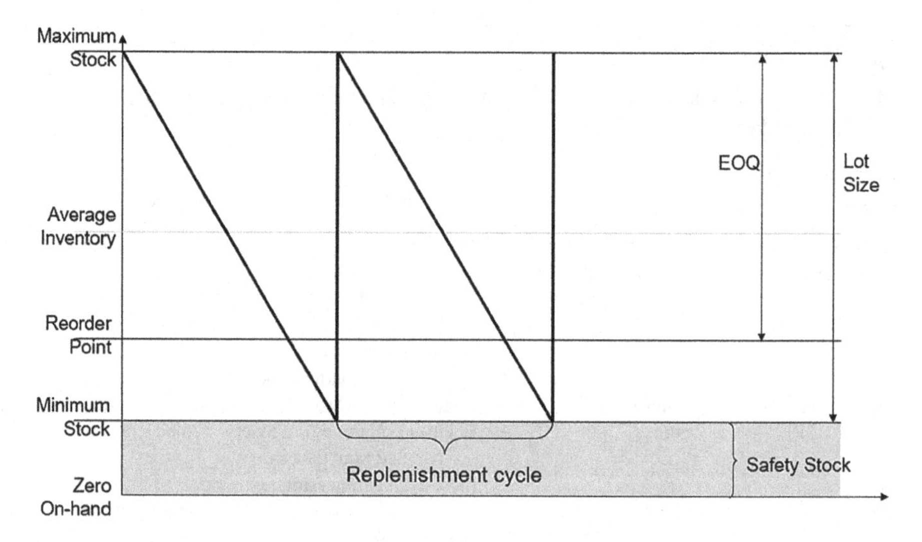


**FIGURE 9.3** Classical sawtooth model of inventory active parts management. (Adapted from copyright 104 MPE Inventory Management ©2020 from Maintenance Parts Management Excellence Training edited by Barry. Reproduced by permission of Asset Acumen Consulting Inc.)

The active part would have values for each item by location: minimum stock level, maximum stock level, lead time expected for the part to arrive from a supplier, safety stock factor for contributing to the inventory service level, a calculated reorder point, a calculated economic order quantity and, if applicable, a part or supplier-specific lot size for parts replenishment orders.

If the expected lead time (LT) is four weeks from a supplier and the designated safety stock level (SSL) is also four weeks (one month), then the reorder point (ROP) would be calculated to be (LT + SSL = ROP or 4 + 4 = 8).

Understanding the off-setting costs can determine a workable economic order quantity.

- What should the order quantity be?
- How do you determine this for each item?
- It's starting to look like a lot of work, even if the calculation is simple.

Every economic order quantity (EOQ) decision, even when done intuitively, is based on comparing three competing cost drivers – the cost to carry the inventory versus the cost to place orders versus business risk. The unit cost of the part and the cost of handling the part (labor and transportation) may have a simple relationship to the economic order quantity. However, the part is likely to be the more significant cost factor dynamic than parts handling cost. Therefore, calculating EOQ *can be done with a table* by parts cost or calculated as shown in Figure 9.4. Note that some organizations would apply a risk factor based on the business risk perceived by the criticality of their business or comfort with the supplier service level. This risk factor is a multiple to the base EOQ calculation in the equation.

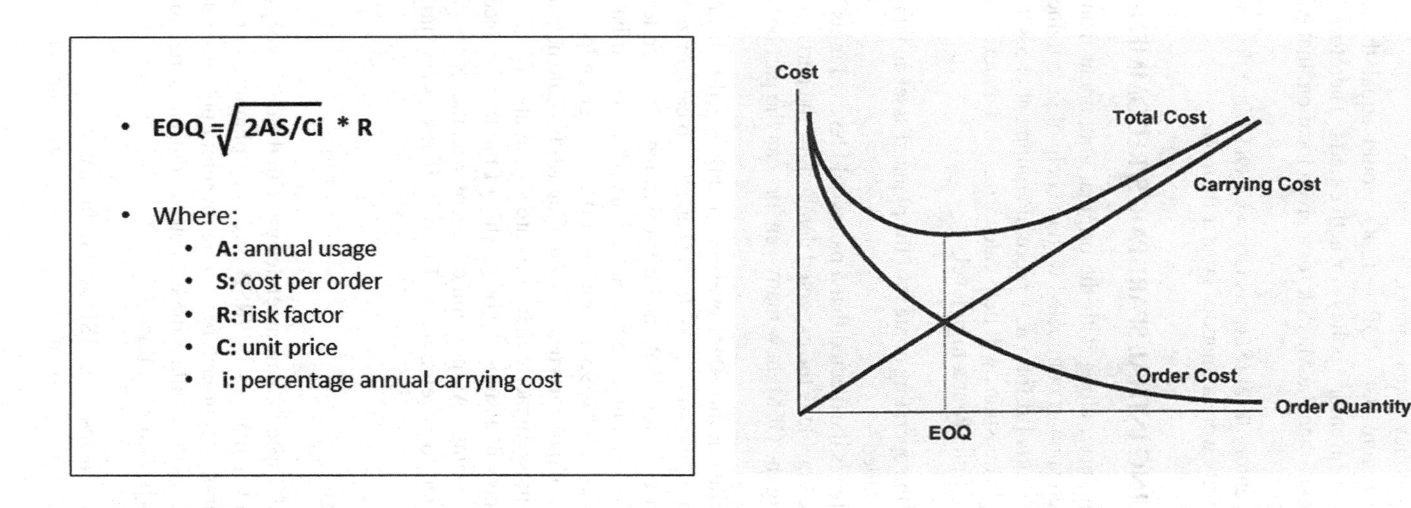


**FIGURE 9.4** The economic order quantity (EOQ) model. (Adapted from copyright 104 MPE Inventory Management ©2020 from Maintenance Parts Management Excellence Training edited by Barry. Reproduced by permission of Asset Acumen Consulting Inc.)

Note that the asset downtime cost is not a variable here. It is expected to be addressed in the ISP/RSP stocking policy component of the overall maintenance parts inventory policy management.

Expanding Figure 9.4, ROP + EOQ would equal the maximum stock level (MaxStk) in the Inventory policy for active parts. Therefore, if a quantity on hand (QTOH) is higher than the MaxStk for a given location, that part/Item would be considered surplus.

Surplus inventory level (Surplus Qty = QTOH-MaxStk). The surplus would exist if surplus quantity were a quantity greater than one.

## INTRODUCING INITIAL SPARE PARTS/RECOMMENDED SPARE PARTS

The design engineer, along with the original equipment manufacturer (OEM) and their committed support contracted with the help of procurement, will work through an initial spare parts list for a new piece of equipment or asset.

A few simple questions will help categorize what materials may need to be strategically stocked quickly at a high level.

- What is the operating context of the planned asset and its impact when functionally failed?
- How often is it expected that a part would be used in its lifecycle?
- How predictable is the expected demand for the part?
- Including the OEM, how many vendors can the part be viably acquired?

Answering these questions, for every item anticipated to be a failing component in a new piece of equipment or asset, you can start to categorize which parts should be included in an ISP. Note that an ISP list will have echelon levels in it. Thus, it can be estimated if the complement-stocked parts should contribute to a site availability level of 50%, 85%, 90%, 95%, or even 100% for a given location.

An analysis of past parts usage for similar assets could help assess the cost of parts usage and business impacts when parts are unavailable.

Maintenance will assume the ownership of the ISP list once the asset is accepted (post-commissioning). At the point of acceptance, the ISP list becomes a RSP list.

The approach to developing an ISP/RSP is discussed further in the chapter on asset lifecycle.

## OPTIMIZER

At the point of a specific new asset design with the design engineer, it may be determined that rather than an ISP/RSP inventory planning complement approach, an inventory optimizer approach for setting inventory policy should be adopted.

"Optimizer" will calculate the expected inclusive parts stocking policy by location echelon while considering:

- All the elements of the ISP/RSP influence
- The actual annual continental parts usage (for the active parts component)

- Multiple other asset reliability and logistics influences set its minimum, maximum, and EOQ levels (Poisson distribution can play a role in calculating the number of spares to maintain an asset at a target service level over some interval of time)

The approach to developing a maintenance parts optimizer policy is discussed further in the chapter on inventory optimization.

## DATE-PROTECTED PARTS

Date protection recognizes that many parts may have become active or less active over time and lose their privilege in the policy. Many of these parts find their way to a stock room because a Maintenance Technician ordered the part to a work order and then recognized that they would not use it and ultimately returned it to the local stock room. To reduce inventory churning between a stock room and then (when deemed surplus due to inactivity) being sent to the central warehouse, date protection can help support a quantity of one item in the originating stock room. Dates protected can be "date last used/disbursed" or "date the part was initiated into stock" for a specific stock room.

## LOCAL POLICY OVERRIDE PARTS TACTICS

Often an enterprise-wide inventory stocking policy will not account for a local need. The site need could be as simple as a service level commitment from a recent asset outage. It may also be to recognize that the distance between one echelon of support and the higher level of support where the desired part is stocked is too far away for a given season or asset operating context cycle. In such cases, some enterprises elect to empower their stock room staff (with the local maintenance management team approvals) to stock a nominal number of additional items in the local stock room. This local override may only be 1–2% of the total value of a local stock room but can go a long way in addressing issues that may otherwise become policy-setting escalations.

## SURPLUS MANAGEMENT IN STOCK ROOMS

There may be very few weeks where a stock room does not show at least one item/part with a surplus to its policy, quantity on hand. Typically, the inventory planner will expect the local stock room staff to return parts to a central warehouse to help keep the inventory levels in line with policy. Some enterprise systems will look across their network for surplus quantities and place a replenishment order on a stock room showing the available surplus quantity if the material and transportation costs to handle such an "excess redistribution" order is viable. Excess redistribution will allow surplus leveling across a parts network and reduce double handling when addressing a stock room surplus to policy.

It is essential to understand that the system used to manage these various Inventory policy influences should first net out the highest supporting policy. For example, suppose the ISP/RSP and optimizer and date protection parameters each suggest a stocking level of one. In that case, the highest stocking level should be one, and replenishment would happen only when the stocked part is used, and the stocking level goes to zero. The highest stocking level would still be one if the Min/Max fields were zero. However, if the ISP/RSP and optimizer and date protection parameters were one or less, the weekly average demand was one, and multiple disbursements happened over a "three-month time period," then the minimum stocking level (and the order point) would be one or greater and the maximum stock level (based on EOQ) would be the order point + EOQ. This maximum stock level represents perhaps six weeks of stock as an order point and another ten weeks of stock (totaling 16 weeks) for the Maximum stock level, thereby netting an order point of six against a netted ISP/RSP, Optimizer, or Date Protection parameter of one. The new order point would be based on the "Active" parts parameters of six (with a maximum stock of 16) and an EOQ of 10. If there were no parts in stock, the first order would be to the maximum stock level of 16.

## MANAGING AUTOMATED REPLENISHMENT

Determining how much to replenish can be determined by systematically reviewing the Inventory policy and determining which policy is netting to be the highest reorder point.

Several methods can be used to determine the timing of replenishment orders of inventory items. Two standard methods are:

- The reorder point system (variable time – fixed quantity)
- The Min-Max system (variable time – variable quantity)

The replenishment orders are triggered if the inventory level reaches a certain point (the reorder point). The only difference between the two systems is the order quantity. A reorder point system uses a predetermined quantity in every order. In a Min-Max system, the order quantity may be varied whenever orders are issued following the rule of "order enough to restore inventory to the Max level."

Running a reorder check against parts used each day and then recalculating the netted leading maximum stock indicator before a weekly replenishment run is believed to be a leading practice.

## SYSTEM PROCESS DIAGRAM FOR INVENTORY POLICY MANAGEMENT (FIGURE 9.5)

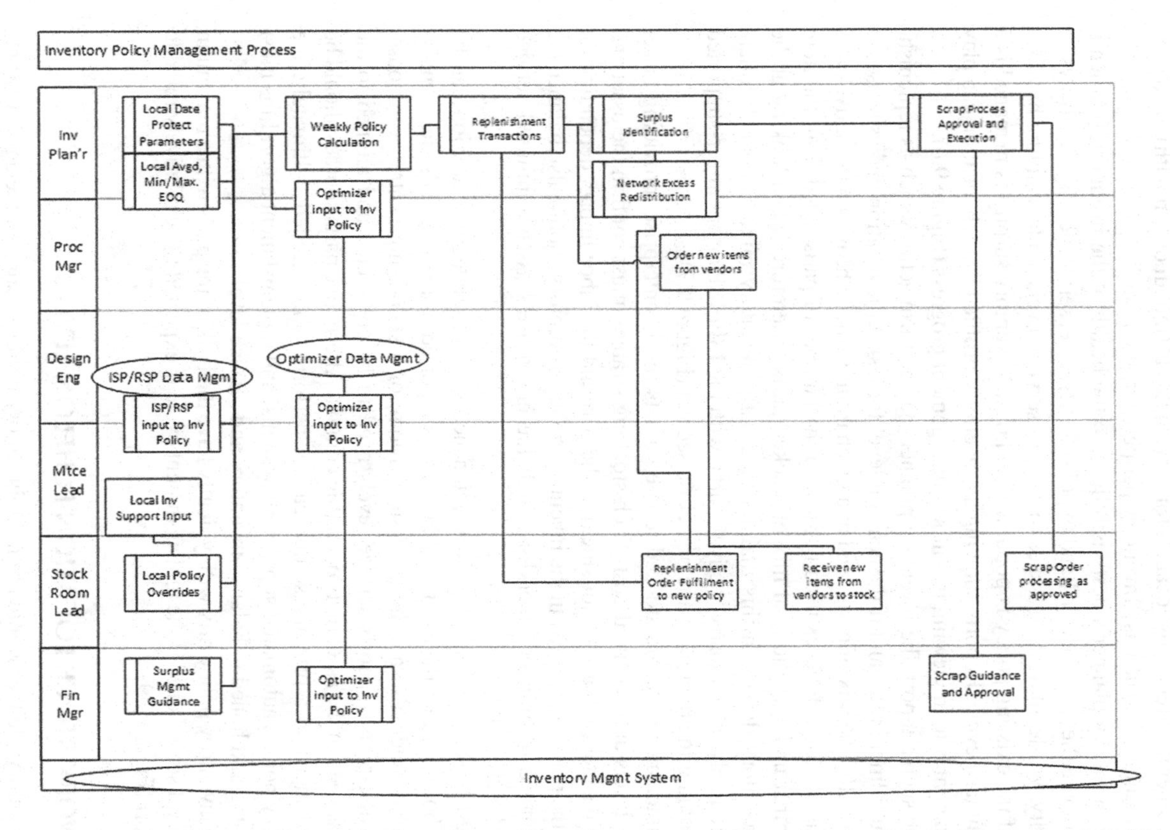


**FIGURE 9.5** System data development diagram for inventory policy management weekly updates. (Adapted from copyright 104 MPE Inventory Management ©2020 from Maintenance Parts Management Excellence Training edited by Barry. Reproduced by permission of Asset Acumen Consulting Inc.)

## INVENTORY VALUE VISIBILITY

Reports showing how many parts order lines were fulfilled versus ordered or how many work oder lines were fulfilled versus needed to be sent to a supplier to complete the order may help understand a stock room's service level. The inventory planner would generate a report by location for service levels (month over month), inventory value calculations, such as "Inventory per percent filled rate."

The inventory planner's best visibility into the health of the inventory is when it can compare the inventory on hand to policy. Therefore, service levels will always be a priority. While addressing issues impacting service levels, understanding by stock room if the existing policy is populated and if the inventory surplus is managed is a constant task cycle. Reports showing an inventory profile by stock location can play a substantial role in viewing inventory management progress (Figure 9.6).

With such a report, the inventory planner may confirm actions such as replenishment fulfillment rates and if the echelon level for a specific location makes sense.

Generally, the mix of parts used in a given month will be heavy on the active parts in terms of stocked pieces used but low in the number of parts listed from the item master that make up the total mix of stocked parts. Therefore, when looking at the total inventory dollars in this maintenance parts inventory mix, the actual dollars used are likely to be mostly "active" parts, with the ISP/RSP or optimizer-supported parts coming in about the same or a close second (Figure 9.7).

Maintenance parts are stocked to keep assets in the enterprise, providing target service levels at an optimal cost. Each stocking strategy must confirm the asset and related business impact of not stocking the part and the labor impact depending on where the part is supported in the inventory policy. Inventory should also balance the maintenance technician's workload with the enterprise's asset-committed service levels (Figure 9.8).

When looking at each asset's maintenance support strategy, a decision can be made about whether the asset can/should be supported from the asset's site, a maintenance technician's van, the local stock room, or a supporting central warehouse. Supporting parts vendor's service levels need to be contracted and aligned to the business need of the asset in its operating context. Alternative parts sources should be factored in when a primary source cannot perform when needed. Ultimately, the inventory policy, automated supporting systems, and replenishment algorithms must be tailored and honed to balance inventory and service levels.

The inventory policy mix will typically have the bulk of the parts with a maximum stock level set to "one" if the replenishment cycle is expected to be relatively reliable and short (Figure 9.9).

## NETWORK STOCK LOCATION ECHELON LEVELS

The inventory planner would align the inventory policy to support the agreed-to service level by stock location. For example, suppose the enterprise has multiple stock rooms across a large geographic area (country). They may accept the "level 1" support, set to 55% parts availability within ten minutes of the asset. Lower stock levels may be tolerated when the enterprise parts network also has a local stock room with

**Stock Room Inventory Profile**

**Inventory On Hand Balance Summary** / **Inventory Policy Summary**

| Policy Categories | Item Lines | % of Total Lines | Pieces in Stock | % of Total Pieces in Stock | $ Value in Stock | % Value in Stock | Lines in Max Policy | % of Total Lines in Max Policy | Pieces in Max Policy in Stock | % of Total Pieces in Max Policy | $ Value in Max Policy | % Value in Max Policy |
|---|---|---|---|---|---|---|---|---|---|---|---|---|
| Active | 1400 | 4% | 39200 | 47% | $ 588,000 | 5% | 1190 | 5% | 33320 | 40% | $ 17,850 | 28% |
| ISP/RSP | 14400 | 38% | 17424 | 21% | $ 4,530,240 | 39% | 10800 | 41% | 13068 | 16% | $ 2,808,000 | 24% |
| Optimizer Rec | 16500 | 44% | 19965 | 24% | $ 5,190,900 | 44% | 12375 | 47% | 14974 | 18% | $ 3,217,500 | 27% |
| Local Override | 1200 | 3% | 1452 | 2% | $ 377,520 | 3% | 720 | 3% | 871 | 1% | $ 187,200 | 2% |
| Date Protected | 1600 | 4% | 2096 | 3% | $ 335,360 | 3% | 1200 | 5% | 1572 | 2% | $ 192,000 | 2% |
| Total | 35100 | 94% | 80137 | 96% | $ 11,022,020 | 94% | 26285 | 100% | 63805 | 77% | $ 6,422,550 | 55% |
| Not in Policy | 2400 | 6% | 3144 | 4% | $ 691,680 | 6% | 0 | 0% | 0 | 0% | $ - | 0% |
| Total | 37500 | 100% | 83281 | 100% | $ 11,713,700 | 100% | 26285 | 100% | 63805 | 77% | $ 6,422,550 | 55% |

**Inventory Surplus Summary**

| Policy Categories | Lines in Surplus | % of Total Lines in Surplus | Pieces in Surplus | % of Total Surplus Pieces | $ Surplus Value in Stock | % Surplus Value in Stock |
|---|---|---|---|---|---|---|
| Active | 210 | 1% | 30170 | 36% | $ 452,550 | 4% |
| ISP/RSP | 3600 | 10% | 6624 | 8% | $ 1,722,240 | 15% |
| Optimizer Rec | 4125 | 11% | 7590 | 9% | $ 1,973,400 | 17% |
| Local Override | 480 | 1% | 732 | 1% | $ 190,320 | 2% |
| Date Protected | 400 | 1% | 896 | 1% | $ 143,360 | 1% |
| Total | 8815 | 24% | 46012 | 55% | $ 4,481,870 | 38% |
| Not in Policy | 2400 | 6% | 3144 | 4% | $ 691,680 | 6% |
| Total | 11215 | 30% | 49156 | 59% | $ 5,173,550 | 44% |

**Note:** Total In policy + Surplus values do not need to equal On Hand Total

12

**FIGURE 9.6** Example of a maintenance parts inventory profile (by Stock room or total for a network). (Adapted from copyright 104 MPE Inventory Management ©2020 from Maintenance Parts Management Excellence Training edited by Barry. Reproduced by permission of Asset Acumen Consulting Inc.)

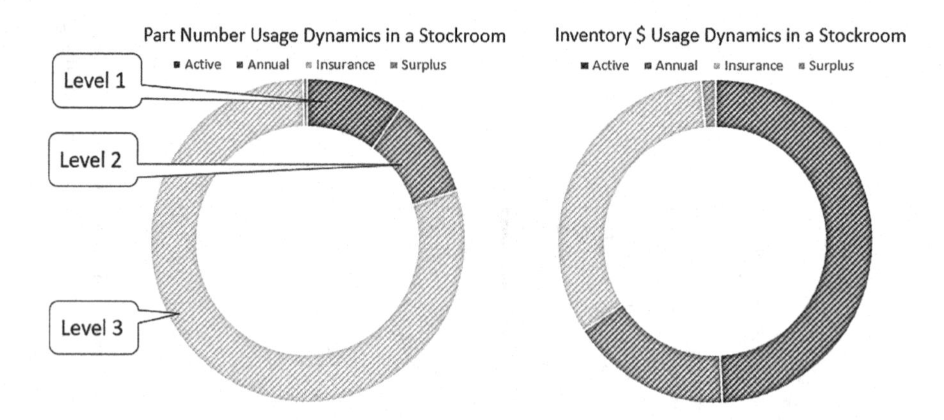


**FIGURE 9.7** Example: Parts usage mix in a maintenance parts stock room. (Adapted from copyright 104 MPE Inventory Management ©2020 from Maintenance Parts Management Excellence Training edited by Barry. Reproduced by permission of Asset Acumen Consulting Inc.)

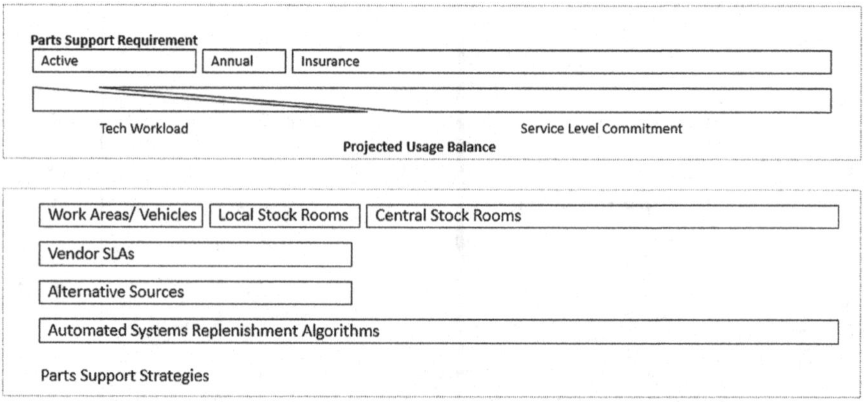

**FIGURE 9.8** Strategies to support parts stocking business dynamics. (Adapted from copyright 104 MPE Inventory Management ©2020 from Maintenance Parts Management Excellence Training edited by Barry. Reproduced by permission of Asset Acumen Consulting Inc.)

a planned 85% parts availability within 20 kilometers of the asset. However, there may be business risk situations where neither service level is appropriate when considering tolerable risk. A 90%, or even 95%, parts availability could be designed to support and tolerate risk, particularly if the next level of support is expected to be a 2–4-hour parts delivery cycle time. The notion of multiple stock room support hierarchies introduces the concept of "echelon" support for an asset from its closest stock room (Figure 9.10).

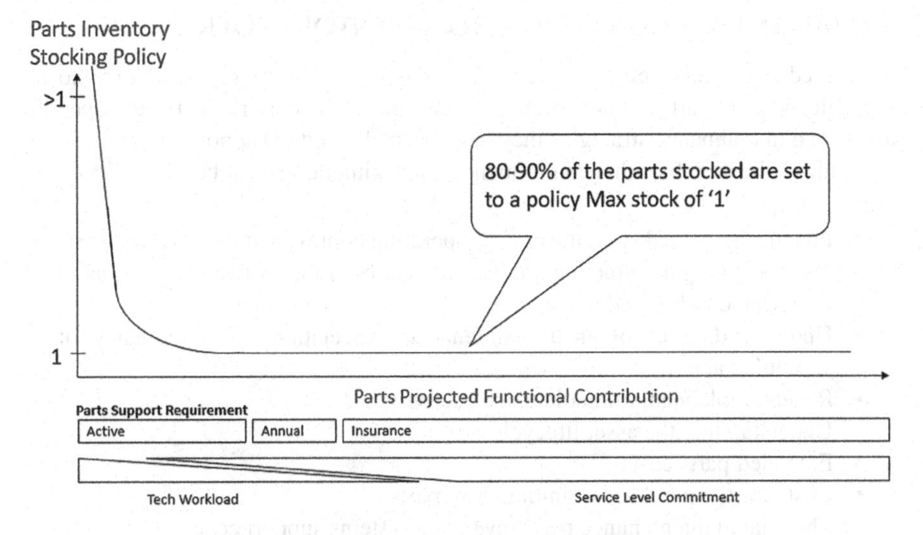


**FIGURE 9.9** Typical inventory policy mix maximum stock level per stock room. (Adapted from copyright 104 MPE Inventory Management ©2020 from Maintenance Parts Management Excellence Training edited by Barry. Reproduced by permission of Asset Acumen Consulting Inc.)

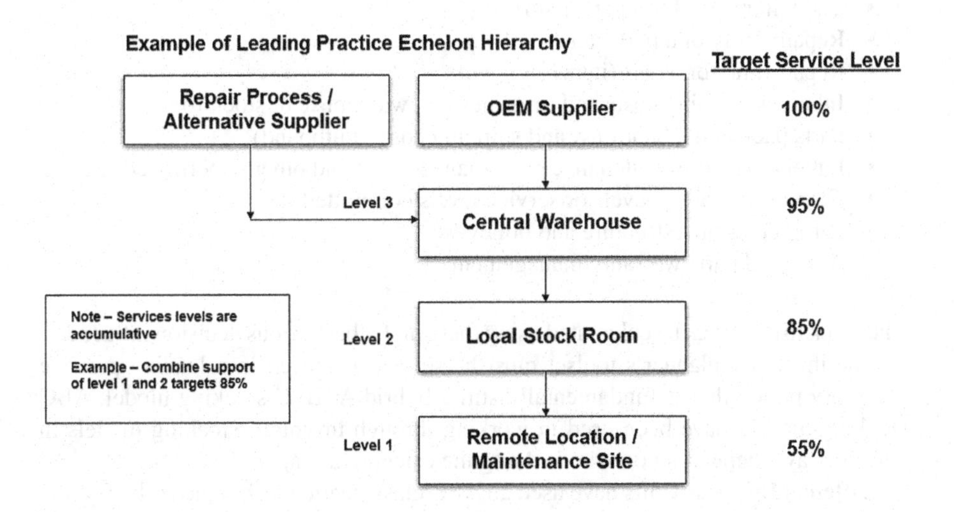


**FIGURE 9.10** The maintenance parts echelon network support model. (Adapted from copyright Figure 6.10 ©2011 from *Asset Management Excellence, Optimizing Equipment Life-cycle Decisions* edited by Campbell, Jardine, McGlynn. Reproduced by permission of Taylor & Francis Group, LLC, a division of Informa plc.)

## VARIABLES THAT CONTRIBUTE TO INVENTORY POLICY

An agreed-to inventory policy in an enterprise is a formal setting of inventory stocking guidance, by part number (item number), for each stock room for the specific assets and maintenance strategies they have been charged to support.

Some of the variables that can influence a stocking policy can be:

- Functional criticality of the assets' operating context and required support
- Understanding the functional failure dynamics of the part and how the asset is expected to be used
- Understanding of planned maintenance expectations and frequency of scheduled activity
- Recent unplanned, immediate need activity
- Understanding the asset lifecycle
- Extended parts costs
- Costs for the purchase of initial spare parts
- The related maintenance parts inventory systems support costs
- Number of locations supported
- The cost of the inventory support equipment, office space, asset support documentation, and training
- Financial influences on the supported asset value and the parts placement
- Surplus management
- Scrap management
- The failure rate for a part (item)
- Repairability of a part (item)
- Repair time for a part (item)
- Inbound shipping costs and timelines to a warehouse or stock room
- Parts packaging, handling, and shipping costs (outbound)
- Labor costs for maintenance parts management and other apportioned fees
- Target availability levels or service levels committed
- Parts repair infrastructure and policy
- Asset and parts warranty management

The stock to carry can be calculated using these and other various decision influences.

The inventory planner's toolset must be set and managed in a leading practice inventory policy that is fundamentally still a hybrid A, B, C stocking model. ABC stocking models have been used in working through inventory stocking models in various ways, depending on who is doing the categorization.

Different Business Units have used an ABC classification technique in the following areas:

- Design engineering:
  - Design engineering parts rationalization efforts are typically focused on Initial Spare Parts designations, and as such, they focus on items with forecasted high usage value rather than on items with a lower value (see the chapter on asset lifecycle planning).

- Procurement:
  - Purchasing activities should concentrate on higher extended usage value (Quantity * Price) items for sourcing and negotiating.
  - More sophisticated supply arrangements are applied to "A" items, while "B" and "C" items can even be considered for outsourcing.
- Inventory planner for replenishment:
  - Sometimes the classification scheme may influence the inventory replenishment control method.
  - It might be more economical to control some "C" items with a simple two-bin system.
  - Scientific management techniques are applied to "A" items.
- Warehouse personnel for Inventory cycle counting:
  - When checking inventory record accuracy using the cycle-counting method, "A" items should be verified more frequently than "B" or "C" items.
- Security:
  - Although the absolute unit price might be a better guideline than usage value, ABC analysis may also be used to indicate which items should be more tightly secured in locked stock rooms

We will discuss the A, B, and C designation for parts support as an echelon support level for the Inventory Planner role.

Table 9.3 was adapted from copyright [Figure 6.7 ©2011] From [*Asset Management Excellence, Optimizing Equipment Life-cycle Decisions*] edited by [Campbell, Jardine, McGlynn]. Reproduced by permission of Taylor and Francis Group, LLC, a division of Informa plc.

Given the business's reliability expectations, these Inventory policy settings would be in place for each level of maintenance parts support, whether it was the central warehouse, a network stock room, or a site location (Figure 9.11).

## PLANNING AND REPORTING INVENTORY RESULTS

The inventory planner will be impacted by the dynamic nature of how parts are ordered and used.

While the planner can set inventory policy, that is only one inventory control influence. Typically, the inventory and created parts availability service levels would be reviewed monthly with the enterprise operations and maintenance executive or, at a minimum, reviewed with finance. Activities that can influence inventory additions (and hopefully improved service levels) can be:

- New ISP/RSP adds
- Parts replenishment
- Parts warranty replacement receipts
- Unusual demand from the enterprise or maintenance technician (perhaps for an asset turnaround or enhancement)
- New parts returns from a customer or maintenance technician
- Surplus parts added from surplus past commissioning activities

**TABLE 9.3**
**Examples of Different Stocking Strategies with Three Levels of Inventory Policy Support**

| Stocking Strategies | ABC | Active vs. Insurance Stock | Criticality Stocking | Optimization |
|---|---|---|---|---|
| Level 1 | Top 80% used items by $ value<br>Less than 10% of stocked items | Top used parts (typically two usages in 3 months)<br>Less than 10% of stocked items | Level 1 is supported through either of ABC or Active Level 1 process<br>Less than 10% of stocked items | Low-value parts that are deemed to be used in asset lifecycle<br>Balanced with stocking and expediting costs |
| Level 2 | The next 15% used items by $ value<br>Less than 20% of stocked items | New part in the past year or at least one usage in the past year<br>Less than 20% of stocked items | High criticality, high usage parts not covered in level 1<br>Can be 20% of parts stocked | Cost-effective stocking of parts expected to be used once a cycle (e.g., year)<br>Can be 60% of stocked items |
| Level 3 | The bottom 5% used items by $ value<br>Can be than 70% of stocked items | Parts deemed to be stocked "Just in case"<br>Can be 70% of stocked items | Lower criticality parts and lower usage parts<br>Can be 70% of stocked items | High-value low, usage parts stocked, often at a consolidation center |
| Comments | Often manual initial stock process with min/max support | Often scientific initial stock process with min/max support | Can have many levels of criticality and echelon support depending on the support network | Considers all costs/impacts in stocking optimization calculations by network location |

Not all of these "adds" will improve inventory service levels, but they all increase the inventory value on the accounting books.

Activities that can influence inventory reductions (hopefully without impacting service levels) can be:

- Parts usage on work orders
- External parts sales
- Parts scrap due to a quality issue (new defective) or a warranty transaction
- A sale of inventory surplus
- The scrapping of surplus inventory that is not expected to be used in the coming years

These disbursements will reduce inventory levels and improve inventory carrying costs, and some directly contribute to inventory service levels (Figure 9.12).

Typically, the inventory- and service-level progress will be tracked and reported to an executive review counsel. For example, a monthly/quarterly report summary

**3 Level Echelon - Distribution Center or Warehouse**

| Inventory Policy Lead | Potential Stock setting for Stock Room Type |
|---|---|
| Active Parts | Max set to 12 weeks of stock |
| ISP/RSP | D Level ISP RSP |
| Optimizer | 95% service level target |
| Date protected | 2 years since last changed |
| Local Override | Manage to 4% of total value |
| Surplus | Manage to 20% of total value |

**2 Level Echelon – Local Stock Room**

| Inventory Policy Lead | Potential Stock setting for Stock Room Type |
|---|---|
| Active Parts | Max set to 8 weeks of stock |
| ISP/RSP | C Level ISP RSP |
| Optimizer | 85% service level target |
| Date protected | 1 year since last changed |
| Local Override | Managed to 2% of total value |
| Surplus | Manage to 10% of total value |

**1 Level Echelon – Asset Stock site (Technician managed)**

| Inventory Policy Lead | Potential Stock setting for Stock Room Type |
|---|---|
| Active Parts | Max set to 2 weeks of stock |
| ISP/RSP | B Level ISP RSP |
| Optimizer | 55% service level target |
| Date protected | 6 months since last changed |
| Local Override | Managed to 5% of total value |
| Surplus | Manage to 10% of total value |

**FIGURE 9.11** Example of how Inventory policy can be applied across the stocking echelons. (Adapted from copyright 104 MPE Inventory Management ©2020 from Maintenance Parts Management Excellence Training edited by Barry. Reproduced by permission of Asset Acumen Consulting Inc.)

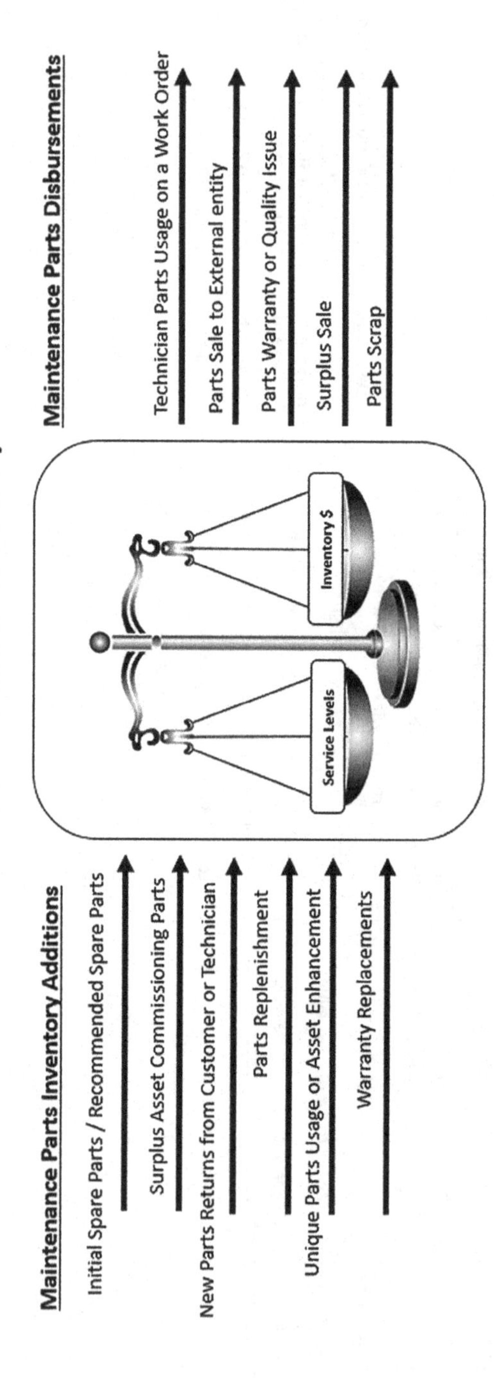


**FIGURE 9.12** The maintenance parts stocking dynamic influences summary model. (Adapted from copyright 104 MPE Inventory Management ©2020 from Maintenance Parts Management Excellence Training edited by Barry. Reproduced by permission of Asset Acumen Consulting Inc.)

**TABLE 9.4**
**Example of a Monthly Maintenance Parts Management Summary Report**

| Reporting Topic | Current Year Target | Year to Date, Achieved | Year-End Estimate | Comments |
|---|---|---|---|---|
| Inventory targets | $100M | $105M | $102M | Target at risk |
| Parts availability levels | 93% | 94% | 94% | |
| Parts reutilization savings | $25M | $18M | $28M | Higher usage Yr/Yr |
| Labor costs | | | | |
| Traffic costs | $4M | $2M | $4.1M | Up due to parts reutilization |
| Inventory scrapped | $4M | $1M | $4M | Needed to make YE target |
| Inventory provisions | $20M | $22M | $19.9M | |

**TABLE 9.5**
**Example of a Monthly Maintenance Parts Management Quality Report**

| Quality Metric | Target | Year to Date Achieved |
|---|---|---|
| Parts availability levels (%) | 93 | 94 |
| Parts sourcing competitiveness | 3% Year/Year reduction | 5% Year/Year reduction |
| Part customer feedback | 90% Satisfied | 90% Satisfied |
| Parts systems availability (%) | 99.8 | 99.9 |
| Parts Ddstribution quality (%) | 99.99 | 99.92 |
| Parts quality (new defective) (M) | $.9 | $.4 |

could contain a status on the business's critical inventory-related metrics and the overall maintenance parts management commitments.

A monthly report would include inventory targets (Table 9.4)

Quality would also be presented and could look like the following (Table 9.5):

Supporting charts and summaries would be provided for each of the metrics above (Reporting topics and Quality metrics). In addition, details would be presented for programs aligned to the Inventory Planner's activities, such as:

- Disposition of stocked parts that were supporting a discontinued asset set in the enterprise
- Top 12 parts that are causing Availability issues (sorted by source vendor or other relevant parameters)
- Parts expediting volumes (year to date and year over year)
- Quality issues caused by inventory availability, defects, or delivery issues
- A new adds chart showing "Year to date" adds versus plan and inventory withdrawals (surplus sale or scrap) versus plan
- Opening and closing inventory levels by month
- Parts usage and lines disbursed by stock room by month
- Service levels, inventory levels, and turnover by key asset types
- Enterprise-wide Maintenance Parts inventory turnover to date and a year-end forecast versus plan

## INVENTORY TURNS

At the highest level, an enterprise finance business unit may determine that inventory usage versus average monthly inventory should be used as a reasonable "year-over-year" metric for Inventory management. This metric is known as an inventory turnover measurement. The inventory turns metric is helpful when other data is lacking.

A maintenance parts inventory does not exist for the privilege of "turning." A maintenance parts inventory exists to optimally support a set of critical assets to the supply chain value of an enterprise.

The inventory planner should be aware of the impact of inventory policy decisions on the inventory turns. Finance wants this insight to justify a direction that does not promote increased service levels and turnover "year over year."

## INVENTORY CARRYING COSTS

The cost of carrying inventory includes the cost of:

- Storage facilities, store personnel for inventory handling
- Transportation between stock locations (if applicable)
- Insurance and tax
- Allowance for obsolescence due to engineering changes
- Allowance for loss due to pilferage or spoilage
- The time value of money (the lost opportunity for alternative investments)

The cost for placing orders includes the whole transaction process, from purchasing to material receiving and the cost of the materials purchased.

## DEALING WITH INVENTORY OBSOLESCENCE

Finance typically receives reports from the inventory planner and monitors expected inventory obsolescence and scrap activities. Scrapping inventory can be an emotional topic for many stakeholders in an enterprise. Maintenance personnel, for example, act like pack rats by nature and see value in things that may have virtually zero probability of needing the part again in the future of the enterprise. A cross-stakeholder agreement on the inventory stocking and support policy should help them be more comfortable assuming that their needs are addressed in the inventory policy. It is always easy to argue that holding on to a part is statistically better than no part if an unexpected demand arises, but which part and at what quantity? It typically costs 20–30% of the annual (monthly average) inventory value to hold the obsolescent part in stock. An obsolete part's annual inventory carrying cost is a high price for something you will never use. On the other hand, an actual tax credit can be gained by scrapping and recognizing the capital loss earlier rather than waiting for a better year.

Finance will be best suited to confirm if the forecasted inventory losses fit into their accounting for inventory provisions. Assuming finance can account for the inventory provision, it is expected that the enterprise would be best served by taking the loss sooner rather than accepting another inventory audit count verification cycle and then scrapping in a later year.

## IN SUMMARY

Inventory planning and policy management are the responsibility of the inventory planner. This chapter is aligned to describe many of the influences and activities this role would manage to drive a successful balance between inventory service levels, inventory levels, and the associated costs. In an ideal enterprise, the inventory planner would perform this role with little need to focus outside his assigned set of stock rooms/warehouses. Regrettably, this is very seldom the actual case.

The inventory planner will also need to:

- Influence total asset lifecycle activities to make sure that the ISP/RSP lists are well identified and sized by stock location echelon if this method is used
- Engage maintenance to ensure they take over the management of RSPs after an asset go-live and inform them when an asset type is discontinued
- Initiate inventory optimization activities when a new program or focus is set on the enterprise-specific Inventory costs and service levels
- Implement rotatable and repairable parts processes to improve parts sourcing options for availability and cost
- Work with procurement to make sure that vendor services can be leveraged where practical to meet the balanced goals of inventory costs and service levels
- Engage HR when the stakeholders need to be trained or prompted to embrace change that will improve the inventory cost and service level balance

Involve finance in all high-cost or service levels impacts supporting the maintenance parts inventory effectiveness

# 10 Technologies Supporting Maintenance Parts Management

## INTRODUCTION TO MAINTENANCE PARTS MANAGEMENT SYSTEM OPTIONS

Few business areas in an enterprise benefit more from optimal data and systems support capabilities than maintenance parts management. By nature, all inventory movement transactions and inventory policy stocking data lend themselves to continual scrutiny by the maintenance parts business unit and maintenance and operations. Likewise, an enterprise's finance group will constantly focus on the inventory values, inventory effectiveness, and the related financial influences on their bottom line as they report business results and forecasts.

For example, a small company's spare part inventory of 100 items could be managed on a spreadsheet. However, most maintenance organizations would not find this valuable and likely have thousands of potential parts it would need over the lifecycle course of a set of assets. In maintenance parts, significant data leverage is required to optimally manage transactions and effective inventory policy. Manual methods would not be efficient or effective in almost all scenarios (Figure 10.1).

Standalone inventory management systems may provide parts activity and in-stock visibility to parts and maintenance staff. Still, they, on their own, are also typically not the answer as they will need to integrate to some subset of an enterprise's:

- Financial accounting systems
- Purchasing systems
- Maintenance systems
- Production planning/scheduling systems
- Sales and distribution systems
- Vendor transportation systems

The above is a shortlist of ideal functional systems the optimal maintenance parts management will need to work with or have within its capability. A standalone maintenance parts system may be difficult to justify without a viable return on investment (ROI).

DOI: 10.1201/9781003344674-10

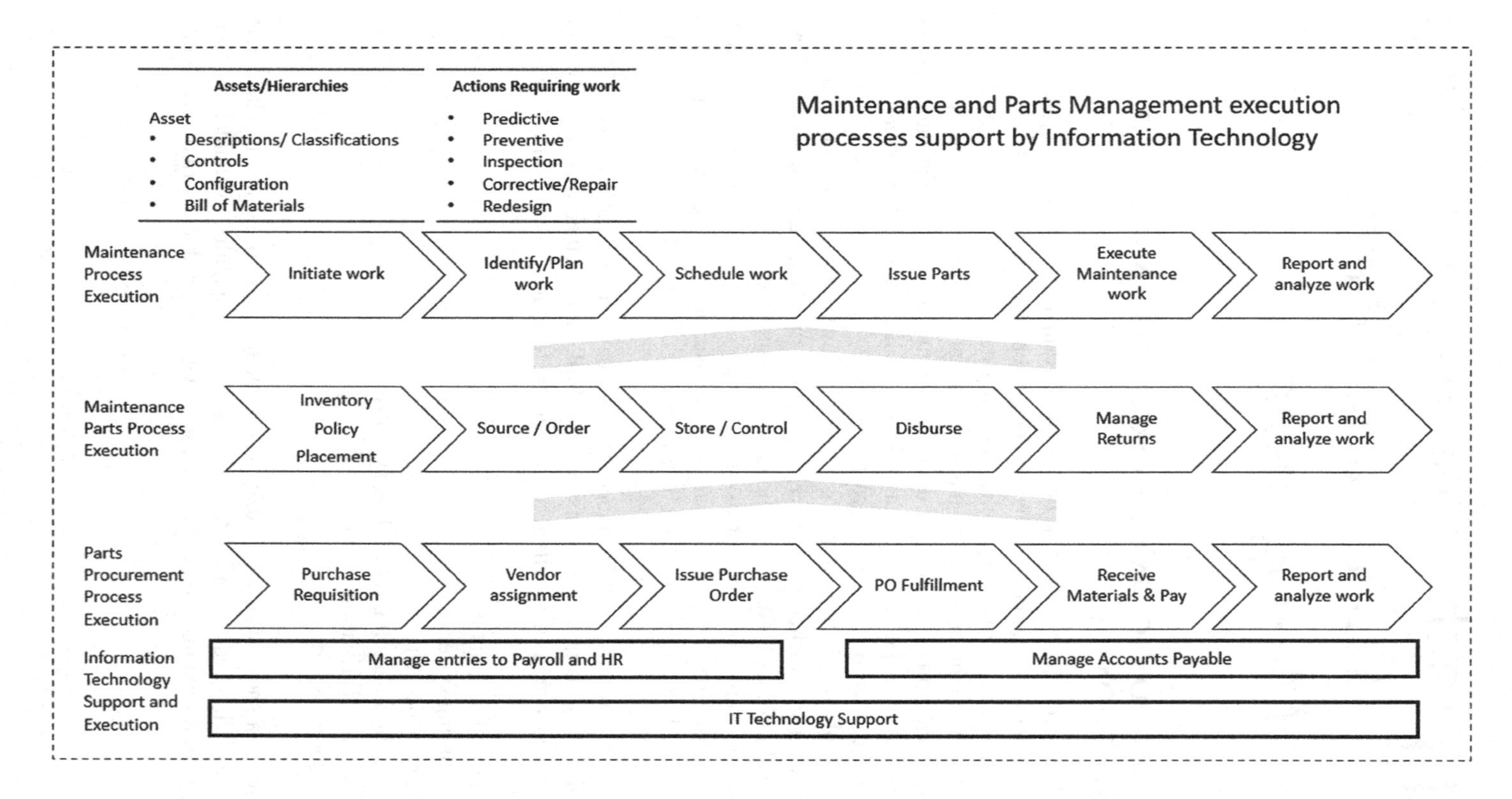


**FIGURE 10.1** Simplified placement of the role for Maintenance Parts Management. (Copyright MPE ©2022 from Maintenance Parts Management Excellence Training edited by Barry. Reproduced by permission of Asset Acumen Consulting Inc.)

In a highly simplified version of a maintenance parts management system, we would see that the parts system supports a role of working with procurement to provide parts to the maintenance technician as part of the maintenance work order process.

Enterprise resource planning systems and enterprise asset management (EAM) systems will offer a pre-integrated solution to many of the above list systems and the ability to integrate the missing elements. These integrated solutions go a long way to get a valued maintenance parts operation working efficiently. However, no ERP or EAM does all that a leading maintenance parts system may need to become optimal even with this ability.

## BASIC INVENTORY MANAGEMENT SYSTEM

The fundamental goal of a maintenance parts operation is to provide the correct part in the requested quantity to the requestor in the committed timeline. Therefore, it needs to be organized and have the forecasted activity to get the right part to the closest stock room when the maintenance technician creates the request. The inventory management system requires an inventory policy capability that will work across the supporting asset's lifecycle, hold and secure the part in the stock room until needed, and track all the inbound and outbound activity by item number.

Fundamental Inventory Management Systems will have the ability to:

- Display item number, description, related commodity data, and attribute data
- Quantities that are on hand, in transit, on order, etc.
- Location (stock room, bin, and overflow locations)
- Supplier (vendor) options and dynamics for each item (and location if different)
- Lead time to fulfill a supplier order (service level agreements – SLA data)
- Demand history by stock room and calculated average demand
- Usage data (if different than demand data)
- A "where used" or "equipment" list that the item supports by location
- Financial data including item cost, replenishment cost, item costing method
- ABC Inventory cycle counts, variance history, and transaction history
- ISP/RSP forecast levels by equipment (asset) type supported and by operating context
- Inventory replenishment policy development and approval
- Some forms of simple analysis tools manage data quality and report on costs, transactions service levels, and internal analysis (e.g., Bin trip analysis) (Table 10.1)

**TABLE 10.1**
**High-Level Capabilities of a Maintenance Parts Management System from an EAM Supplier (IBM)**

| General Capabilities | Materials Management | Purchasing | Contract Management |
|---|---|---|---|
| • Reports and KPI management | • Item Master, Service Items | • Purchase requisitions and Purchase Order management | • Purchase contracts |
| • Mobile technologies for order picking, receipts, and cycle counting | • Tools management | • Order revision management | • Lease/Rental Contracts |
| • Integration to: | • Inventory balance visibility | • Shipment Receiving | • Labor rate contracts |
| • Financial systems | • Inventory transaction visibility | • Invoice management | • Master contracts |
| • Maintenance systems | • Costing method options (average, standard, LIFO, FIFO) | • Request for Quote management | • Warranty contracts |
| • Asset (equipment) ISP/RSP strategies by asset/location | • Consignment inventory | • Supplier master data management | • Software contracts |
| • Vendor transportation systems | • Commodity groups and codes | • Terms and conditions | • Pay Rate management |
| • Vendor-consigned inventory systems | • Classifications, Specifications, Attributes | • Currency and exchange rate management | • Payment schedule management |
| • Supplier systems | • Network and Stock room location management | • Financial Chart of Accounts tracking | • Service catalogs |
| • Warranty approval systems | • Intercompany transfer | • Cost management | • Service groups |
| | • Reservation management | • Receipt tolerance | • SLA Management |
| | • Kitting | | • Service Incident Management |
| | • Picklists | | |
| | • Cycle counts | | |
| | • Inventory Policy Management | | |
| | • Minimum, Maximum, EOQ algorithms | | |
| | • Automated replenishment | | |
| | • New parts returns | | |
| | • Used parts returns | | |
| | • Repairables/Rotables management | | |

### Inventory Reservation Types

Critical parts orders planned for a near-future date, such as the next two weeks, may benefit from locking in the required parts against the planned work order. In these cases, an inventory reservation capability helps secure the part and allows some visibility should a true emergency arise before the part is planned to be used. Inventory item reservations can come in three types (Hard, Soft, and Automatic):

- Hard Reservation:
  - Set by inventory planner
  - A requirement for a part by a specific date, usually tied to a work order
  - The requested balance is on hand
- Soft Reservation:
  - Set by inventory planner
  - A requirement for a part, not date-specific and usually tied to a work order
- Automatic Reservation:
  - Can be hard or soft reservations
  - Set by maintenance parts system based on selected criteria and run daily

Stock location types should be flexible enough to allow inventory to be accounted for and managed as:

- Vendor inventory/location
- Stock rooms in the enterprise
- Maintenance Technician locations such as a remote location or a service vehicle
- Courier held inventories
- Holding locations for capital spares or kitted parts for planned work orders

Inventory transfers between stock rooms should be available in the system to facilitate Transfer requests that may be automatically created due to a replenishment demand from a weekly calibration of inventory needs in a network location or manually created as needed (ad hoc).

Leading maintenance parts management systems will feature visibility to the essential data elements of the inventory planning, financial, safety, and parts transaction processes. A maintenance parts inventory system should be relatively easy for the inventory planner and the material handlers. Navigating their system, they should be able to quickly locate essential inventory data for their location and across their enterprise (or network), if needed. As an example: Inventory related data screens could be in the format of displaying:

- Stock rooms (and locations)
- Inventory item master data (including commodity codes and condition codes)
- Tools data
- Inventory usage(disbursements), transfers, and receiving data

It should be easy for the system to display essential item data when looking deeper into the Inventory specific data. An example of the essential item data is:

- Item number and description
- Storeroom and site location
- Unit of measure (quantity of X or meter or volume)
- Expirations or shelf-life dates (if applicable)
- Safety data (e.g., hazardous parts, special transportation instructions)
- Unique item receiving instructions (e.g., inspection requirement or tolerance)
- Primary stocking (bin) location (and alternates, if applicable)
- Type of part (rotating, tool, consumable, etc.)
- Critical financial data (e.g., applicable tax codes, item costing method, Standard cost, weighted average cost, last receipt cost, FIFO, LIFO)
- Inventory adjustment histories (period to date, current balance, physical count)
- Inventory balance dynamics (quantity on hand, quantity in transit, quantity on order, quantity reserved for committed work orders, quantity staged
- Date last used, date first added to the system
- Inventory policy tactic (active, date protected, ISP/RSP protected, local over-ride protected) and a netted policy inventory reorder level, minimum stock level, maximum stock level, and economic order point
- Condition code data could display as well – if applicable
- Usage (disbursement) history data could be displayed for the past year and the previous two to three years if needed

The ability to see if a specific item is in stock in another stock room within their network should also be easily accessible. A part referral to the stock room with the available part could be generated when required.

The system should manage and track Item movement transactions from an inventory location and track asset movement activities for non-inventory items. Inventory (Item) transactions can typically be:

- Inventory issues
- Inventory returns
- Inventory transfers (to another stock room)

The transaction record should have an auditable trail data set that would include:

- Transaction unique identifier
- Item number, item description, the quantity disbursed
- Issue type (Issue to usage, transfer or return, etc.)
- Relevant stock room and location (to and from) data, including work order number if applicable
- Relevant financial accounting (general ledger codes, costing method) data

Non-inventory (asset) transactions can typically be:

- Asset moves from one location to another (an asset can include rotating repairable assets if designated – that is carried as an item in inventory)

## INVENTORY POLICY MANAGEMENT AND REPLENISHMENT ALGORITHM MANAGEMENT

Inventory management systems aim to create a service-level solution that enables the complete maintenance parts supply chain, from suppliers to network stock rooms and warehouses to maintenance technicians and network customers (if parts are sold).

Ultimately, the inventory planner would optimize the inventory of a multi-echelon distribution network to reach best-in-class customer service at the lowest possible cost. A system designed to minimize manual efforts drives exception management, strategic simulations, and automatic optimization.

Each item in each stock room location should display the inventory policy influences for each item listed against that stock room. The data elements should be:

- Whether the part is enabling automated replenishment to be calculated
- Inventory policy tactic (active, date protected, ISP/RSP protected, local over-ride protected) and netted policy inventory reorder level, minimum stock level, maximum stock level, and economic order quantity
- Visibility to all quantity influences for an item (part) in a specific location (quantity on hand, quantity on order, quantity in transit, quantity in available for repair, quantity reserved or staged for work orders, etc.)
- Visibility to an item (parts) assembly hierarchy and direct substitution
- Monthly and annual inventory planning by region and globally
- Network visibility and automated stock referral capability (with parts locator)
- Scenario simulations
- Unique demand and seasonality considerations
- Repairable parts and rotatable inventory planning
- Multi-echelon
- Excess re-distribution
- Reporting & KPIs
- Source vendor (supplier) name, description, and item name, number, and description in the vendor's catalog
- Manufacturer vendor name, description, and item name, number, and description in the vendor's catalog
- Algorithm data sources for:
  - Netting inventory policy
  - Calculating new stocking support levels
  - Calculating replenishment transactions

The system should be able to hold or integrate the source data for:

- Demand forecasting
- ISP/RSP netted settings by asset type (equipment supported) and location (see chapter on asset lifecycle)
- Optimizer settings by item and location (see chapter of asset optimization management)
- The data elements used to calculate a revised inventory policy each week or each item by location (see chapter on inventory policy management)
- Automated stock replenishment

## MANAGING ROTATING AND REPAIRABLE ITEMS IN INVENTORY

A "rotating" item will have an Item number, inventory value, and current balance in inventory and can also be maintained as an asset in the asset management system. A rotatable item will not be entirely consumed as it remains an asset on the financial accounting books. However, it will typically replace another rotatable item when needed in an operational operating context. When being addressed as an asset, the same item is interchangeable and will have a unique location number and, in some cases, a unique asset (or serial) number (Figure 10.2).

Repairable parts and rotatable parts, when applied in maintenance, replace the defective part in operational use. Figure 10.2 that either the used repairable or rotatable part will be tracked as a "used part" back to a used parts disposition depot. The parts will be stored at the Used Parts depot as an "available for repair" status until needed to be repaired. When demand for repaired (like new) parts is evident at the central warehouse, the used but "available for repair parts" will be sent to a repair depot (or a third-party repair vendor) to be repaired. The third-party vendor will repair, test, and package the repaired part for reuse as a repaired (like new) part and ship it to the central warehouse to be received into stock.

The maintenance parts management system will need to manage the tracking of the movement of each item through each step of this process. Rotatable parts will already be serialized, and the repair process and quality will likely be tracked using that identifier. Likewise, repairable parts will have a unique repair tracking identifier added to them to enable post-repair quality tracking. Related KPIs for this process will also need to be managed by the supporting systems.

Finance will treat repairable parts, particularly rotating items, by how their value is calculated when repaired and returned to stock. Some organizations may elect to assess a percentage of the total new value for repaired or rotatable parts (e.g., 80% of the average cost of new). Other organizations may elect to rationalize that the returned part or rotatable item has a residual value and may apply an algorithm with the cost of repair plus 25% of new as the value set to return a repaired part to stock (e.g., 25% X average new + cost of part repair). For simplicity, the repair costs may be averaged against all repairs of a specific repairable item in such a calculation.

Ideally, the maintenance parts system can manage the related data and transactions to make this viable activity easy to execute and report for the financials, quality, and critical business metrics.

Repairable parts and non-repairable parts can be considered "non-rotating" parts. They are simply the regular parts in the maintenance parts system. At a high level, they typically would be:

- Managed with an inventory bin location and inventory quantity and balance
- Consumed to a work order
- Managed as a spare part with a spare parts policy
- Optionally purchased and managed in "lots"
- Optionally be "condition" tracked if repaired, returned from the original asset commissioning activities, or some other approved process
- Optionally be kitted for planned maintenance activities
- not uniquely serialized tracked like rotating items unless worked through the repair process as a repairable part

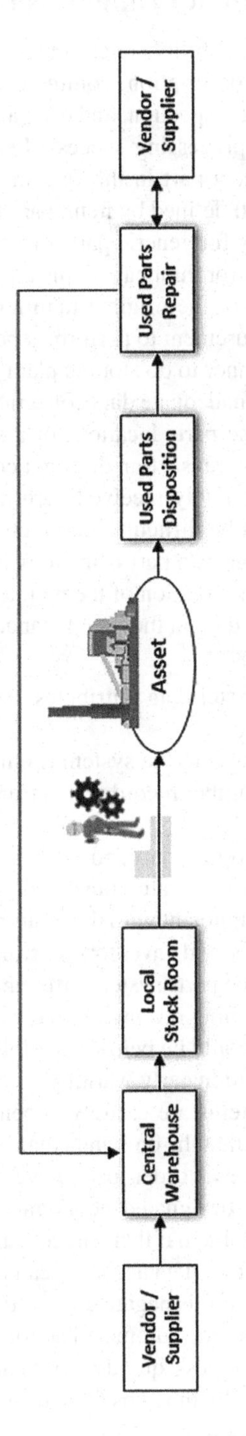


**FIGURE 10.2** Example of rotatable and repairable parts high-level process flow. (Adapted from copyright 101 MPE Introduction ©2020 from Maintenance Parts Management Excellence Training edited by Barry. Reproduced by permission of Asset Acumen Consulting Inc.)

## THE IMPORTANCE OF DATA ACCURACY IN PARTS MANAGEMENT

Parts (item) data accuracy and availability are critical to a leading maintenance parts system. Understanding the item description, commodity group, attributes, what to look for when doing a part receipt inspection, and even a system-stored picture could be considered essential to parts processing success. In many cases, a unique item number would be applied to a new part in the system with a cross-reference to a supplier's part number. When well defined by item, part attributes allow the procurement team to search more broadly for generic parts that meet the required standards. Attribute descriptions also search for duplicate items that may already be in stock or become a recognized substitute for the original item requested. Commodity codes and attribute data will allow procurement to perform "spend analysis" and "sourcing analysis" and the stock room planner to do storage planning.

Limited data requirements or inadequate data collection will lead to an ineffective parts operation. Time is lost if the parts locations of a specific stock room are not accurately executed or the receive transaction does not control putting the part away in the designated location. Inaccurately received parts will be the beginning of an inventory integrity issue that will be difficult to correct. It will create lost time and expedited parts if the system suggests a part is in stock but cannot be found. Such a simple misstep also degrades the reputation of the maintenance parts operation. Two widespread causes of insufficient data in the maintenance parts system are:

- Poor control of the item master data (attributes, description, vendor cross-reference, etc.)
- Flawed transaction reporting into the system (item receipts, disbursements not reported, wrong item number recorded in a transaction)

In many cases, an inventory system's on-hand dollar balance will not match the financial systems over time. This item value mismatch can be caused by inheriting flaws in the processes and their alignment with the systems. It will also be influenced by how finance manages invoices and inventory costing versus how your system manages. Finance and maintenance parts system differences should be reviewed and reconciled quarterly. When performing inventory count verifications, variances in the on-hand balances to the system should be below 2% (dollar value). Above 2% could suggest a system or process execution early warning.

Automated replenishment systems are entirely dependent on the data provided. For example, if a system suggests that the on-hand balance is zero and the order point is two, the system will likely create an order to whatever the system declares is their maximum stock level or by the designated economic order quantity. This auto-replenishment process will be a problem if there are actually "10" pieces on hand, but the system cannot see this. Therefore, data accuracy can have a significant impact on optimal parts management. The result is negative when the data is incorrect but optimistic when the data is correct, and the automated actions are effective.

Training will help maintain process quality for material handling staff. Stock room access control will also help Technicians from missing the reporting of an item disbursement they forgot to declare during off-hours.

The leading stock rooms understand the relationship between item transactions and the system records. One misstep in receiving (as an example) can cause a need to do variance counts, inventory reconciliations, system inventory adjustments, receive unneeded replenishment, expand a bin location to accommodate excess stock, and explain why the inventories are inflated and the variances are high. Receiving accuracy and timeliness may be a discussion for the warehouse management team, but it is critical for system data accuracy.

## ITEM PARENT–CHILD RELATIONSHIP TRACKING

The leading systems will have the where-used asset data for a specific item. For example, the "where-used" data field would be helpful when adding new initial spare parts (ISP) supported parts and becomes critical when an asset is at end-of-life so that surplus parts can be declared and disposed of equitably.

Similarly, significant parts components will have parent/child relationships acknowledged in a leading maintenance parts system. For example, a diaphragm on a reciprocating compressor may not uniquely be available as a stocked part. Still, understanding that the item is a sub-component of a larger item called "compressor" could help the maintenance technician and the material handler support a key "asset down" scenario (Table 10.2).

A leading maintenance parts system will be able to create parent–child relationships starting with a parent or the higher-level rotating item. They should also be able to associate these parents with other related items. Typically you would purchase and receive the whole assembly with this capability or apply the received item assembly structure later. Purchasing would leverage the item master record in a system and serialize the child records as part of the receiving process. In addition, the maintenance parts system should be able to manage a specific parent–child set of data attributes.

**TABLE 10.2**
**Example of an Item, Parent–Child Relationship**

| Relationship | Level | Example (Options) |
|---|---|---|
| Parent | 1 | Compressor |
| Child | 2 | Rotary<br>Reciprocating |
| Child | 3 | Rotary<br>• Screw<br>• Lobe<br>• Vane<br>• Ring<br>• Casing<br>Reciprocating<br>• Piston<br>• Diaphragm<br>• Casing |

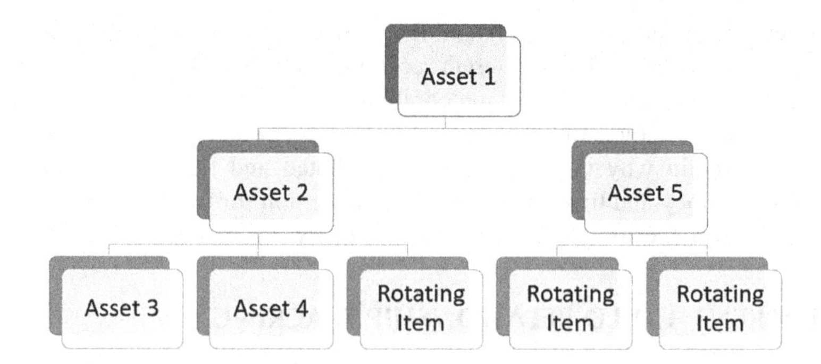


**FIGURE 10.3** Example creating item assembly structures. (Copyright MPE ©2022 from Maintenance Parts Management Excellence Training edited by Barry. Reproduced by permission of Asset Acumen Consulting Inc.)

For example, applying an item assembly structure to an existing asset: rotatable items can be linked to a sub-assembly, and non-rotatable items can be added to a spare parts list. A leading system would also allow easy movement of assets in a hierarchy or lock changes if needed (Figure 10.3).

## COMMODITY CODE AND CONDITION CODE CONSIDERATIONS

Many items in a stock room may have some commonality that could help work through transactions with:

- Companies (vendors)
- Purchasing
- Tools
- Service items
- Any item in the item master

Commodities will likely be grouped as either items, tools, or services. Within these commodities, commodity codes will help to standardize:

- Different commodity sets
- Commodity hierarchies across several levels

The US Department Standard Classification of Transported Goods and the United Nations Standard Products and Services Code (UNSPSC) could be applied here.

As an example, The US Department Standard Classification of Transported Goods would organize a commodity set as follows:

44 Office Equipment Accessories and Supplies (Segment)

10 Office Machines and their supplies and accessories (Family)
11 Office and data accessories
12 Office

15 Mailing supplies (Class)
16 Office supplies
17 Writing instruments
18 Correction media
19 Ink and lead refills
  01 India Ink (Commodity)
  02 Lead refills
  03 Pen refills

There may be scenarios where a part received is stocked but is not at the full (like new) condition. In these events, the part may be stocked in a unique location, but often it is identified and stocked with other parts. Finance will want to treat the value of these parts differently than a "like new" part. For example, codes may suggest that a new part is valued at 100%, but a repaired part is valued at 50% and a slightly used part at 80%. The ability to declare the condition when receiving these parts into inventory will help track the inventory values for such a scenario. The Inventory management system should distinguish that they now have X like new parts in stock, Y repaired parts, and Z slightly used parts as their inventory mix.

## EVOLUTION OF MAINTENANCE PARTS SYSTEM REQUIREMENTS

Maintenance parts systems recognize the need to be competitive in the market. This competitive need is particularly true for EAM solutions that are uniquely focused on asset maintenance and the supporting processes (e.g., procurement, parts management, vendor service level management).

Leading solutions will have improved vital areas in the part's transaction management, integrate with automated storage and retrieval systems (ASRS), provide deeper analytics, and further manage and automate the minimum, maximum, and economic order quantity capabilities.

Some of the gaps recently addressed or being addressed are the ability to:

- Handle used parts returns, parts repair processes, and planning for "like new" parts to be returned to stock in a pull order process
- Do item-related lifecycle history tracking
- Handle the accounting issues inherent in a work order or field service operation
- Handle warranties for equipment and maintenance parts
- Maintain a historical record of the demand and generate rolling average demand algorithms
- Handle a "parent/child" item relationship to a bill of material (BOM) of a specific asset as needed
- Track maintenance technician orders and backorders through multiple tiers of inventory stock rooms
- Manage lot-sizing quantities in the system
- Track "moving" inventory located in technician vehicles

- Track "moving" inventory located at parts repair vendors
- Track "used/defective" inventory located in the waiting-for-repair stock room
- Manage a network of stock rooms with hierarchy echelon support parameters
- Calculate weekly a new inventory replenishment policy based on multiple parameters maintained in the system by stock room location
- Generate daily and weekly replenishment orders based on multiple parameters
- Leverage blockchain to support order and item tracking audits
- Create wave-picking lists to optimize the end-of-day picking effort

## MANAGING TOOLS IN A MAINTENANCE PARTS MANAGEMENT SYSTEM

Most leading maintenance parts systems will also be able to manage tools and the unique activity related to securing, repairing, managing, and calibrating tools. Like inventory, there likely will be rotating and non-rotating tools to be managed. The tool would have an item number and stock location and be disbursed to a work order for the maintenance technician to manage while out of the stock room. Again, like inventory, the tools system would have the ability to:

- View tool availability
- Manage tool status for the stock room
- Issue tools
- Manage tool balances
- Make inventory adjustments (as needed)
- Perform vendor analysis and other related KPI metrics (Table 10.3)

All tools management systems should have the flexibility to assign an hourly usage rate to be charged to a work order with proper accounting setup capability in the maintenance parts system

**TABLE 10.3**
**Stocked Versus Rotating Tools System Requirement Considerations**

| Stocked Tool Functionality | Rotatable Tools Functionality |
|---|---|
| • Tools can be "reserved" to a specific work order | • Managed rotating assets |
| • Tracks the location of a tool | • Can reside in a stock room and therefore issued and returned |
| • Resides in stock room until needed | • Maintenance plans can be created against the use of a rotatable tool |
| • Issues to a work order (job) | • Calibration can be tracked against the qualification of a rotatable tool |
| • Returned after the work order (job) | |
| • Can be cycle counted as part of inventory controls | |

## INVENTORY INTEGRITY MANAGEMENT (CYCLE COUNTING)

Cycle counting is the process of continually validating the accuracy of the inventory in the maintenance parts system by regularly counting a portion of the inventory on a daily or weekly basis so that every item in your inventory is counted several times in a given annual cycle.

ABC N cycle count designations

- All items in a stock room are to be counted and verified at least once a year
- ABC concept spreads the counting workload over the year
  - A Parts – Higher extended value parts
    - Counted more than five times a year
    - Typically, "Insurance" spares or high-value/unique items
    - Target counts every 30 days
  - B Parts Medium extended value parts
    - Counted 3–5 times a year
    - Target counting every 60 days
  - C Parts – Low extended value parts
    - Counted less than three times a year
    - Target once a quarter
  - N Parts
    - Do not count parts
    - Not inventoried parts, extremely low value, and high quantity parts

The system should help designate a Financial approved cycle counting tactic for ABC, N parts. Including confirming negative balance options. It should also be able to:

- Auto creates inventory adjustment transactions as part of the inventory variance reconciliation processes
- Create variance report detail and summaries for approval from finance as needed

## FINANCIAL TRACKING OF INVENTORY TRANSACTIONS IN THE MAINTENANCE PARTS MANAGEMENT SYSTEM

The system should have the ability to manage different inventory accounting methods such as:

- Standard
- Weighted average costing (WAC)
- "First-in, first-out" (FIFO)
- "Last-in, first-out" (LIFO)

Inventory receipt calculation capability

- An example would be to perform a receipt of 10 items to a stock room:
  - The current inventory quantity on hand is increased by 10 at $15 each:
    - The costing method is "average"
    - The current quantity on-hand balance is 6 at an average cost of $12 (for a total of $72)
    - With the received additional 10 pieces
    - The new Inventory quantity on-hand balance becomes 6 + 10 = 16
  - The new average cost becomes:
    - ([Quantity on hand X average cost] + [Added piece count X purchase price])/(new Quantity on hand)
    - ([6 pieces X $12)] + [10 pieces X $15])/16
    - ([$72] + [$150])/16 = $13.88 each

Inventory Issue transaction capability

- An example would be to issue 10 items from a stock room to a work order:
  - The costing method is "average":
    - The average cost is $13.88
    - The current quantity on-hand balance is 16
  - With the 10 pieces being disbursed:
    - The new inventory quantity on-hand balance becomes 16–10 = 6
    - The new inventory on-hand level becomes
      (new quantity on-hand) X (average cost)
      (6) X ($13.88) = $83.28

Inventory adjustment transactions

- Inventory adjustments will be required when an inventory count determines a variance between what is counted in a specific stock room and the system quantity on-hand balance
- If you reduce the balance down by a quantity of 4 and each item costs $13.88, then the inventory value should be reduced as a result of the inventory adjustment by 4 pieces X $13.88 = $55.52

Inventory adjustment implications

- Care must be taken to ensure that inventory adjustments have been investigated and approved before executing them in the maintenance parts system
- All transactions against a specific stock room create an expense/credit accounting that will go directly to the enterprise's bottom line
- An issue or transfer out will decrease the stock room's value
- A receipt or return will increase the stock room's value
- A physical count adjustment does nothing to the stock room's value

The system should manage unique capital spares for the financial accounting of how finance may treat them. Various-sized capital projects need to manage parts in the system but treat the capital project demand as "exceptional demand" when calculating weekly demands for inventory policy calculations.

## AUTOMATED PROCUREMENT LINKS TO SUPPLIERS

A recent shift in the role of procurement is to set up the ability for an automated transaction between the maintenance parts management system's procurement capability and the supplier's catalog and order systems. This allows procurement to have their staff focus more on "supplier relationship management" and less on the actual "procurement transaction." Electronic commerce capabilities have been in place for over a few decades. However, a maintenance technician could punch out to a supplier catalog, generate a purchase requisition that would be auto-approved in many scenarios, and auto-generate a "purchase order" directly to the supplier. This type of interconnectivity is emerging more and more to make the process more efficient. Buyers now buy the products of the catalogs offered by the vendors through the portals of vendors.

Now e-procurement with third-party-based middleware web portals can host the e-catalogues of the suppliers and facilitate the selection and purchasing document creation (PR/PO). Furthermore, the open catalogues interface (OCI) is a business to business (B2B) mechanism. A buyer's system can "punch out" to link to a third-party tool via the internet for supplier product view and selection (Figure 10.4).

Leading organizations install an open catalogues interface (OCI) solution/system capability that will readily interface with their own procurement/maintenance parts management systems and key supplier's order management and catalog systems via the internet. Suppliers make this type of transaction available to promote customer loyalty. Such end-to-end interaction between businesses (B2B) would feature:

- Purchase requests capability
- Supplier marketplace and feature-rich catalogs
- Purchase orders
- Intelligent sourcing
- One-click goods receipt/confirmation
- Mobile framework capability
- Web services framework for integration with ERP/EAM systems

With OCI, buyers can see the product ranges of different vendors and their prices and choose to select and purchase. For the buyer using this type of system, the perceived benefits of this type of process tool could include:

- Cross catalog search across all suppliers
- One view of all supplier content
- Punch-out integration
- Dynamic product check based on rules
- The potential for internally unique fields and input boxes
- Supplier diversity attributes
- Price and lead time comparisons

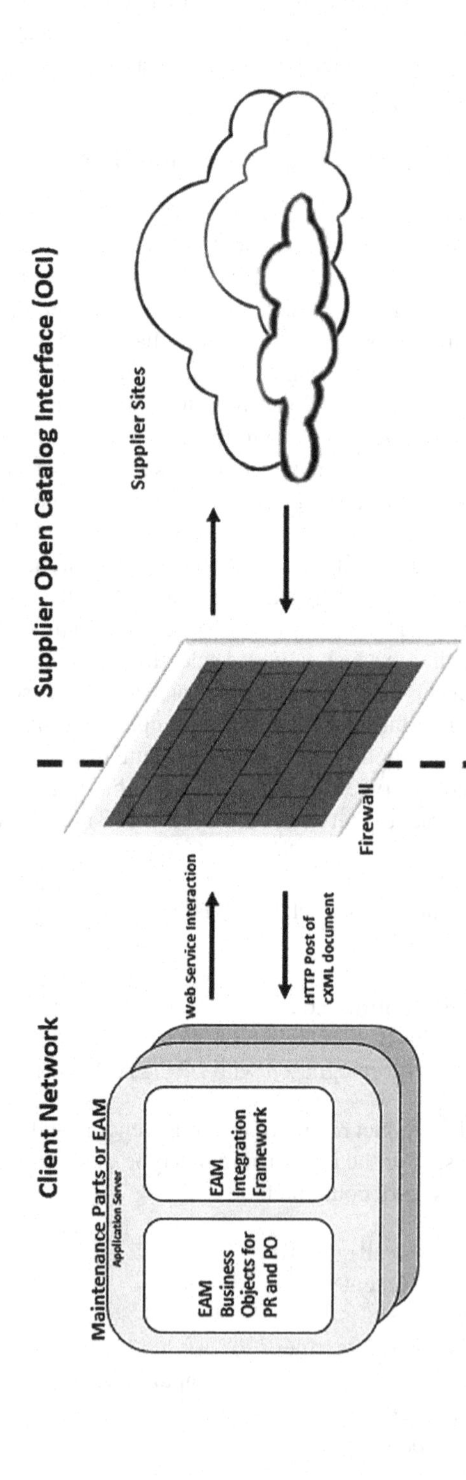


**FIGURE 10.4** Example of a simple Open Catalogues Interface (OCI) to a Maintenance Parts Procurement system. (Adapted from copyright 104 MPE Inventory Management ©2020 from Maintenance Parts Management Excellence Training edited by Barry. Reproduced by permission of Asset Acumen Consulting Inc.)

Third-party vendors also offer services to facilitate multiple supplier views for the same item by acting as the consolidated catalog input for many common suppliers.

## PROCUREMENT BLOCKCHAIN CAPABILITIES

Blockchain is a new relatively new term. It can add security and accuracy to a parts order transaction, which could be particularly helpful if sourced from a foreign country.

A Blockchain is a distributed ledger structured in fixed sets (blocks). Blockchain comes with:

- A header, which would contain a reference number, timestamp, and a link to a previous block in metadata
- Content to validate a set of digital assets and instruction statements (e.g., transaction, quantity, and the relative stakeholder's critical location data)

It would be linked to the previous block to maintain history, making it verifiable and independently auditable.

Blockchain can help trade, logistics, and shipping when freight logistics involve many different parties (e.g., manufacturers, forwarders, shippers, customs agents, and insurers). These parties use different systems to track shipments with potentially contrasting or conflicting operating or business targets.

An IoT-enabled blockchain was used as a shared ledger to record shipping containers moving through a logistics system and automatically update the "smart contract" and blockchain through an AI cloud solution.

Specific to supply chain transactions such as maintenance parts or production supplies, it helped to provide:

- Greater transparency of shipment progress to improve shipping efficiency
- Greater trust since all transactions are indelibly recorded
- Greater accuracy and lower cost
- Ability to optimize and automate business processes
- Confidence in a future vision that would enable "freight autonomy"

## INTEGRATING AUTOMATED STORAGE AND RETRIEVAL SYSTEMS OR VERTICAL LIFT MACHINES INTO A MAINTENANCE PARTS SYSTEM

Automated storage and retrieval systems (ASRS) and vertical carousels have a place in high-volume parts operations. The benefits can be taking advantage of the full height of a warehouse ceiling space for storage and reducing the labor it takes to put parts away, do cycle counts, and pick parts. In one example, installing a total of 17 carousels (or vertical lift machines - VLM) was justified because of their ability to densely store a large volume of smaller parts and eliminate an approximate 40,000 square feet of warehouse floor space. The consolidation to high-density storage in the warehouse generated additional floor space to manage large maintenance turnaround projects at a large facility. Material handling labor savings were also identified in the business case. There is little doubt that labor savings would be experienced for higher volume (wave) picking orders, but VLM may slow down immediate need (emergency) picking time. Integrating the VLM systems to the maintenance parts system needed to consider the following transaction activities as part of this justification:

- Item master, Item/location assignments and updates, cycle count activities, negative confirmations (discrepancies), order number alignment with other lines to be picked on the same order, pick efficiencies and parts put away efficiencies

Integrating would need to consider six bi-directional interfaces and one single-direction interface plus two reports: system reconciliation report and cycle count report.

Critical decisions would be taken to confirm how they plan to work detailed scenarios such as:

- Inventory discrepancies:
  - Where the system suggests a quantity on hand of 3, the request is 4, and the pick quantity is 3, a negative return confirmation is 1
  - Where the system suggests a quantity on hand of 3, the system says 1
- Stock location communication:
  - Primary and overflow locations
  - Which bin is the primary pick location
- Misplaced part in the carousel. Much more difficult to find than in a manual storage system
- Cycle count data formats

Ultimately a risk versus benefit decision would need to be made on the criticality of a part stored in an automated system if the item was urgently needed and the location VLM unit is down due to a power, system, or software failure.

# 11 Asset Lifecycle Alignment to Inventory Spare Parts Management

## INTRODUCTION TO ASSET LIFECYCLE

An asset management lifecycle includes managing the asset from a business strategy (where the assets' function (value) is accounted for) through the asset design stages, into the constructing and commissioning stages in preparation for the operating stage and supportive maintenance and modifications that may be expected throughout the asset's operation and inevitably the asset de-commissioning.

Before committing to the enterprise, the asset lifecycle should also include an expected de-commissioning strategy and cycle to be well understood and documented as part of the asset's business case.

Total asset lifecycle management (ALM) includes a lifecycle for each part (or item) in the composition of each asset in an asset lifecycle. It should include considering an inventory stocking policy from parts' acquisition to the disposal or disposition of its eventual parts. Asset lifecycle management would include costs for the initial asset purchase and related initial spare parts purchases required to support an asset installation, repair, replacement, upgrade, and movement to and from repair facilities (if applicable). The removal from service, dismantling, and disposal should be considered when designing and costing out the total cost of ownership (TCO) (Figure 11.1).

The asset lifecycle management lead role (including declaring the initial spare parts requirements) would typically be performed by the design engineering group in a large facility or a product manager in a field service organization. In either event, the decisions, while owned by this role, would require input and support from other business units in their enterprise, including finance, IT, the maintenance parts organization, operations, and maintenance. For the sake of this chapter, we will assign the role to the design engineer.

## THE DESIGN ENGINEER'S ROLE IN SUPPORT OF A PARTS INVENTORY STRATEGY

The maintenance parts inventory support strategy should focus on minimizing the direct parts costs while maximizing the planned value from the supported asset.

The design engineer would support the early asset lifecycle activities, including the asset's business case and support strategy. They would lead, or at least coordinate,

DOI: 10.1201/9781003344674-11

**FIGURE 11.1** An example of Asset Lifecycle Management stages. (Adapted from copyright Figure 1.3 ©2011 from *Asset Management Excellence, Optimizing Equipment Lifecycle Decisions* edited by Campbell, Jardine, McGlynn. Reproduced by permission of Taylor & Francis Group, LLC, a division of Informa plc.)

the asset design stages from conceptual design through detailed design and actively support the asset construction, initial testing, commissioning (with operations and maintenance), and declare a de-commissioning strategy and approach for the asset's end of life. In many organizations, the design engineer may play an active role in how the asset performs against the original business case, but more likely, they hand over the responsibility of the asset performance when starting the operations stage to the operations and maintenance business units.

When a trusted supplier is providing an asset, the design engineer needs to confirm that:

- The requirements have been appropriately communicated to the supplier
- The resulting product (asset) is delivered to the supplier as requested

Let's suppose essential data is passed to the asset management operating team (operations and maintenance). The design engineer can ask the supplier to provide the required information before the commissioning stage in the early design stages. As suggested in Figure 11.2, the required asset creation support effort reduces once the asset has been commissioned. However, the asset support effort increases for the asset management team from commissioning to the assets' end-of-life de-commissioning.

The same figure shows a high-level timeline of which business units will be needed to support an effective asset management lifecycle program. For example:

- Finance is required across the entire lifecycle
- Design engineering is most active from early analysis through to the end of asset commissioning
- Procurement would need to be active from the detailed design stages through to the end of the operating stage (but could also be involved earlier with engineering)
- Inventory management would need to be active from the detailed design stages through to de-commissioning
- Maintenance and operations may be needed in the detail design stage if an early RCM program is used – in any event, they would be needed during the asset commissioning to the start of the de-commissioning stage

Figure 11.2 also illustrates when much of the essential data must be exchanged from one business unit to another. As suggested earlier, this data is essential for the maintenance and operations teams to succeed during the commissioning and operation stages. Likewise, for assets that may be sold after the end of life or de-commissioning (such as aircraft), the maintenance history data is expected to be part of the asset's sale if the asset recovery value is to be considered optimal during an end-of-life asset sale.

## ASSET LIFECYCLE SERVICE CYCLES BY ASSET INDUSTRY

At a high level, the asset lifecycle (ALC) products can be organized into categories of products aligned to the length of their service lifecycles, as shown in Table 11.1.

Companies with short, mid-length, and long-service lifecycle products have different goals and objectives when running their spare parts businesses.

The data transfer from the original equipment manufacturer (OEM) is essential in creating an efficient and successful asset lifecycle execution.

More significant assets will require maintenance and parts support, and these sizeable long-service lifecycle products have been organized into two categories.

Plant-specific assets comprising assets in enterprises such as:

- Refineries, pipelines, exploration and production, petrochemical & oilfield services, power generation, T&D, water production, distribution or airlines, and

Product-specific assets comprising assets such as:

- Equipment manufacturers, aircraft and train manufacturers

For plant-specific assets and product-specific assets, the asset lifecycle steps for commission, operate/maintain, enhance, and de-commission are at a high level, the same. However, the focus steps before construction and commission

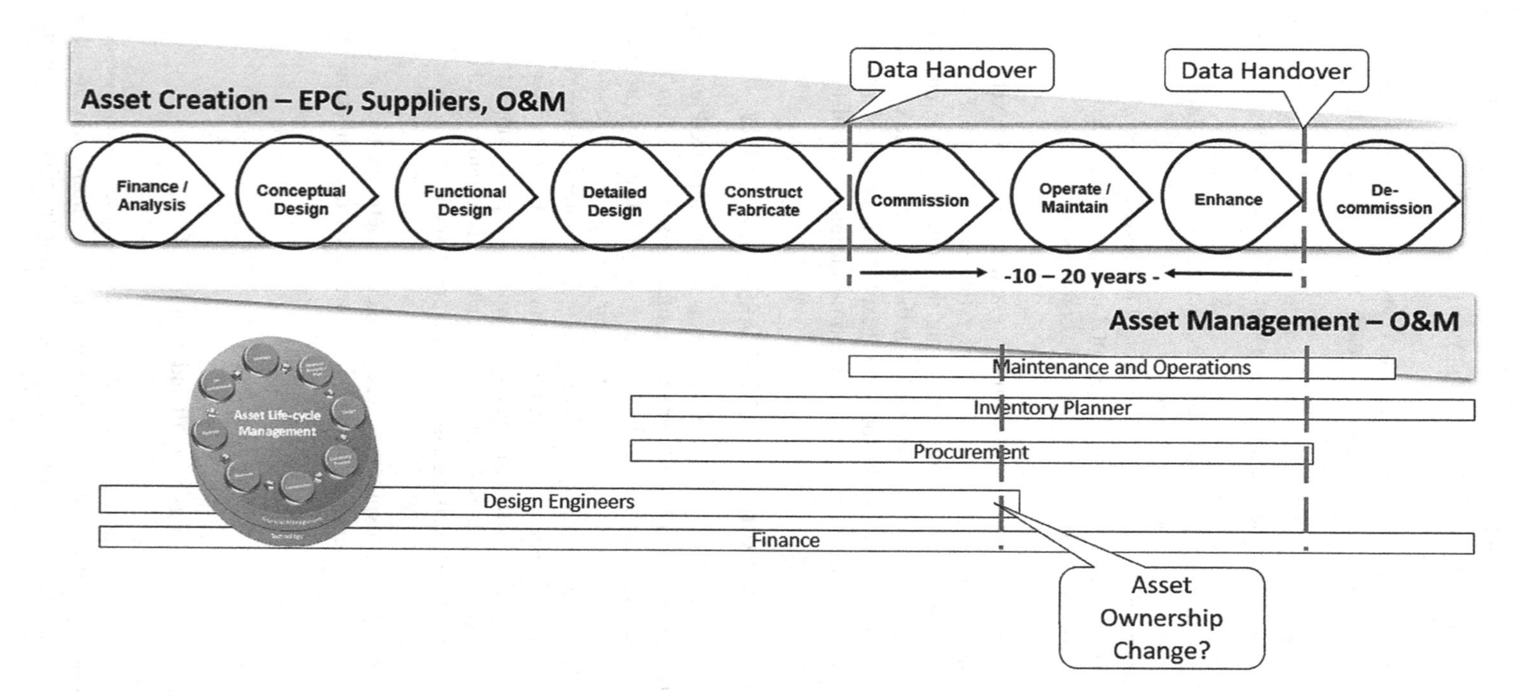


**FIGURE 11.2** An Asset Lifecycle Management focus in Inventory Management. (Adapted from copyright 101 MPE Introduction ©2020 from Maintenance Parts Management Excellence Training edited by Barry. Reproduced by permission of Asset Acumen Consulting Inc.)

**TABLE 11.1**
**An Example of Asset-Service Lifecycle Groupings**

| Product Service Lifecycle | Lifecycle Range | Product Type Examples |
|---|---|---|
| Short lifecycle | 0–3 years | Personal computers, consumer electronics, semiconductors, mobile devices |
| Mid-length lifecycle | 3–10 years | Appliances, cars, medical devices, copiers, business servers, business office telecommunications equipment, consumer durables |
| Long lifecycle | Ten plus years | Aircraft engines, buildings, heavy machinery, industrial equipment, telecom, infrastructure |

will likely differ as the marketing and procurement approaches demand different preparation steps (Figure 11.3).

An example of the expected steps across the asset lifecycle for a plant-specific asset type is shown in Tables 11.2–11.4.

Complex asset management requires detailed design information to be kept up to date. The interfacing of various equipment types requires the equipment's receiver to create and manage detailed designs of each type. Often, specific information is required to operate and maintain from design. The OEM can provide spare parts lists and bill of material (BOM) lists to their clients as examples of how data transfers can help the equipment receiver execute their asset lifecycle more effectively. In addition, some manufacturers offer operating support services (e.g., aircraft and aircraft engine manufacturers) to support their products, client loyalty, and market share.

An example of essential information to collect from an OEM early in the lifecycle and before the committed purchase may be to:

- Collaborate on how the asset design continues to be enhanced over the operate and maintain lifecycles
- Confirm the expected equipment downtime (perhaps MTBF and other data points)
  - What is the production and maintenance cost of downtime?
  - What is the expected ratio of planned versus unplanned maintenance?
  - What parts need to be stocked close to mitigate downtime?
- Confirm how parts will be provided throughout the asset lifecycle – sold or supported through the equipment warranty
- Confirm what the OEM's experience has been in parts usage with like equipment at other clients by getting their recommended spare parts lists and fulfillment service level commitments
- Confirm if the design specification is owned by the customer or equipment manufacturer
- Confirm that the transfer of data needed to operate and maintain the equipment is provided from the detailed product design information by the equipment – how is this done? Is it timely? Is it accurate?
- Confirm that there will be a customer collaboration group in placc that is responsible for essential data transfer and customer feedback

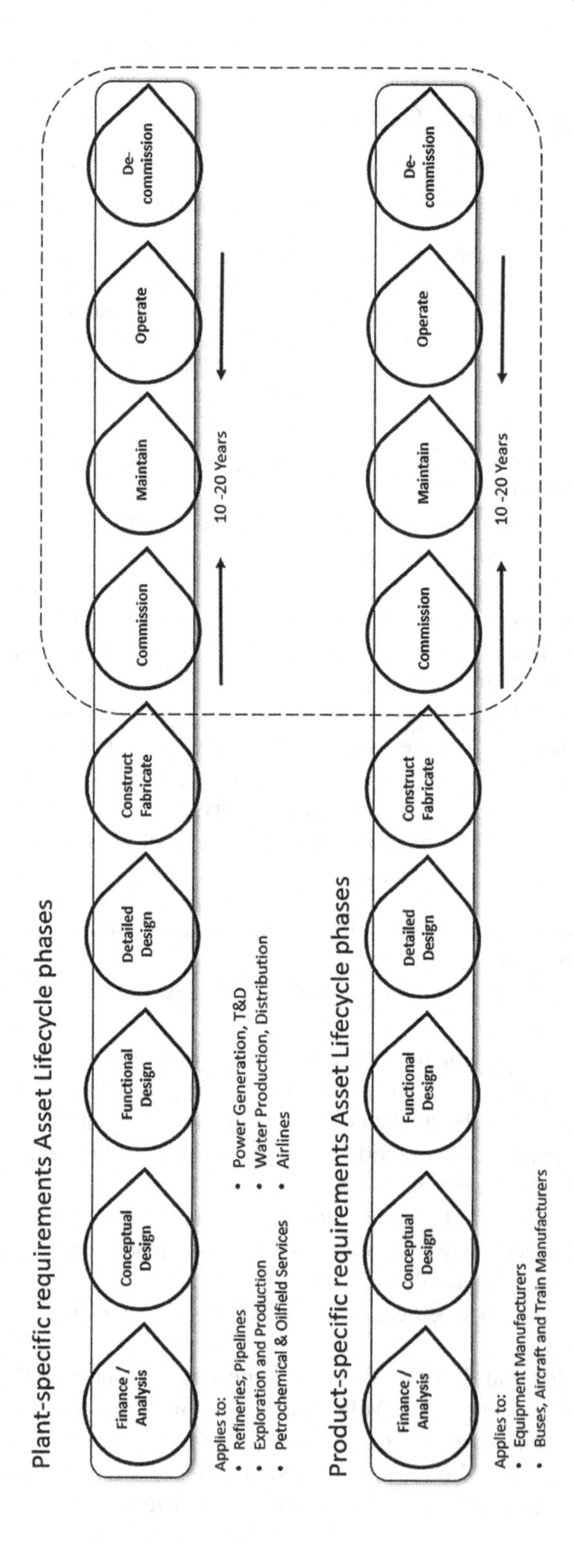


**FIGURE 11.3** Asset lifecycle examples across multiple Industries. (Adapted from copyright 104 MPE Inventory Management ©2020 from Maintenance Parts Management Excellence Training edited by Barry. Reproduced by permission of Asset Acumen Consulting Inc.)

**TABLE 11.2**
**Example of First Five Phases of ALC for Plant-Specific Asset Types**

| Financial Analysis | Concept Design | Plant Design | Detailed Design | Construct |
|---|---|---|---|---|
| • Business case | • Plant Design concept | • Standard Equipment<br>• Custom Equipment | • Specify Configuration<br>• Qualify<br>• Procure<br>• Set Total Cost of Ownership | • Assist/oversee solution build and test<br>• Collect early spare parts lists<br>• Create End-of-Life plan |

**TABLE 11.3**
**Example of First Five Phases of ALC for Product-Specific Asset Types**

| Market Analysis | Product Design Concept | Product Design Requirements | Detailed Design | Construct |
|---|---|---|---|---|
| • Develop Market position | • Create Products and Value Propositions concepts | • Confirm products, design, and value propositions | • Tailor product to market/client needs<br>• Set Total Cost of Ownership | • Build and test the product to client order<br>• Collect early spare parts lists<br>• Create End of Life plan |

**TABLE 11.4**
**Example of the Last Four Phases of ALC for Both Plant and Product-Specific Asset Types**

| Commission | Operate and Maintain | Enhance | De-commission or Post EOL |
|---|---|---|---|
| • Jointly work through product testing<br>• Product data transfer<br>• Spare parts lists and priorities<br>• Documentation<br>• Training<br>• Operating manual | • Execute Asset logs<br>• Maintenance/ Repairs<br>• Manage usage and warranty<br>• Maintain as-built design/BOM<br>• Implement Spare parts policy | • Modify/upgrade as the business evolves<br>• Update configuration<br>• Update Spare parts policy | • Execute equipment "end-of-life" plan<br>• Facilitate a buyer for the used asset<br>• Coordinate the disposition of the equipment<br>• Coordinate the downsizing and disposition of stocked parts support |

## ASSET LIFECYCLE COSTING

Asset lifecycle costing accounts for the asset's economic life and includes costs beyond the initial acquisition costs. Additional costs can be:

- Maintenance parts management spares support costs
- Operating costs (personnel, facilities, energy spend)
- Product supply chain costs (transportation, material handling)
- Maintenance costs (customer service, field/supplier factory maintenance)
- Test and support equipment costs
- Technical data costs
- Software support costs (operations and maintenance and maintenance parts management
- Training costs (of maintenance and operations staff)
- Asset disposition costs (de-commissioning) (Figure 11.4)

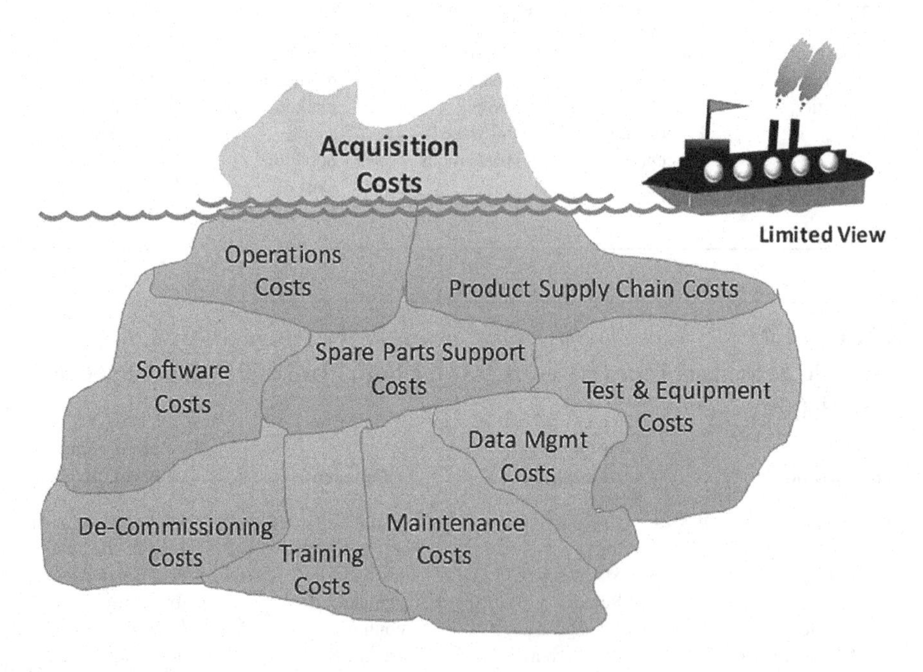


**FIGURE 11.4** Modified asset lifecycle cost iceberg. (Adapted from copyright Figure 12.31 ©2011 from *Asset Management Excellence, Optimizing Equipment Lifecycle Decisions*) edited by Campbell, Jardine, McGlynn. Reproduced by permission of Taylor & Francis Group, LLC, a division of Informa plc.)

## WHAT ITEM NUMBERS SHOULD BE IN THE PART SYSTEM

Asset component data in maintenance parts management is essential in helping to understand what parts should be set up in their systems well before the asset is expected to start commissioning. All parts that are "reasonably likely" to be expected to be used in the asset's lifecycle should be in the maintenance parts system. The design engineer should also provide a list of initial spare parts that should be in stock to support the asset stages, from commissioning through the operating stages to the de-commissioning (end of life) stage.

## INITIAL SPARE PARTS

Once provided and in place, the ISP list can be renamed the "recommended spare parts" (RSP) list once the maintenance team has formally accepted ownership of the ongoing maintenance support and the asset and has completed the commissioning stage.

Spare Parts are a significant portion of the asset lifecycle costing considerations for the design engineer. While the asset is expected to bring value to the enterprise, the support costs (including the parts usage and parts stocking costs) need to be factored into the asset lifecycle costing along with the many other factors such as:

- Research and development
- Manufacturing and installation
- Operations and maintenance and
- Asset de-commissioning

The formal description of the steps needed to create a set of spare parts to support a new asset includes:

- Establishing the asset maintenance plan (leveraging input from multiple stakeholders such as design engineer, RCM, maintenance/operations, etc.) and determining the most effective maintenance strategy:
    - Field replaceable unit (could also be a rotatable part), predictive maintenance, preventive maintenance, corrective maintenance, detective maintenance re-design asset system context
    - Determine parts support requirements for supporting the agreed-to maintenance approach/plan
- Establishing an initial set of spare parts to complement the existing parts support in the parts network (to support the new asset), including:
    - Establishing initial spare parts stock requirements
    - Forecasting parts usage or demand
    - Determining the location of the inventory (distance from the supported asset)
    - Maintaining/modifying network locations
    - Procuring the initial spare parts inventory
    - Fulfillment of work order requests as they arise
    - Reclaiming network surplus/excess as it arises

The spare parts process ends with withdrawing the support for the installed asset.

An ISP List or RSP list is expected to complement the enterprise inventory management policy. Each asset will be supported by multiple sets of dynamic influences, including:

- Local parts disbursement volumes (highly active or activity in the past year)
- The ISP/RSP settings or optimized inventory program overrides
- Local history date-related parameters
- Surplus management policies and activities

The inventory planner would be expected to embrace the ISP input from the deesign ngineer as one of many inputs on how it will achieve its business goals when setting the actual inventory policy to support an enterprise complement of assets. Once past the commissioning stage of the asset lifecycle, the ISP will become an RSP, and the input owner would likely change to the maintenance team.

An initial spare parts list to support the maintenance parts management and policy can be derived from multiple sources. It can come from:

1. The maintenance and operations team's insights and parts history when looking at previously installed similar assets (best guess)
2. The insights the design engineer gains working with the OEM supplier's experience and data
3. A co-developed RCM program executed early in the asset's design cycle

Depending upon the maintenance and asset acquisition strategy, parts acquisition costs may vary. They can include the design, evaluation, engineering, and project management costs.

Initial investment costs vary when comparing repairable versus non-repairable designs and operational (field) maintenance tactics. Likewise, the costs of the initial spare parts lists will differ significantly depending on the selected maintenance strategy. For example, let's suppose asset component parts are deemed to be swapped in the field and to perform centralized parts refurbishment. The repair and supporting logistic costs must be considered in the asset's total cost of ownership calculation.

Suppose the asset components are not "field repairable" (locally or in a central facility). In that case, the mitigating stocking strategy must be factored into the maintenance strategy and the related part usage costs. In addition, the unique tools required to provide maintenance support will need to be accounted for to assist with asset troubleshooting, repair, and testing of the asset and its components during a maintenance activity.

The planned total cost of ownership against an asset should include asset disposal costs and the likelihood that the asset-specific parts costs can be recovered through a viable market sales activity after the asset lifecycle ends and the asset is de-commissioned.

## A MAINTENANCE AND OPERATIONS TEAM'S INSIGHTS AND PARTS HISTORY APPROACH TO ISP/RSP DEVELOPMENT

Many organizations develop their ISP lists from their insights from existing equipment. For example, if you have a fleet of vehicles and decide to move to a new fleet (perhaps from the same manufacturer), you could elect to use a spare parts list based on the experience and lessons learned from the existing fleet. Factoring parts for seasonal issues and accidents may be particularly useful for your operating context if expected to be the same. If the ISP/RSP is expected to complement other inventory stocking strategies such as a Min/Max EOQ algorithm for particularly active parts and a date parameter for annual usage, this could provide a relatively practical and straightforward approach.

Targeting high parts availability for the support of a critical asset should be influenced by the reliability calculations for the asset in its operating context. High availability (reliability) targets and failure rates will drive the models to a potentially high number of stock parts. "Item criticality," expected usage, and the predictability of the usage will directly influence how a "successful parts" sparing list should be generated.

Many asset failure trends can be modeled using Weibull distribution, but most use an exponential distribution model for simplicity. Modeling various options, including "repair on failure" or planned time-based replacements, can be done using the same models but varying the inputs. As an example:

- A time-based replacement can be influenced by a mean time between failure (MTBF) rate, which is calculated using the replacement frequency only
- A repairable (first-time fixed) rate may be used to calculate a repair success frequency or asset attrition rate.

It is paramount to have a parts support strategy to address unscheduled, immediate demands for parts to service critical random maintenance events.

If the asset or specific part failures become more common, the failure frequency may make the failure event more predictable, and then perhaps a preventive action may make sense. However, if the frequency increases, the maintenance organization may be challenged to consider whether the technical maintenance skills are at the best level. An asset or a higher assembly replacement (versus repair) tactic may need to be leveraged.

Failures with wearing mechanisms, cyclical stress, or thermal reversals are more likely to lead to time-based (preventive) restoration or replacement tactics without condition monitoring.

A Poisson distribution is a statistical distribution showing the likely number of times an event will occur within a specified frequency. The Poisson distribution is commonly used to calculate the number of spares to maintain an asset over time. It is used for independent events that occur at a constant rate within a given frequency, in other words, the cycle time required to repair a failed component. A formula to approximate the Poisson equation can be used to model the spare parts needed to be stocked over one or multiple years to achieve the desired availability. This calculation can benefit the Design engineer looking to understand the total cost of ownership of a new planned asset system.

Consider insurance stock (just in case) scenarios for parts that are not expected to be highly used or predictable. For example, an ISP minimal philosophy could force a stocking policy quantity of one. This ISP is often the case to keep the maintainers and business risk managers comfortable addressing potential issues.

## CREATING AN INITIAL SPARE PARTS LIST WITH THE HELP OF OEM DATA

Many sparing decisions are made based on the OEM's recommendations. While the OEM's input is critically needed, they typically do not have an intimate understanding of the planned operating context of the new asset to be installed. If only the OEM recommendations are taken as the inventory policy, then:

- The local stock room will discover that it is holding many parts that will never be used or at least underused
- In many cases, it has not recommended enough of the parts to be stocked that are used multiple times in a three-month cycle

OEM ISP guidance can only make for unhappy stockkeepers, maintenance technicians, operations managers, and Finance managers as none of their KPIs are being served well by the maintenance parts inventory.

Original equipment manufacturers make their reputation on the reliability of their products. However, the risks associated with only adopting the OEM input can create sub-optimal results given that:

- The OEM seldom actually operates their assets and, as a result, cannot provide the spares insight most needed
- The OEM may be influenced by mitigating their own risk rather than mitigating clients' risk in the client operating context
- The OEM makes more money the more you buy their parts as spares inventory
- An enterprise stocking more parts than needed is more likely to serve the OEM than the enterprise
- Not accepting installation surplus parts as a return to the OEM often influences the enterprise to stock the part, often when no effort is made to forecast it will ever be used

OEM spare parts recommendations should be taken with caution. A rigorous enterprise-driven ISP approach is the best foundation for support throughout the enterprise's asset and spare part lifecycle.

The design engineer, along with the OEM and their committed (contracted) support, and with the help of procurement, will work through an initial spare parts list for a new piece of equipment or asset.

A few simple questions will help categorize what materials may need to be strategically stocked quickly at a high level.

What is the operating context of the planned asset?

- How critical is it for the operation?
- If the asset or component became unavailable, what would the immediate adverse effect be on production, safety, or the environment?
  - What would the business impact be?
  - What are the risks and probability that this defect could happen?
  - How secure is the enterprise value chain if this defect happens?

How many times is it expected that a part would be used?

- What is the forecasted annual parts dollar usage?
- Does this represent a high cost for the organization?

An analysis of past parts usage for similar assets could be helpful here to assess the cost of parts usage and business impacts when parts are unavailable.

How predictable is the expected demand for the part?

- Is the part expected for predictive maintenance, preventive maintenance, or unexpected unscheduled and immediate need maintenance orders?
- Is it used individually or in multiples or sets (such as a bill of materials kit in a maintenance activity)?

Including the OEM, how many vendors can the part be viably acquired?

- Is the part linked to the vendor/OEM warranty?
- Is the part a commodity part and readily available locally?

Answering these questions for every item expected to be a (part) component in a new piece of equipment or asset can categorize which parts should be part of an ISP.

Note that an ISP list will have echelon levels in it. For example, would the suggested parts stocking complement support an availability level of 50%, 85%, 90%, 95%, or 100% for a given location?

Most enterprise operations may desire 100% parts availability but would likely stock parts for a lower support echelon. An echelon level lower than 100% better manages the inventory costs across the asset's lifecycle. It avoids purchases to inventory that will be later scrapped when the asset gets to "end-of-life" and reduce the related annual inventory carrying costs.

## INITIAL SPARE PARTS CATEGORIZATION APPROACH (FIGURE 11.5)

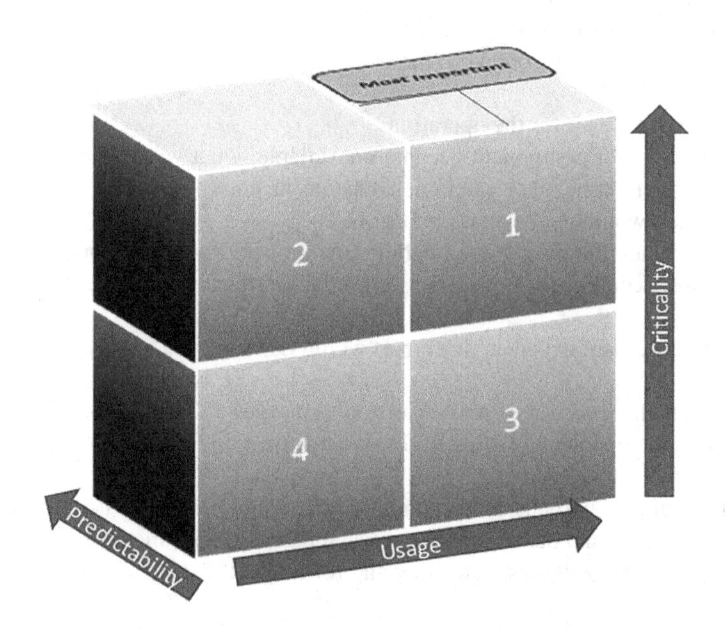


**FIGURE 11.5** Maintenance parts initial spare parts categorization approach. (Adapted from copyright 104 MPE Inventory Management ©2020 from Maintenance Parts Management Excellence Training edited by Barry. Reproduced by permission of Asset Acumen Consulting Inc.)

Each part is categorized, and each cell can represent a different support strategy. To identify the categories in simple terms, we will number the categories one through four.

Highly critical parts with high expected usage (Category 1) and highly predictable demand could be considered most likely to be added to an ISP list and kept throughout the asset's lifecycle. They represent the highest strategic importance in impacting lost production and support costs. Demand predictability contributes to the likelihood of the asset's productivity when these parts are in place early in the asset's lifecycle. The maintenance support strategy should focus on the sustainment and continuity of a new asset, so this is particularly important to get in place before the asset commissioning and early production stages. Once the asset is actively used, an active parts support strategy may likely complement (override) the ISP list and stock higher quantities as experience dictates. Active parts management is discussed in this book's earlier Inventory Planning chapter.

High criticality and low usage parts (Category 2) with a high expectation (predictability) of demand hold a lower expectation of parts usage cost but are still critical to the business and should be part of an ISP list presented to the business from the

Design Engineering group. This ISP approach represents a simple way to ensure crucial parts are in the ISP stocking policy, replenished when used, and should suffice here. Because parts usage volumes are expected to be low, the stocking level can also be low but should be above zero. The supporting ISP strategy would typically be to stock a quantity of one of these parts with a process to replenish as they are disbursed from the local parts stock room to ensure support for unexpected new demand.

Low criticality and high usage parts (Category 3) with expected high predictability should be identified and set into an ISP at a higher echelon.

For example, suppose it is estimated that a 90% availability can be reached with all the Category 1 parts and some level of Category 2 parts. In that case, none of the Category 3 parts should be expected to be included in the ISP list for a location looking to reach 90% availability.

This simple method may deviate when some higher-category parts are costly. It may be determined that lower category parts with significantly lower costs could make or more likely exceed the desired 90% availability level if traded in a designated ISP list.

Category 3 and 4 parts would not likely be part of an ISP list. However, parts with expected criticality and statistical usage could also be considered (particularly the low-cost items) to complement the ISP lists when Category 1 and Category 2 items have been assigned. The targeted availability level still cannot be achieved.

A demand for a "not stocked" part (not in stocking policy) should be expected periodically, mainly if the target fulfillment support strategy is less than 100%. On the other hand, having a usually "stocked" (in stocking policy) part not available from the local stock room should be less likely but will happen. Therefore, some stock-outs should be considered acceptable in the short term.

The parts that are simply unpredictable in their demand (assuming low criticality and low expected usage) would not be considered in the ISP lists and are left for the inventory planner to manage (see Chapter 9 – inventory planning).

Low criticality and low usage parts with high predictability typically would not have special support efforts unless needed to make up a higher stock room availability commitment statistically. There can be significant variation in the parts required, influenced by the context in which the asset is being used and maintenance technician's experience. Many of these low criticality and low usage parts would not be in stocking policy. These exceptions to stocking policy would also be left for the inventory planner to manage (see chapter 9 – inventory planning).

This is a simplified summary of creating an initial spare parts list. However, as mentioned earlier, the inventory planner would be expected to embrace this input as one of many for achieving their business goals when setting the inventory policy to support an enterprise's complement of assets.

## CREATING AN INITIAL SPARE PARTS LIST FROM AN RCM ANALYSIS PROGRAM SET OF DATA

How a component or whole asset fails will influence the best maintenance tactic to mitigate asset failures in a specific operating context and when parts will be needed. Therefore, RCM disciplines should ideally be used to determine the appropriate maintenance action (run to fail, predictive, preventive, detective, or re-design) depending on the expected failures. For example, random failures cannot be predicted or prevented and cannot be eliminated using a time-based preventive replacement maintenance approach.

An RCM analysis is an asset-specific discipline that confirms the most effective mitigating maintenance tactic for a specific asset (in its operating context) and the related asset's risk and reliability business impacts. It is typically facilitated with the help of the maintenance, operations, and design engineering staff and line supervisor, all trained and ideally experienced in the assessment discipline. Once the asset functionality and how it may functionally fail is defined, a failure modes and effects analysis (FMEA) is facilitated to document each failure, the affected risk to the business, and how an applied repair-strategy cost dynamic happens under normal operating circumstances. An RCM team would then determine the best maintenance tactic to be engaged by documented failure mode. The documented mitigating tactic would have identified the type of expected maintenance mitigation (inspection, predictive, preventive, repair, re-design), the frequency (if applicable), and the skills and parts needed (where applicable). Using an RCM approach will influence the parts that should be stocked and whereby understanding how the specific asset:

- Drives value to the enterprise
- Functionally works and could fail in its operating context
- Failure impacts the business costs and risks
- Failure could be mitigated
- Frequency of the planned maintenance tactic mitigation action

Getting this set of questions (and data) right takes an enterprise's committed community that will often include the supplier and perhaps the customer (operator or end-user). The reason is that no one business unit will fully understand this data's effect. Only by working together will the most effective maintenance strategy and initial spare parts list be created (Table 11.5).

**TABLE 11.5**
**No One Business Unit Knows All the Criteria for Initial Stocking Parts**

| Business Unit/Team | Insights for Maintenance Parts |
|---|---|
| Original Equipment Manufacturer (OEM) | Manufacturer's spare parts list |
| Design engineer | Asset Lifecycle costing approach |
| Maintenance tech | Business, environmental, and safety impacts |
| Inventory Manager | Number of "like" assets supported |
| | Number of "like" parts supported by the location |
| RCM team | Operating context |
| | Maintenance history and approach |
| RCM team with Inventory Manager | Parts delivery and service level expectations |
| Procurement Lead | Parts procurement and repair strategies |

With the design engineer in the lead role, they can facilitate a reliability-centered maintenance assessment for the new asset to:

- Declare the asset operating context
- Proactively define functions, functional failures, failure modes
- Do failure modes effects analysis(FMEA)
- Define maintenance strategy
- Define planned and unplanned/emergency maintenance scenarios that drive parts demand

In an ideal environment, the RCM analysis will identify the asset and operating context, document the desired functions from the assets, and how these functions can fail (failed state). With the failed state known, an RCM team would identify the root cause using an FMEA approach and document the failure state's consequence for each identified and "reasonably likely" failure mode. The RCM process prescribes whether a predictive, preventive, corrective (run to failure), detective (inspection) or Re-design task action is best to mitigate the risk of the failure mode. Next, the task, the resource (technician type), parts required, the required tools, and the task frequency are determined. Here the tools and parts information can be input for a part stocking policy. RCM is the best way to create an "accurate and mutually agreed to" ISP list (Figure 11.6).

RCM can support multiple stages in an asset lifecycle process. Figure 11.8 demonstrates where RCM can ideally be performed early in the concept design stage. Then, after some feedback to the design team, again be applied after the detailed design is completed. Working at these stages allows for the RCM team to identify the most effective maintenance tactic for each identified failure mode, and with each failure mode – maintenance tasks, staff requirements, training requirements, infrastructure support processes, and maintenance parts requirements can be identified and calculated in terms of their contribution to the business risk of a non-functioning asset. With the parts needs defined from an RCM analysis, the RCM team with the design engineer can determine an ISP list to be implemented to support the commissioning through to de-commissioning stages of the asset lifecycle (Figure 11.7).

RCM can also be applied to existing (already installed) assets during the operation and maintenance. However, the earlier an RCM analysis can be applied in the asset lifecycle, the better the opportunity to optimize the asset lifecycle costs, including the related maintenance spare parts costs.

## ISP/RSP DATA ELEMENTS

With the output of an RCM analysis, the Design engineer can set up a progression of data for:

- Planned maintenance
- Unplanned maintenance
- Emergency maintenance
- Field replaceable spares (centrally repaired) maintenance

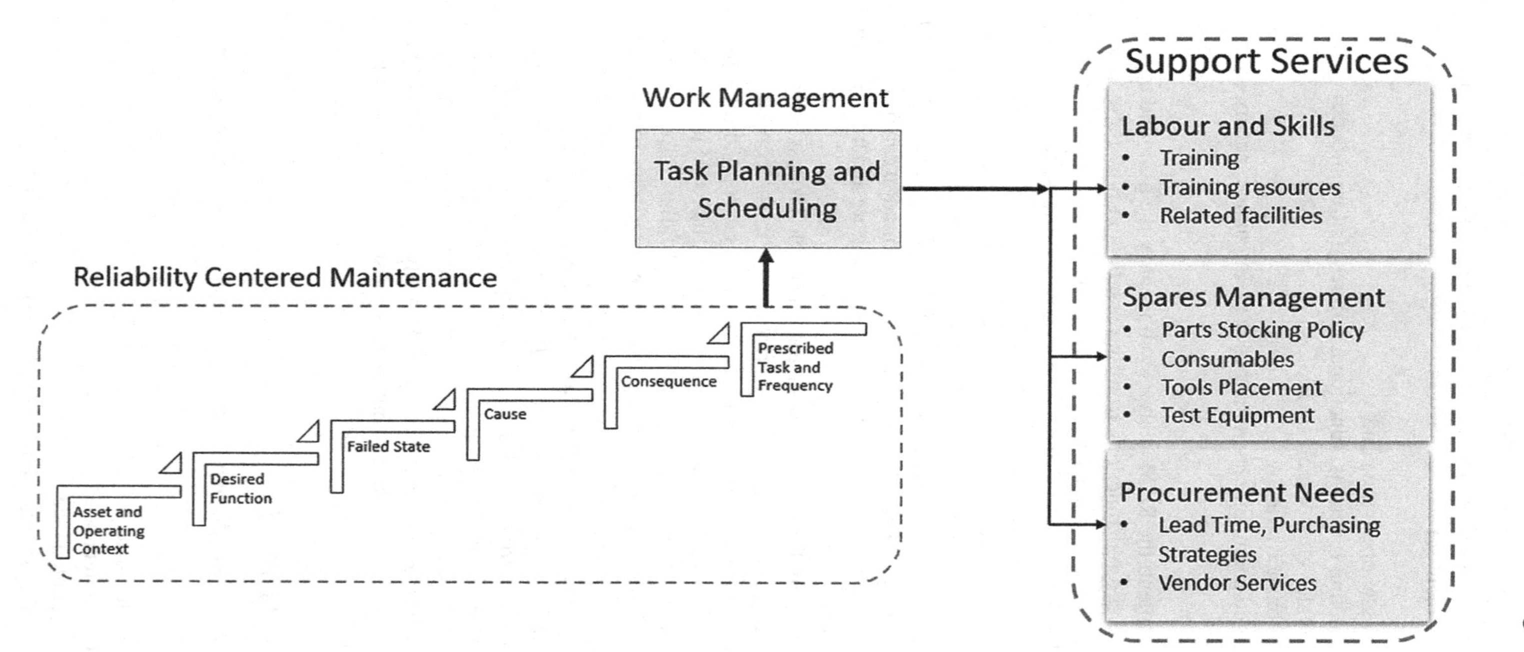


**FIGURE 11.6** Simplified RCM analysis process. (Copyright MPE ©2022 from Maintenance Parts Management Excellence Training edited by Barry. Reproduced by permission of Asset Acumen Consulting Inc.)

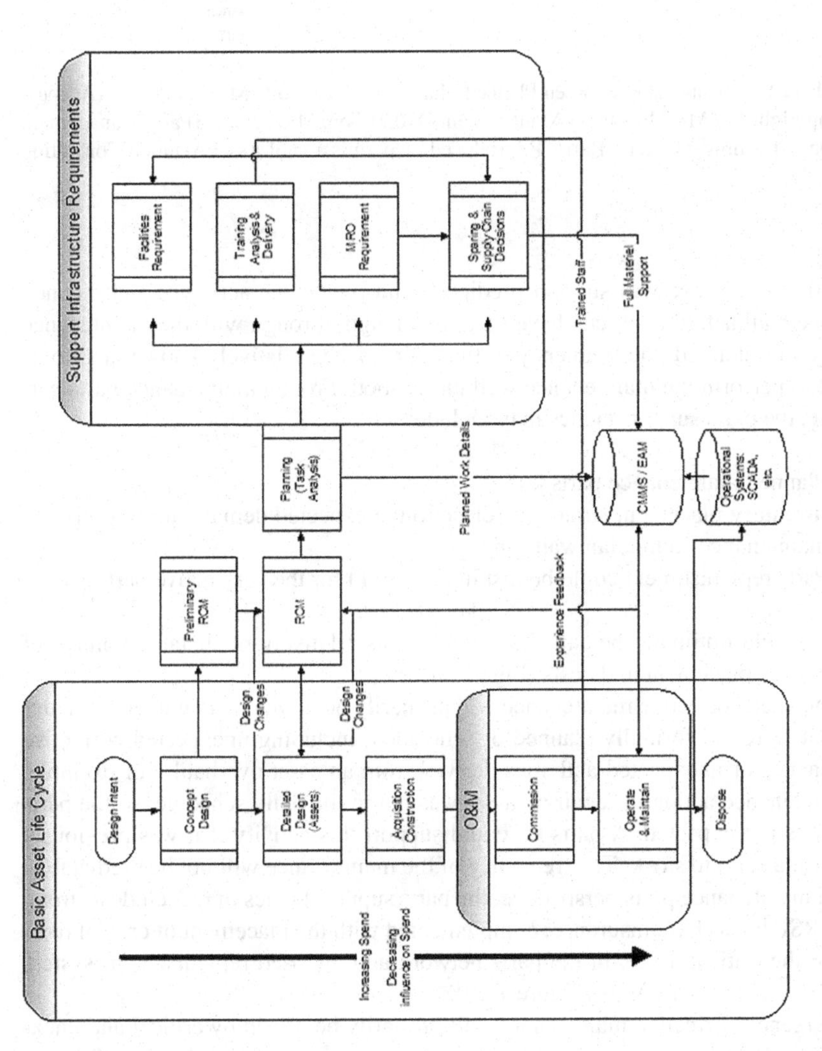


**FIGURE 11.7** Asset Lifecycle Management stages with RCM steps included. (Adapted from copyright Figure 2.5 ©2011 from *Asset Management Excellence, Optimizing Equipment Lifecycle Decisions* edited by Campbell, Jardine, McGlynn. Reproduced by permission of Taylor & Francis Group, LLC, a division of Informa plc.)

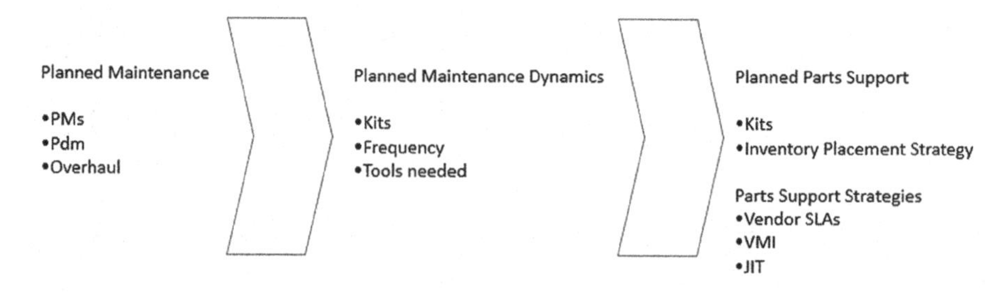


**FIGURE 11.8** Relationship between Planned Maintenance and Parts Management. (Adapted from copyright 104 MPE Inventory Management ©2020 from Maintenance Parts Management Excellence Training edited by Barry. Reproduced by permission of Asset Acumen Consulting Inc.)

Planned maintenance will support predictive and preventive activities and planned asset overhaul activities. It can have part kits thought through with the maintenance strategy well defined, the frequency of the maintenance relatively known, and tools needed to perform the maintenance well understood. From a maintenance parts perspective, the parts support tactics may include:

- Planned maintenance parts kits
- Inventory placement strategies (close to the expected demand to minimize maintenance technician wait time)
- Parts replenishment could be just in time (JIT) for the very active parts

Vendors could optimally be called to manage this relatively predictable volume of parts activity placement and costs (Figure 11.8).

Unplanned corrective maintenance will primarily be driven by maintenance activities that were not formally planned or scheduled, including unexpected corrective maintenance or unexpected activities derived from an asset overhaul. For example, the maintenance action to repair an asset was not in the daily schedule, so the parts usage was not expected. A parts kit could support this activity if it was previously defined and set into stock. The frequency of the maintenance will not be predictable. From a maintenance parts perspective, the parts support tactics may include a strong ISP or RSP list as base insurance, complemented with the placement of crucial parts close to the critical assets in the parts network and a closed replenishment system with dynamic reorder points (Figure 11.9).

Emergency corrective maintenance will primarily be driven by critical and unexpected activities from an asset failure. To support this type of activity, an ISP list of critical spares provides essential insight to support the expected service levels of the assets. "Where used" and parts history data will complement the foundational ISP/RSP list in the maintenance parts system. If the unexpected failure needs a relatively active part, the maintenance parts system for managing active parts will likely support this demand. The required maintenance frequency for a specific asset failure will

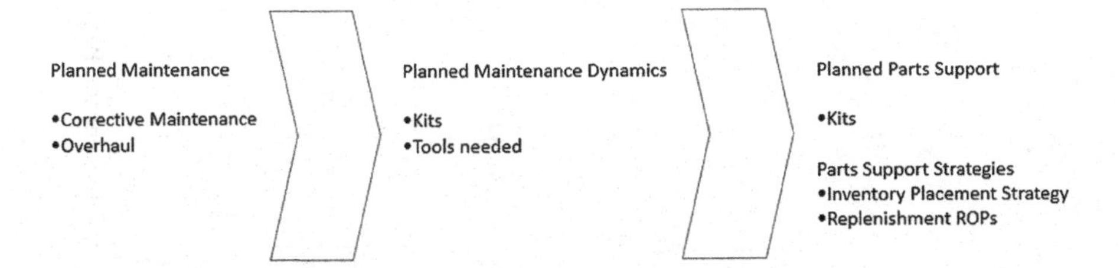


**FIGURE 11.9** Relationship between Unplanned Maintenance and Parts Management. (Adapted from copyright 104 MPE Inventory Management ©2020 from Maintenance Parts Management Excellence Training edited by Barry. Reproduced by permission of Asset Acumen Consulting Inc.)

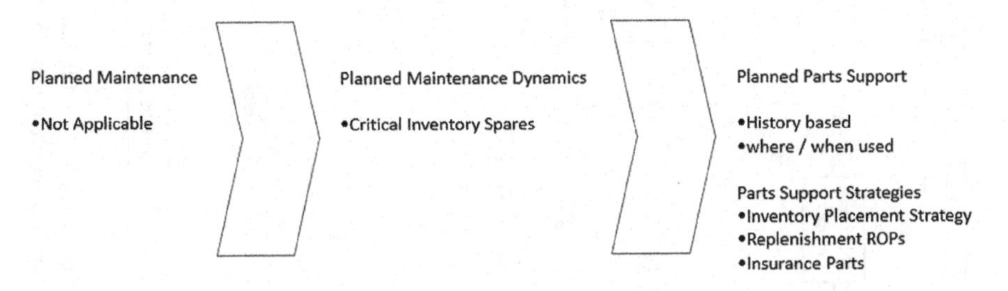


**FIGURE 11.10** Relationship between Emergency Maintenance and Parts Management. (Adapted from copyright 104 MPE Inventory Management ©2020 from Maintenance Parts Management Excellence Training edited by Barry. Reproduced by permission of Asset Acumen Consulting Inc.)

not be predictable. From a maintenance parts perspective, the parts support tactics may include a strong ISP (or RSP) list as base insurance, complemented with the placement of crucial parts close to the critical assets in the parts network and a closed replenishment system with dynamic reorder points (Figure 11.10).

Field replaceable maintenance tactics are often applied to mass-volume assets expected to be installed. The repair demand will also be expected to be at a predictable frequency. Given this, the replacement parts (asset) that would be stocked will likely have a volume that will drive a min/max and EOQ and, as such, will be more influenced by the recent historical demand than an ISP strategy.

## SYSTEM PROCESS MODEL FOR ISP/RSP DATA CREATION (FIGURE 11.11)

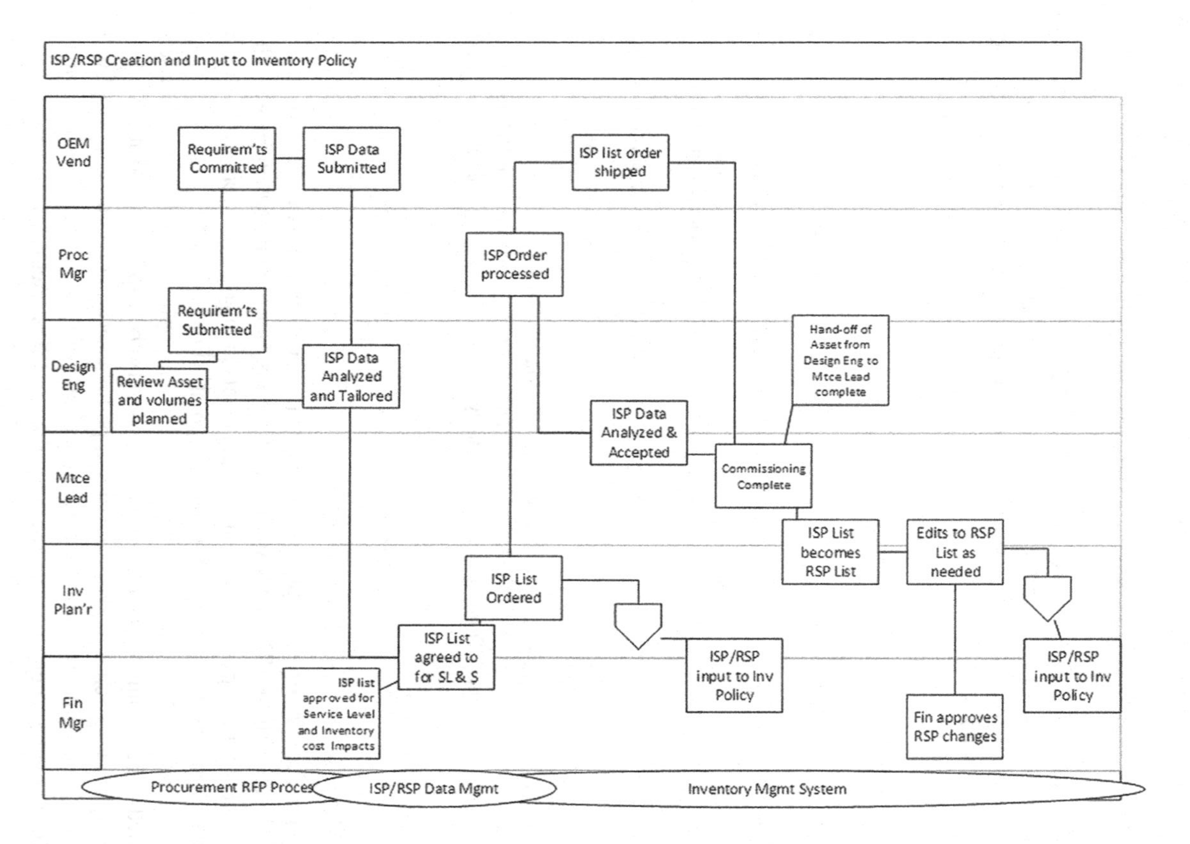


**FIGURE 11.11** ISP/RSP data creation process model. (Adapted from copyright 104 MPE Inventory Management ©2020 from Maintenance Parts Management Excellence Training edited by Barry. Reproduced by permission of Asset Acumen Consulting Inc.)

**TABLE 11.6**
**ISP/RSP Data Element Considerations**

| Generic ISP/RSP Data Description | Comments |
| --- | --- |
| Part Number/Item number | As listed in the maintenance parts management system |
| OEM Item Number | As provided by the vendor, if different than above |
| Where used | Asset/equipment the part is supporting |
| Item costs | |
| Criticality Factor | |
| Predictability Factor | |
| National Usage factor | |
| Stocking echelon designation | |
| Technology code | Electrical, mechanical, etc. |
| Stock control factor | Allowed for stock rooms versus satellite installation stock |
| Item complement code | Unique part for the asset (can be swapped to a less critical area of asset) |
| Stocked/not stocked part | Parts forced not to be stocked (Cost, health and safety issues, etc.) |
| Date ISP last updated | Yyyymmdd format |
| ISP/RSP owner code | Engineering/maintenance, etc. |
| ISP/RSP method used | Like equipment/OEM collaboration/RCM/Other (name it) |
| OEM Collaborator name | If applicable |
| ISP Approver | Executive name |
| Finance Approver | Executive name |

ISP/RSP data elements should be captured and managed by the ISP/RSP owner. A partial list of data considerations is provided in Table 11.6.

One of the final considerations for integrating the inventory spare parts management with the asset lifecycle is recognizing that the inventory stocked for each piece of equipment (asset) will rise and shrink as the number of a specific installed asset increases or shrinks in the enterprise's footprint. Thus, while the inventory is seldom set to an annual usage level, the asset install base will likely drive parts usage early in the asset lifecycle and perhaps later in the asset lifecycle. Overall, however, it is expected that the need for parts will decrease as assets are de-commissioned. Ultimately, when the last asset of an asset type is no longer in use, the supporting maintenance parts complement should also be set for (disposition) sale or scrap. In any event, stocking parts for assets no longer in the enterprise is poor (Figure 11.12).

Regrettably, many organizations still do not have a workable systemic way to track why parts are in policy (e.g., what ISP/RSP list are they part of). If the ISP/RSP lists are well established and maintained, these lists should allow the Inventory

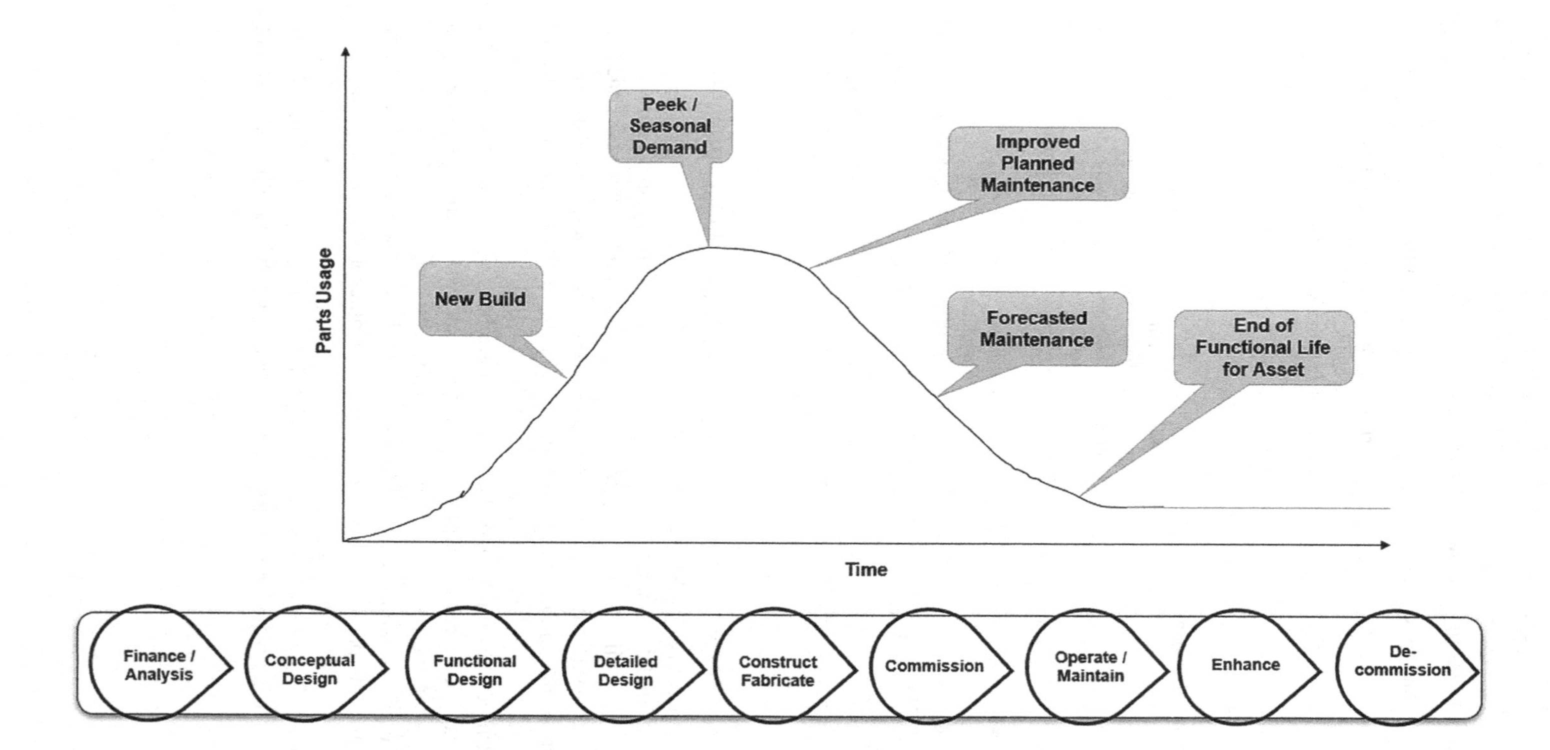


**FIGURE 11.12** ISP/RSP data element considerations. (Adapted from copyright 104 MPE Inventory Management ©2020 from Maintenance Parts Management Excellence Training edited by Barry. Reproduced by permission of Asset Acumen Consulting Inc.)

Planner to execute a policy change by removing the supporting ISP/RSP list – equipment that is no longer in the enterprise nor identified as equipment inventory surplus. The early identification of inventory surplus will enable the Inventory Planner to consider inventory cost recovery options beyond just scrapping inventory surplus. Selling surplus parts inventory would go a long way in helping to manage the overall operating costs of inventory and the asset's total cost of ownership.

# 12 Inventory Optimization Tactics in Spare Parts Management

## INTRODUCTION TO INVENTORY OPTIMIZATION

The term "Optimizer" as an inventory policy tool has been introduced in previous chapters. Indeed, when optimizing your inventory tactics, an Optimizer tool can provide great insights into what part should be stocked and what location of a maintenance parts network to meet a specifically targeted availability (service) level. Thus, an Optimizer tool would ideally provide optimal service levels and spare parts inventory across a managed network for maintenance parts support.

This chapter will discuss the components that could be in such a tool. In previous chapters, we suggested that Optimizer, if available, could control a significant part of the network inventory. However, the Optimizer tool and data required are not always dynamically available. The Optimizer tool is considered a complement to an overall maintenance parts strategy and on its own: Optimizer is a maintenance parts policy tactic.

## WHY PARTS ARE STOCKED

Some inventory management fundamentals should be discussed before going too deeply into inventory optimization. For example, understanding why parts are needed and should be stocked is a critical fundamental philosophy and culture ideally shared across the enterprise (Table 12.1).

Maintenance parts should be stocked locally to support immediate needs like urgent or emergency orders. All other parts can be acquired and brought in (perhaps "just in time") to align with specific work order requests. The rule of thumb should be that:

- Planned inventory requirements can be effectively passed to suppliers or at least higher in the inventory support network
- Unplanned requirements require significant planning to predict effectively – ISP/RSP, active and annual parts usage tracking can help here for immediate need activities
- New assets require asset lifecycle data to assess initial support (ISP/RSP) effectively

DOI: 10.1201/9781003344674-12

**TABLE 12.1**
**Types of Parts Requests and Should Local Stock Support Them**

| Activities Maintenance Parts Support | Comment | Are Local Spares Needed? |
|---|---|---|
| Routine/running repairs | All of this work can be planned & scheduled | Some |
| Emergency repairs | Some can be planned, but none are scheduled | Some |
| Shutdown repairs and projects | All of this work can be planned & scheduled | No |
| Project work | All of this work can be planned & scheduled | No |
| Proactive maintenance | All of this work can be planned & scheduled | Some |

## INVENTORY POLICY NETTING

Inventory policy netting should occur each week as part of the dynamic inventory policy confirmation and after the previous week's demand has been considered in the new active parts Min/Max, EOQ calculations. Order levels are recalibrated weekly by location with:

- New order points and maximum stock levels were created and compared to available quantity in each location
- Orders are identified where the available quantity is below order point
- Surplus inventory levels are identified by location
- Automated system actions that search and direct new replenishment orders on stock room locations show surplus quantities on hand before generating purchase requests and ultimately purchasing orders to outside vendors. This automated process redirects stock room surplus to support new replenishment orders to requesting stock rooms is called "excess redistribution."

Excess redistribution happens when:

- Financially justified to procure internally from surplus
- Inventory transfers from one local stock room to another are allowed in the supporting Maintenance parts management system (Figure 12.1)

When focused on "service levels," the Inventory policy would assume the highest of the stocking policy declarations if the following conditions are met:

- The new min/max is suggesting a maximum stock of "five"
- The date protect (annual) policy suggests a maximum stock of "one"
- The ISP/RSP policy suggests a maximum stock of "one"

The new inventory policy for that part in that location would "net" to the highest protection level, "five."

Inventory deemed surplus would remain as "0" (assuming the quantity on hand is below "5") (Figure 12.2).

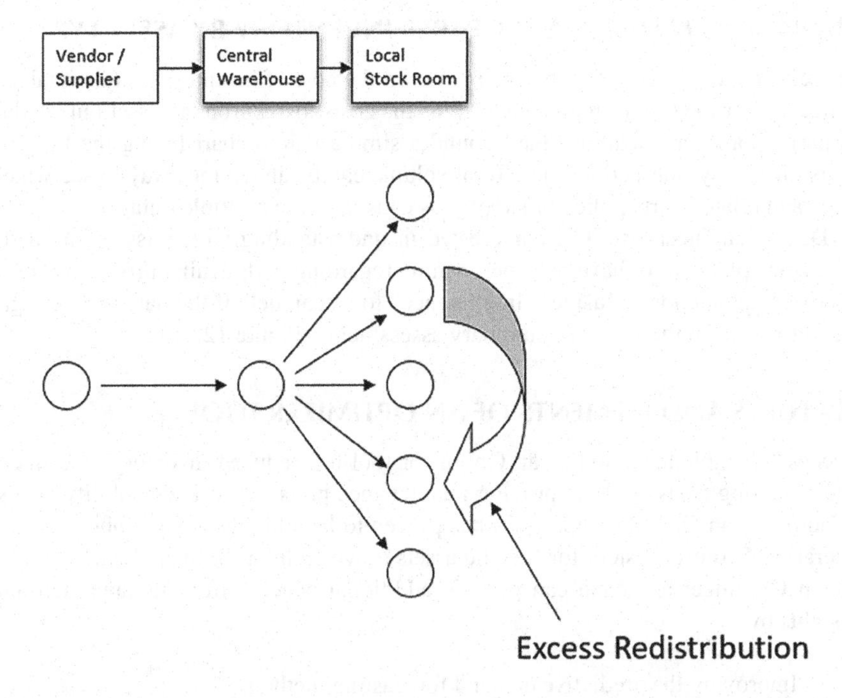


**FIGURE 12.1** Excess redistribution: Managing surplus in a parts network. (Adapted from copyright 104 MPE Inventory Management ©2020 from Maintenance Parts Management Excellence Training edited by Barry. Reproduced by permission of Asset Acumen Consulting Inc.)

| Stocking Strategies | ABC |
|---|---|
| ▪ Level 1 | ▪ Top 80% used items by $ value<br>▪ Less than 10% of stocked items |
| ▪ Level 2 | ▪ Next 15% used items by $ value<br>▪ Less than 20% of stocked items |
| ▪ Level 3 | ▪ Bottom 5% used items by $ value<br>▪ Can be than 70% of stocked items |
| ▪ Surplus | ▪ Surplus to policy supported by ABC above as well as down level parts |

Example:

- Active Inventory =5
  - Leveraging Safety Stock, Order Point, Max Stock, EOQ, Average Demand algorithms
- Annual Activity (by location) =1
  - Leveraging Date Last Used, Date Stocked
- Initial Spare Parts Recommendations =1
  - Leveraging input from Manufacturer, Design Engineer, Maintenance Tech, Inventory Manager, or RCM Team
- Netted 'Max Stock' Stocking =5
  - is set to the highest level of the three stocking strategies if our goal is a high service level
  - Surplus is deemed to be Available Quantity minus Netted Max Stock

**FIGURE 12.2** Inventory policy netting concepts. (Adapted from copyright Figure 6.7 ©2011 from *Asset Management Excellence, Optimizing Equipment Life-cycle Decisions* edited by Campbell, Jardine, McGlynn. Reproduced by permission of Taylor & Francis Group, LLC, a division of Informa plc.)

## MANAGING MAINTENANCE PARTS INVENTORY BY ASSET TYPE

There is always an expected balance between the cost of inventory and the maintenance cost or the cost of maintenance plus the costs/losses from an asset outage due to parts. Some organizations have found a simple Pareto chart listing the top 100 parts missed by transaction lines. Total volume can be an obvious way to see which parts and related parts policy or supplier is causing these multiple outages.

Design engineers would be interested in understanding which assets and asset types they planned for have been performing regarding parts availability, parts costs, and user satisfaction. "Business Intelligence" tools can help if the data can be organized to perform these simple summary assessments (Figure 12.3).

## FEATURES AND ELEMENTS OF AN OPTIMIZER TOOL

Successful implementation of an Optimizer tool has reduced inventory investment and operating costs while improving maintenance parts network availability levels. In addition, an Optimizer approach has proven to be a highly flexible planning and operational control system for specific assets to which it has been applied.

An Optimizer approach can provide additional benefits from its approach and insights by:

- Improving the predictive demand forecasting method
- Accounting for how a part is used on a specific asset and its commonality
- Accounting for the multi-echelon structure
- By enhancing cost-service trade-offs within the network echelon (Figure 12.4)

When creating a typical ISP/RSP list, the target service availability level may be set to 85% for a local stock room when complemented with an active and annual parts

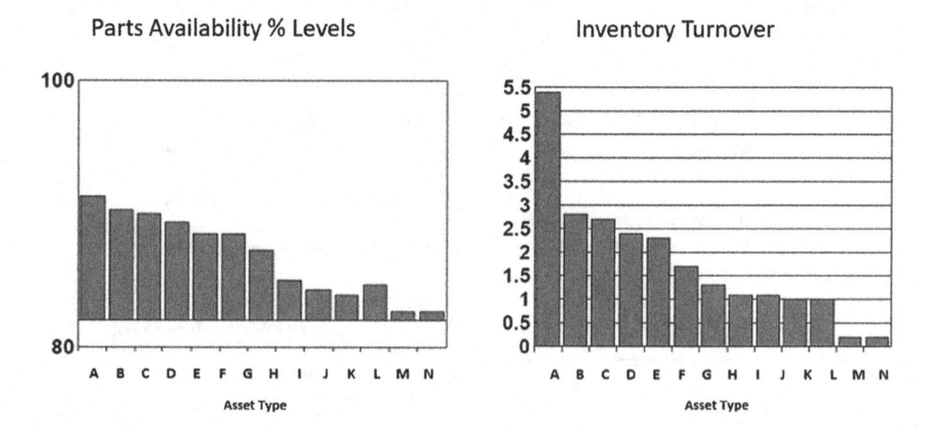


**FIGURE 12.3** Asset type analysis of inventory service levels and inventory activity. (Adapted from copyright 104 MPE Inventory Management ©2020 from Maintenance Parts Management Excellence Training edited by Barry. Reproduced by permission of Asset Acumen Consulting Inc.)

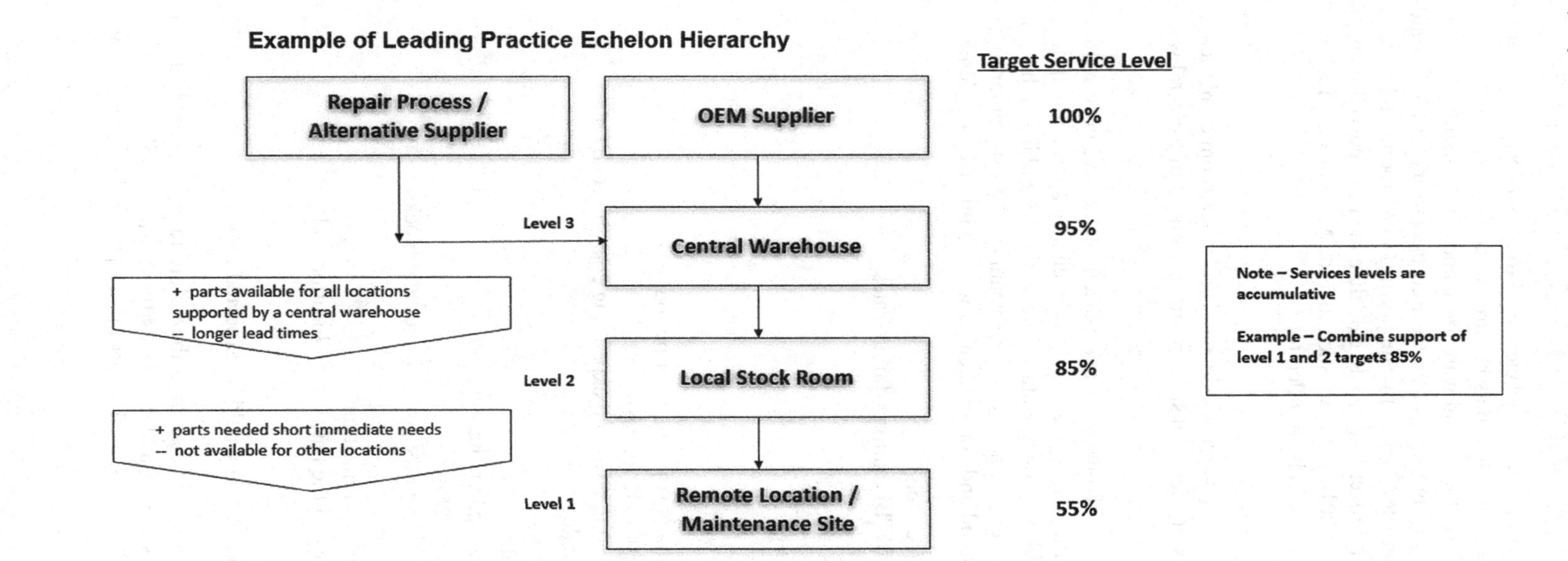


**FIGURE 12.4** Leading practice echelon support with optimizer. (Adapted from copyright Figure 6.10 ©2011 from *Asset Management Excellence, Optimizing Equipment Life-cycle Decisions* edited by Campbell, Jardine, McGlynn. Reproduced by permission of Taylor & Francis Group, LLC, a division of Informa plc.)

policy. Still, you can set the target service level to 85% or 87% with an Optimizer. It will consider all the available inputs required to achieve the level set for the stock room and the level set for the central warehouse and supplier support as the network is defined. As shown in Figure 12.4, the desired target service level could be 55% availability at the asset site, 85% at the supporting stock room, and 95% at the central warehouse. An Optimizer tool would calculate the most cost-effective inventory placement in the network, by echelon, such that the Maintenance Technician would expect that their needs would be fulfilled to:

- 55% from the asset site
- 85% of the time from the asset site and local stock room combined
- 95% from the asset site, local stock room, and supporting central warehouse combined.

Creating an optimal set of inventory and placement of inventory requires significant data. Understanding the applied asset maintenance approach will help leverage asset configuration and component data, failure histories, and install base into a more effective inventory policy setting. First, the dynamics of a list similar to an ISP/RSP list need to be understood and complemented with additional data elements such as:

- Install base by location
- Failure history/MTBF/national failure rates
- Service strategy
- Part function
- Distribution costs
- Inventory costs
- Lead times
- Parts costs
- Inventory holding cost per unit per period
- Setup cost for normal replenishment
- Emergency shipping costs and expediting costs by echelon
- Operation schedules
- Repair BOMs/Kits
- Asset configuration
- Asset component (part) vitality
- Vendor source dynamics
- Local overrides (inventory policy forces and blocks)
- Budget constraints
- Network echelon optimization
- Current inventory in the network by location and
- Churn control

An Optimizer system solution would be expected to be a:

- Forecasting system that estimates the failure rates of individual part numbers in each asset type and combines these known failure rates with the "asset location" to estimate the part failure probability distributions to the closest stocking location

- Decision system to manage the multi-echelon stock-control dynamics as described earlier
- An interface with the existing maintenance parts inventory policy management system to align and provide a platform for the inventory planner to manage the inventory optimizer tool
- An interface with the existing maintenance parts inventory policy management system to align and provide a platform for the inventory planner to manage other asset support inventory policies (Figure 12.5).

Utilizing technology for inventory placement policy and dynamically managing it each week benefits the traditional active, annual, ISP/RSP approach (Table 12.2).

Benefits to an Optimizer system have been observed to be:

- Reduction in the inventory required across the network
- Improvement in stock room availability levels and, more often, higher combined service levels delivered from the supporting stock room network
- Improved inventory planning and forecasting capability
- Improved understanding of the impact of parts operations on the Maintenance Technician and Operations
- A dynamic flexibility capability in responding to changing business requirements
- Reduction in the number of touches required by the material handler in the stock room.

The overall parts logistics problem was complex! It is important to note that transitioning to an Optimizer system can be a complex endeavor to manage as often there could be:

- Tens of millions of part-location combinations
- Tens of thousands of asset-location combinations
- Maintenance parts system control parameters that must be updated frequently (weekly) in response to the dynamic changes in the installed base, enterprise metric demands, and operational demands
- The risk associated with the success of the transformation and the supporting system that is considered vital to an enterprise's daily operations and can have a significant impact on its future reputation and revenues
- Resistance to change by the affected employees within the enterprise, given that the existing inventory policy control system is functioning and sophisticated

Some global enterprises, or multinationals, may have an Optimizer tool. Still, it cannot support all of the assets and related maintenance parts it could have in its scope of responsibility. Even these larger organizations will be asked to optimize their inventories regardless of the tools they may have to manage their maintenance parts supply chain more effectively. In most organizations, the pursuit of continuous improvement will drive a challenge to continually work to minimize inventories and material handling costs while driving improved asset support service levels.

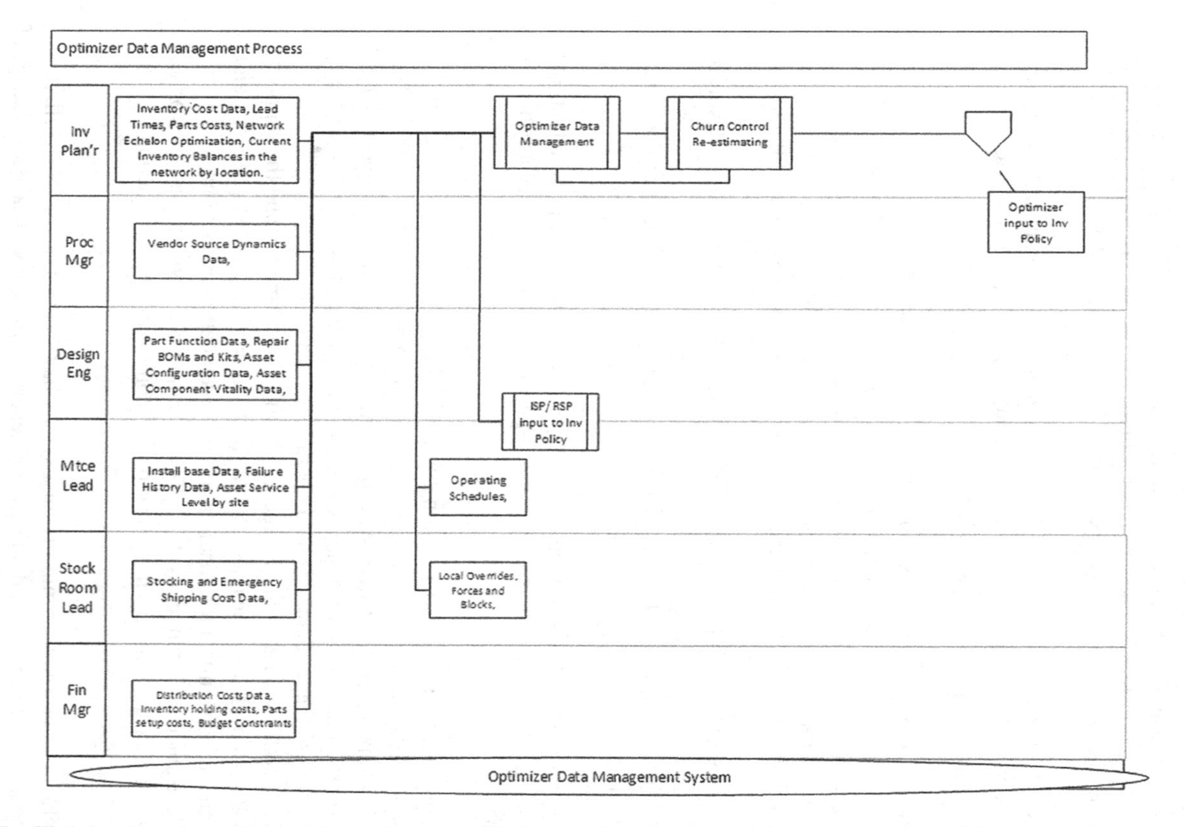


**FIGURE 12.5** High-level process diagram for an Inventory Optimizer data development system. (Adapted from copyright 104 MPE Inventory Management ©2020 from Maintenance Parts Management Excellence Training edited by Barry. Reproduced by permission of Asset Acumen Consulting Inc.)

**TABLE 12.2**
***Observations from the Optimizer State-of-the-Art Inventory Control System Initially Developed by IBM and the Wharton School of Business***

| Standard Inventory Policy Programs | Optimizer |
|---|---|
| Service levels by part number and location | Service levels by part number, asset, location, or echelon |
| Mean Time Between Failure (MTBF) estimates | Actual national failure rate database to calculate MTBF from |
| Local stock room demand | Local asset installation base by location |
| One location system | True multi-echelon system |
| Features: | Features: |
| • Manual Churn control | • The system calculated Churn control |
| • Manual forces and blocks | • Service part type (how used) |
| | • System-driven Inventory forces and blocks |
| Suboptimal total cost | Least total cost/higher network service levels |

## COST OF MISSED PARTS AVAILABILITY (FIGURE 12.6)

The target is constantly improving service levels while reducing the cost of stocking the supporting maintenance parts inventories. The impact of not having the correct quantity of parts at the right place and at the right time can impact the maintenance technician's time and reputation for provided services. Field technician impact costs may be all you are interested in if your business is a field service business. However, for most infrastructure and plant maintenance operations, the loss of the assets available to produce at the desired volume and produce product quality is the actual cost of not having the part available when immediately needed. *The cost of the lost maintenance technician time will be addressed later in this chapter.*

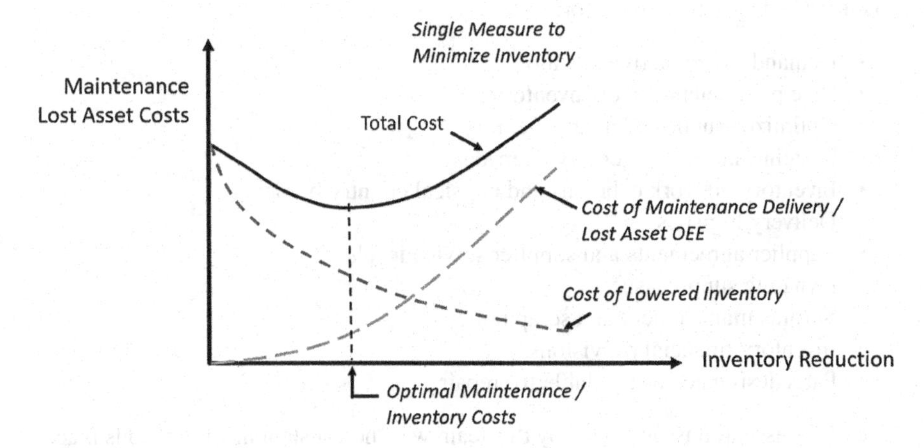


**FIGURE 12.6** Optimal Stocking Strategic Policy to achieve Optimal Maintenance and Lost Asset Impact costs. (Adapted from copyright 104 MPE Inventory Management ©2020 from Maintenance Parts Management Excellence Training edited by Barry. Reproduced by permission of Asset Acumen Consulting Inc.)

## STAKEHOLDERS OF AN EFFECTIVE PARTS OPTIMIZATION PROGRAM EXAMPLE (CASE STUDY)

Almost every enterprise constantly looks for ways to improve this optimal stocking strategic policy balance with the cost impacts. For example, the inventory planner may adjust the lead time and safety stock levels for some very active parts or adjust the date-protected parts and local overrides. Likewise, the maintenance manager (or design engineer for assets still in design) may look to adjust the ISP/RSP levels up or down to address this balance. Often, the opportunities to tune the inventory to be more optimal will create a need for all the stakeholders to get together from process supplier to client (maintenance technician) if a workable solution is expected to be agreed to and properly executed. Figure 12.8 maps out the business elements of a global initiative considered to improve service levels while reducing inventory carrying costs. The stakeholders included:

- Inventory planner management from multiple countries
- Finance representation from each country impacted
- IT support from each country impacted
- Design engineering/maintenance management representation for critical focused assets
- Vendor/supplier representation (OEMs, logistics providers, internal and external repair vendors)

The challenge is to maintain or improve the service levels while reducing the combined inventory plan levels (on the financial books) by 5% within the calendar year (10 months). While this may not seem like a large number, this is on a base inventory of hundreds of millions of dollars, and after a well-managed and well-executed optimal inventory plan has been in place for over a decade. Therefore, finding an additional 5% while improving service levels was a significant challenge to the enterprise.

The existing inventory network infrastructure and policy supported multiple countries and system support for:

- Demand-driven active inventory
- Date parameters-driven inventory
- Optimizer-supported asset inventory
- System-managed inventory overrides
- Inventory network echelons and physical country borders
- Delivery logistics
- Supplier agreements and supplier service levels
- Dynamic supply issues
- Surplus management and scrap policy
- Inventory financial provisions
- Parts designated as available for repair

One of the essential issues faced by this team was the constant inventory adds placed into the inventory mix. These could come from new ISP/RSP adds and maintenance technician returns of parts ordered, unused, and returned to inventory for credit.

A competitive forces model shows the high-level dynamic of the current operating program in Figure 12.7.

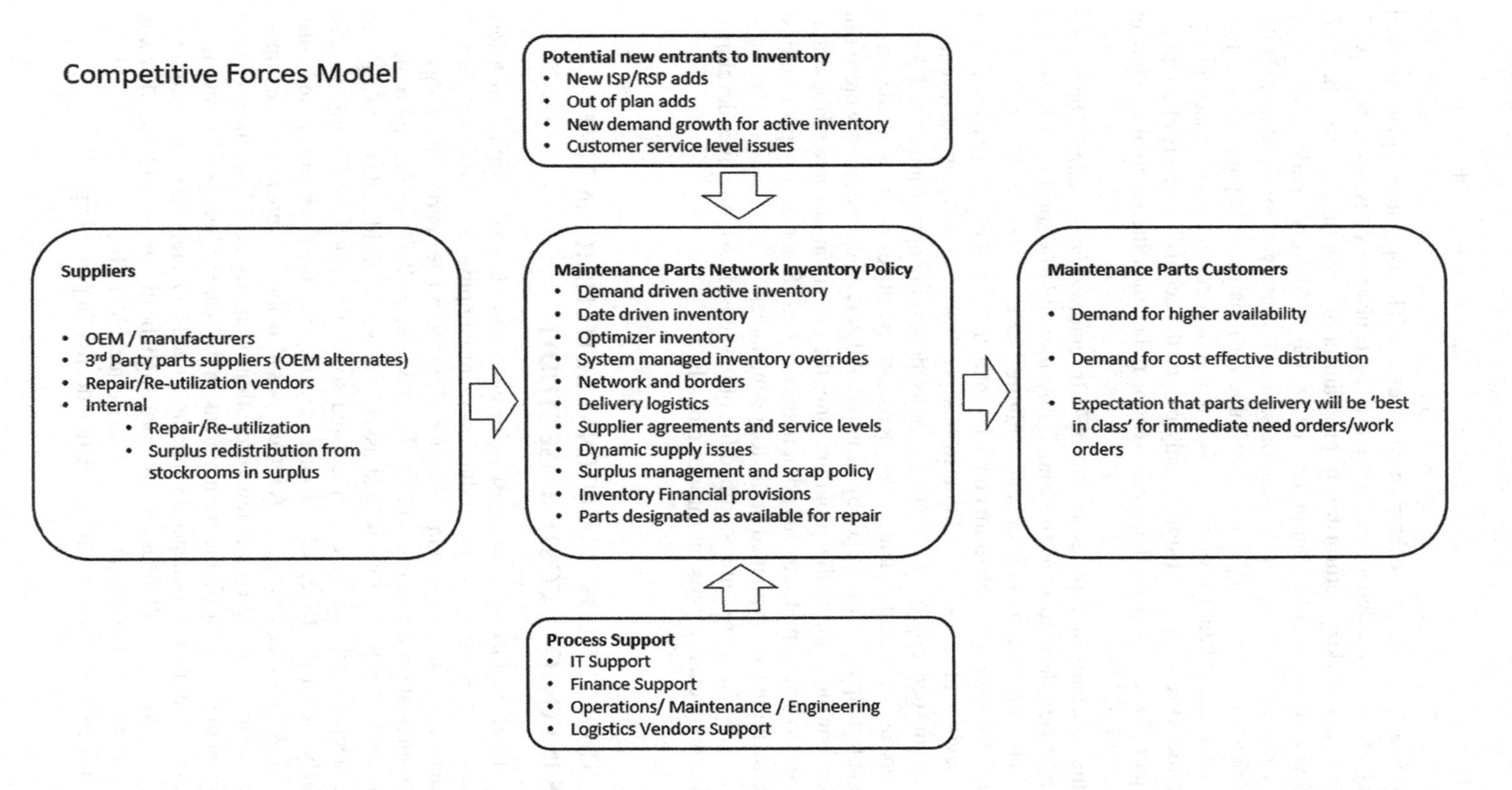


**FIGURE 12.7** An effective Parts optimization program requires support from multiple business units. (Adapted from copyright 104 MPE Inventory Management ©2020 from Maintenance Parts Management Excellence Training edited by Barry. Reproduced by permission of Asset Acumen Consulting Inc.)

A high-level inventory model was created when mapping out the brainstormed initiatives landscape from this group. Note that "new inventory adds" include ISP/RSP adds and Maintenance Technician parts returns. This organization also tracked used parts returns and called the collected, used parts inventory "available for repair" if the part had a working refurbishment program in the plan or in place. The parts repair would be done either "in-house" or by the third-party vendor and (when repaired and packaged "as new") would be sent to inventory as replenishment if the replenishment order was outstanding. Available for repair (AFR) parts were not on the books for this inventory initiative, but it was a focus for the Finance group. However, the AFR parts could significantly reduce the acquisition costs of a replenishment part if sourced from the repair vendor rather than the original equipment manufacturer (OEM).

Both the new parts inventories and the AFR Inventories were needed. They could scrap parts when deemed defective and viably unfixable or surplus to forecasted needs beyond a specific set of parameters (Figure 12.8).

The initiatives from this program created a challenge target inventory reduction of 5%. This revised plan was in addition to that year's already established inventory plan. The team spent approximately two months identifying opportunities and defining the scope and expected benefits. Early agreed-to initiatives were executed as soon as practical to help make the target by year-end. All agreed-to initiatives were set in place after four months into this program. Only the execution of some of the initiatives was left to be completed at this point in the program. Some of the opportunities/initiatives identified are explained in the following tables

Tables 12.3–12.6 represent a modified list of the considered initiatives developed to optimize maintenance parts inventories globally.

## THE COST OF THE LOST MAINTENANCE TECHNICIAN TIME VERSUS INVENTORY COSTS (CASE STUDY)

In Figure 12.3, we discussed the cost of inventory versus maintenance. In a few instances, companies or specific production plant campuses looked at the cost of having the maintenance technician "down tools" and its related parts expediting costs when an immediate need part was ordered and was unavailable as needed from the local stock room. If the cost of the lost asset time were considered, the cost graph would be steeper. Still, for the example discussed in the following Figure, we will assume only technician time, related management time, travel expenses, and parts expediting costs are on the cost line. A review was also done to understand the impact of their recent inventory actions. A few perceptions were captured from the review as they believed they were suffering from a break in the balance between the inventory service levels and the maintenance technician's needs. Overall, the perception was that the new corporate challenges on inventory might be disconnected from what was happening in the workplace. The maintenance technician believed they could not perform their role efficiently with the existing decline in parts support.

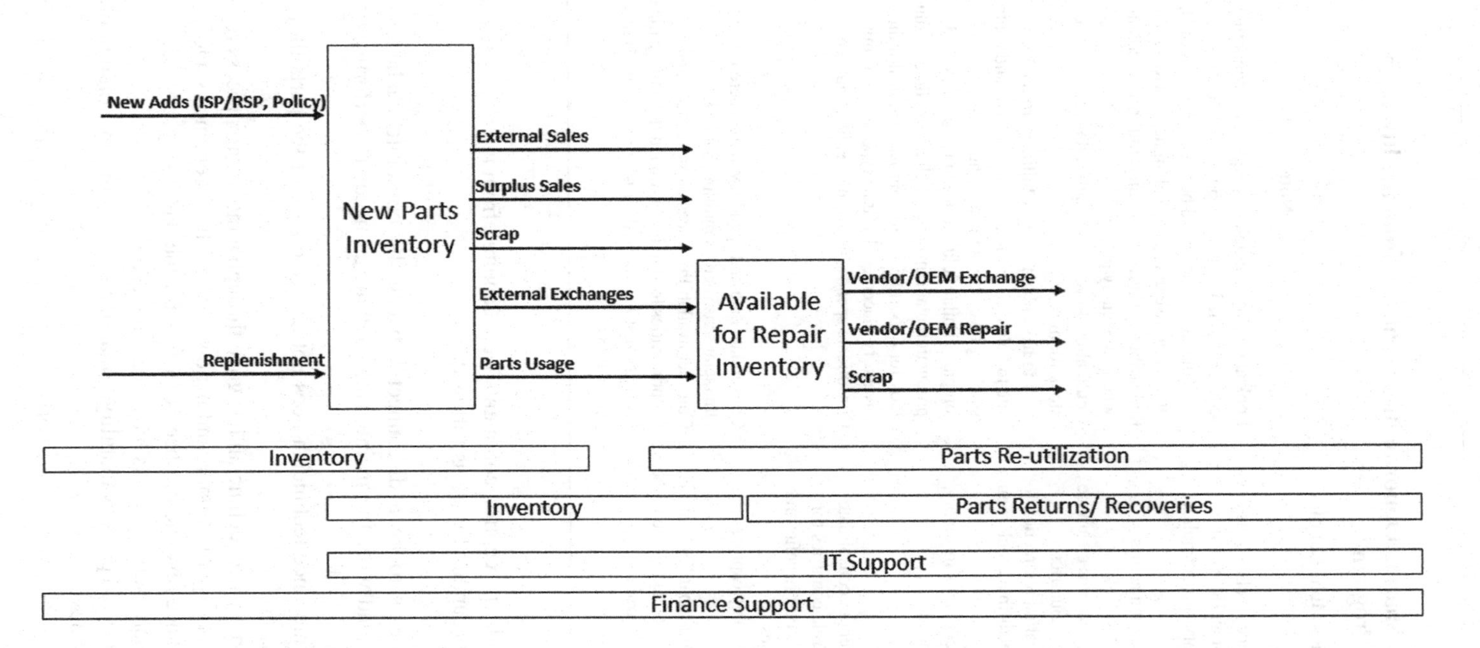


**FIGURE 12.8** Mapping an Inventory Optimization Program. (Adapted from copyright 104 MPE Inventory Management ©2020 from Maintenance Parts Management Excellence Training edited by Barry. Reproduced by permission of Asset Acumen Consulting Inc.)

**TABLE 12.3**
**Multi-Country Sharing Opportunities Identified from the Inventory Optimization Program**

| Potential Enterprise Multi-Country Sharing Actions | Notes |
|---|---|
| Challenge the service levels versus inventory levels of an ISP versus Optimizer Inventory support policy | Look at country examples by equipment supported to confirm if the Inventory Optimizer algorithms need tuning or if the ISP policy could be better managed in Optimizer at a lower inventory cost and higher service level |
| All surplus redistribution across countries | Redistribute surplus (in-country and) across countries where financially practical |
| Work on improving customs clearance cycle time between countries | Would allow lower safety and lead time stock in every stock room |
| Parts order prioritization redirection for non-immediate needs in a Stock room | Work to have planned work fulfilled from a more central site to balance workload in stock rooms and centralize the stocking inventory and handling costs |
| Review each business unit's inventory, and parts usage spend | Compare similar field service units, utilities, and production plant inventory to business unit revenue to look for opportunities to improve technical training, and inventory placement by asset type/business unit |
| Look for procurement to consolidate spend across multiple countries for the basket of goods cost benefits and logistics efficiencies | Consolidate spend where it makes financial sense |
| Improve vendor warranty capture and execution | Consolidate data and spend leverage where it makes financial sense across multiple countries |
| Review "available for repair" inventories to allow parts repair to take place in the most cost-effective location | Consider availability issues on some parts and tax opportunities to ship lower-valued defective parts to a country that may have their repair operation and use the parts as new. |

The stakeholder perception comments were captured from interviews at the time. Some comments captured were as follows:

- "There seems to be a disconnect between the corporate challenges on inventory and what the maintenance technician needs to perform the job efficiently."
- "The maintenance technician does not have the base of the parts they had five years ago."
- "The maintenance technician has gone through a mourning process regarding the loss of parts support and is now more at the acceptance stage."
- "Maintenance revenue has been constant for the past five years, but inventory and parts service levels have been reduced."
- "With improved parts availability, a maintenance technician would improve their call backlog."
- "The decline in maintenance parts service levels may not have hit their revenues yet."

## TABLE 12.4
## Inventory Policy Opportunities Identified from the Inventory Optimization Program

| Potential Enterprise Inventory Policy Actions | Notes |
|---|---|
| Review the policy for stock room policy "overrides." | Review the policy for local policy overrides and confirm how each stock room is performing within their inventory policy |
| Create visibility to work order staged or in-transit parts | See how this should complement the considered on-hand inventory balances in the replenishment algorithms |
| Look at lead times by vendor | See if a safety stock/lead time by vendor data set would create a more optimal set of active parts levels |
| Challenge repair vendors for Just in Time (JIT) fulfillment | Track repair vendors and supplier's ability to support fulfillment aligned to our EOQ actual orders and not over ship quantities ordered (pull orders versus push orders) |
| Close stock rooms with low volumes that can be supported elsewhere | Where stock rooms are across a campus or city, perhaps only one stock room can support all |
| Open high-echelon stock room | Consider increasing the service level for a more centralized stock room as part of a strategy to close out remote inventory locations and low-volume stock rooms |
| Parrado, the inventory request "misses" | Do a root cause on why the service levels are not being addressed |

## TABLE 12.5
## Parts Repair/Re-utilization Opportunities Identified from the Inventory Optimization Program

| Potential Enterprise Parts Repair/ Re-utilization Actions | Notes |
|---|---|
| Confirm sharing of parts vendors and processes where parts repair and re-utilization have been successful | Look at country/regional examples by equipment type and model and confirm the quality and cost benefits would fit. Create an international database of parts that should be captured for potential re-utilization |
| List skills by each country for parts repair and re-utilization | Look to understand if a vendor or internal skills transfer opportunities could help with parts re-utilization growth |

A multi-pronged approach was initiated to understand the current inventory performance versus when it was believed acceptable (about five years prior) and the relationship between maintenance cost when a part is unavailable versus when a part is available. In other words, what is the additional cost of maintenance when the Technician must leave the site of a non-functioning asset when parts cannot be made available and then must return later when a part finally comes available?

The inventory efficiency proved to be performing well against the business revenue streams the inventory was supporting. This inventory efficiency metric was confirmed by comparing the annual Equivalent Asset Maintenance Revenue (EAMR) to

**TABLE 12.6**
**Inventory Financial Handling Opportunities Identified from the Inventory Optimization Program**

| Potential Inventory Financial Handling Actions | Notes |
|---|---|
| Review how parts warranty fulfillment is being treated in inventory | Consider repaired parts to be returned to inventory as new with the burden of the cost of repair (and associated logistics) plus a percentage of new (e.g., 20%) as credit for the used part carcass |
| Create visibility to work order staged or in-transit to technician work order parts | Treat the parts in transit to a work order as expensed versus inventory. Best to understand the dynamics and generally accepted accounts practices (GAAPs) |
| Review how available for repair inventory is being treated in the inventory | Typically, it would be below the new parts inventory line but still in the overall inventory at perhaps 20% of new |
| Challenge how inventory provisions (reserves) are being funded and released | Critical when an inventory strategy change could generate a need for different inventory and identify unplanned surplus in a given period |
| Confirm what additional costs are burdened into the inventory number | Costs to ship a part to the first warehouse/stock room from the vendor are typically burdened by inventory costs. Other costs such as shipping between stock rooms or to a technician and material handling and warehouse costs are treated as part of the inventory carrying costs |
| Look to like companies for how they treat inventory financial accounting to see if other opportunities exist | Some may be depreciating their inventory rather than scrapping it. Best to understand the dynamics and GAAPs |

the average inventory level over the previous year. As a result, the Inventory to EAMR ratio improved from 1:5.7 to 1:8.3. Inventory turnover also performed well (improving from 1.2 turns per year to 1.8 turns per year). Still, the Inventory service level indicators (parts availability and parts procurement time) decreased. Lowered parts availability was the reason for the poor perception from the maintenance technicians and the low employee satisfaction ratings received (Table 12.7).

From looking at the inventory levels and the cost of delivering the critical maintenance metrics, it was clear that the optimal maintenance to inventory costs ratio had been passed (see Figure 12.9). This organization reduced the parts inventory, resulting in additional maintenance costs and bad employee morale among the maintenance stakeholders and the maintenance parts management staff. The asset management revenue was there to support additional support costs. In this case, the costs were an additional burden given to the maintenance technician and not being complemented with well-placed maintenance parts inventory.

An annual corporate directive created the root cause of the inventory and maintenance time mismatch on Inventory level targets to improve the enterprise balance sheet and turnover outcomes and year/year financial market analysts' expectations. This directive was a financially driven business influence to improve annual inventory turnover by 15%.

**TABLE 12.7**
**Example: Year over Year Inventory Audit Review (5+ Year Lookbacks)**

| Operating Metric | Baseline Metric | Delta After 5 years | Delta | Operating Metric |
|---|---|---|---|---|
| Equivalent Asset Maintenance Revenue (EAMR) | $ XXX M | 120% | Increase | Equivalent Asset Maintenance Revenue (EAMR) |
| Inventory | $ XX M | 85% | Lower | Inventory |
| Turnover | 1.2 | 1.8 | Increase | Turnover |
| Inventory: EAMR Ratio | 1:5.7 | 1:8.3 | Increase | |
| Country Availability Level Achieved | 92% | 87% | Lower | |
| Parts Procurement Time | 2 Hours | 2.9 Hours | Lower | |
| Stock Room Availability Level | 85% | 78% | Lower | |
| Maintenance Technician Satisfaction level | 80% | 75% | Lower | |

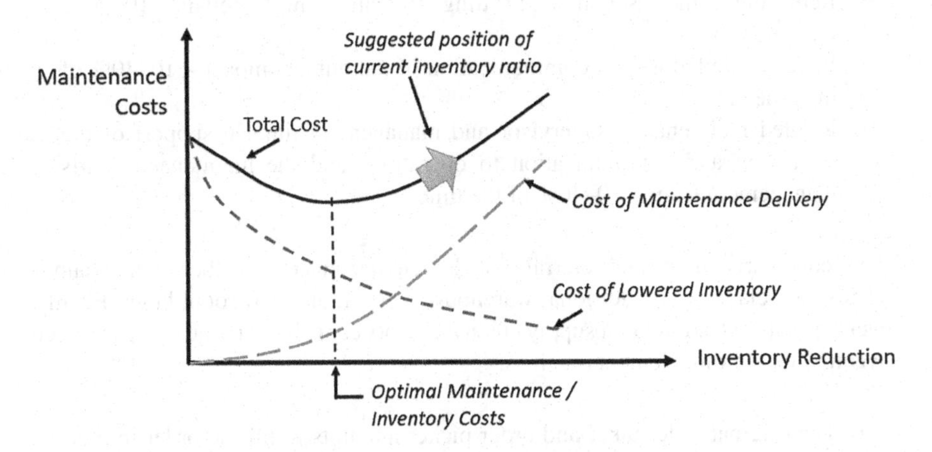


**FIGURE 12.9** Study suggested that current inventory reduction had impacted maintenance costs (well past Optimal target). (Adapted from copyright 104 MPE Inventory Management ©2020 from Maintenance Parts Management Excellence Training edited by Barry. Reproduced by permission of Asset Acumen Consulting Inc.)

With the root cause identified, it was suggested that inventory performance targets should be set based on their local market operating metrics such as inventory/revenue ratios or a similar performance measure such as a return on assets (ROA).

It was also recognized that corporate directives should be aware of and consider geographic market similarities and differences, operating environments, and culture needs to maximize efficiencies and garner the maximum yield of operating as an integrated business where practical.

The response to this corporate directive at this cycle was that *"in the existing business model and operating environment, there are few remaining inventory reduction initiatives that will positively impact operating profitability."*

A high-level analysis of identified inventory reduction initiatives suggested that further reductions must be selective so that the resulting parts referrals due to the low availability of parts are minimized. In addition, the approximate margin ratio of each $1000 in reduced inventory must not result in more than one parts availability miss.

This study also looked at two primary cost contributors when a part was unavailable. First, the maintenance technician had to leave the asset non-functioning due to a part not being available from the local stock room. In a parts outage and part order referral to a supplier or other internal stock room, the first cost element introduced into the supply chain is the cost associated with non-productive maintenance technician tasks. Second, the cost of a maintenance technician leaving the site and returning and the related parts expediting costs due to a local parts availability miss. Maintenance technician elements considered in the study included maintenance technician elements such as:

- Escalation efforts are assumed to happen – 50–80% of the time
- Leaving the asset site (tools down and packed up) – 100% of the time
- Returning to the asset site and setting up again from the repair – 100% of the time
- Initiating and closing a communication of the outage impact – 10–40% of the time and
- Related maintenance supervisor and management time in support of the escalation and communication to operations and the maintenance parts management team – 10–40% of the time

The second parts outage and referral cost element introduced into the supply chain is the cost associated with additional warehouse, carrier, and part room labor. Finally, inventory support and related supply chain elements considered in the study included Inventory expediting elements such as:

- The original order taker and order picker attempts to fill the order in their stock room – 100% of the time
- A role to research where the part may be in the stock room by checking the system and receiving areas (if applicable) is performed
- The picker researches the most time/cost-effective location to refer the parts order to and creates the order referral (to a supplier or other internal stock room)
- The arrangements to expedite the order (via a courier or next flight out) will be coordinated
- The costs of the overnight courier or the following flight airline fees plus the travel to and from an airport (if applicable) need to be considered
- The asset's local stock room material handler to receive the expedited part and inform the maintenance technician
- The parts will be shipped to the asset site or picked up by the maintenance technician

These costs vary depending on the maintenance organization, the criticality and urgency of the parts outage, and the business impact of the asset downtime. The carrier's main costs can be particularly if a next flight out or even a charter aircraft is deemed required to meet the business need. The Maintenance costs here ranged from $100 per incident to $250 per incident, depending on the study. The stock room material handling costs were in the range of $100 to $200 per incident. The additional carrier costs could be $50 to $5000 depending on the business supported and the criticality of the non-functioning asset.

For the studies viewed, the cost of a parts outage scenario ranged from $400 to $800 per incident.

## BENCHMARKING INVENTORY AND SERVICE LEVELS (CASE STUDY)

Comparing inventory baselines and benchmarks can be difficult. Comparing one utility's approach to maintenance parts management is often like comparing apples to oranges. Likewise, comparing an OEM's maintenance parts management to a competitor can prove problematic as they will likely look at their business with a different history and strategic imperatives. Comparing the same enterprise organization to where they were five years prior (as suggested in an earlier Figure) can be challenging as the business direction, imperatives, and focus evolve. Table 12.7 is an

**TABLE 12.8**
**Dynamics of Maintenance Parts Management Costs Can Change Year over Year**

| Maintenance Parts Cost Environment | Year X | Year Y | |
|---|---|---|---|
| Inventory Level ($ M) | $ XX | 85% | Year Y inventory is 85% the value of X |
| Inventory Annual Turnover | 1.2 | 1.8 | |
| EAMR Supported | $ XXX | 120% | Year Y inventory is 120% the value of X |
| Inventory to EAMR Ratio ($ M) | 1: 5.7 | 1: 8.3 | |
| Maintenance Parts Cost Elements | Year X | Year Y | |
| Cost of Inventory Reserves ($ M) (Provisions) | 9 | 8.1 | Provisions down due to lower average inventory |
| Warehouse and Inventory Management Costs ($ M) | 25 | 28 | Number of orders and transactions up to match the EAMR activity |
| Inventory Systems Support Costs ($ M) | 3.7 | 3.5 | System costs are relatively steady |
| Total Managed Costs ($ M) | 37.7 | 39.6 | |
| Parts usage (and Parts Sales) ($ M) | 72 | 92 | |
| Total Maintenance Parts Costs ($ M) | 109.7 | 131.6 | |
| Parts Re-utilization Credits | 14 | 20 | Up due to higher parts usage |
| Parts Sales Credits (Profit) | 2 | 2.5 | |
| Surplus Sales Recoveries | 1 | 2 | Up to supported mix changes |
| Total Maintenance Parts Inventory Credits | 17 | 18.5 | |
| Net Maintenance Parts Costs | 92.7 | 107.1 | |

excellent attempt to show how one can make a baseline comparison. In that example, the business focus evolved from doing field maintenance parts support to doing field maintenance parts support and supporting a new enterprise market offering of field maintenance outsourcing. The old metrics did not recognize the inventory ratio to this new revenue stream in field maintenance outsourcing. Therefore, a new metric called "Equivalent Asset Maintenance Revenue" (EAMR) needed to be separated to include all maintenance-related streams inventory was supporting.

If the enterprise looked to sell the parts they were carrying to business partners or other third parties as a parts-specific revenue stream, that revenue (and profit) also needs to be considered in the baseline comparisons. An example of a model used to compare some enterprise organization's parts "value" is shown in Table 12.8.

Actual parts usage will be the highest cost influence on a maintenance parts management operation. This cost is owned and influenced mainly by the operations and maintenance community. This metric is still something that the maintenance parts management team should be tracking as it drives (along with the ISP/RSP policies). The inventory management-related costs dynamically influence what steps could be taken to optimize the inventory levels and placement while working to deliver improved, or at least optimal, maintenance parts service levels.

# 13 Preparing for Uber Change in Maintenance Parts Management

## CHALLENGING THE "STATUS QUO" IN MAINTENANCE PARTS MANAGEMENT

Every organization must challenge the current operating model and the paradigms that seem to be influencing the maintenance parts management operation on a cyclical basis. Challenging the status quo of this operation could promote and celebrate an outstanding or supreme new model in which the organization can work toward driving higher parts service levels and reducing the related parts management costs.

The need to "re-think" and execute change in maintenance parts management can come from many sources:

- A business merger or acquisition
- A business imperative to become significantly more cost-effective while increasing service levels
- A significant need to recalibrate the current processes and the resources required to deliver optimal inventory management services
- Significantly new equipment technology solution to install and support
- Changes in the enterprise's market or competition
- Changes to other business units in the enterprise
- Changes to the systems planned to support the enterprise – including the maintenance parts processes
- Significant dissatisfaction with the costs or service levels currently being received
- Changes to how finance accounts for capitalized inventories
- Employee morale

With a need for significant change, it is important to carefully understand what issues need to be addressed and what initiatives will best leverage its people, funds, and resources. Some form of "assessment of the current issues and analysis" is recommended to ensure effective use of resources. When the problem and solution are perceived to be known, the enterprise can be guilty of the "ready, fire, aim" sequence of problem resolution. Getting an agreed understanding of what the solution and success will look like is critical to getting the leadership and affected staff aligned to support the committed initiatives to achieve the new model. "ready, aim, fire" is the sequence that will help all the stakeholders support the change continuum (See Figure 5.4 Change enablement continuum.)

DOI: 10.1201/9781003344674-13

Methods for facilitating re-think of maintenance parts management processes can come from:

- An overall quality/strategic development approach (see Figure 13.2)
- A maintenance parts management maturity assessment (see Figure 13.1)
- A design thinking set of workshops (See Figure 3.5)
- A formal International Organization for Standardization (ISO) process documentation and quality management disciplined approach
- A classic process re-engineering set of processes and disciplines

Three methods for identifying current issues, opportunities, and potential initiatives have been introduced in previous chapters.

This chapter will focus on challenging the current parts execution models, a more classical process re-engineering approach, and the expected value (outcomes).

Each approach has been used multiple times in many different maintenance parts management organizations. The first three approaches often leverage a third-party consultant to facilitate the process and timelines. One of the first three is often used to drive the significant change needed in a specific timeline. The classical approach involves the internal stakeholder staff more directly as they are invited to re-think maintenance parts management as leaders and participants. In addition, it will provide them the training and experience that likely will help them with future challenges in the enterprise throughout their career. These initiatives will expect major change through significant service level improvements, cost efficiencies, and improved employee morale.

Maintenance parts management assessment uses predefined questions, stakeholder interviews, and workshops to identify and prioritize issues, opportunities, and initiatives. This process was discussed in more detail in Chapters 2 and 3.

Chapter 3 discussed the high-level process for facilitating new overall strategy development. For example, re-thinking parts processes could follow the same strategy steps, as shown in Figure 3.2 process flow but is often more internal in focus and can be facilitated in fewer steps.

Design thinking workshops immerse the people/entities affected by the current business model execution and what change may look and feel like to those affected. This approach also focuses on defining the client's requirements, how the current program functions, how the future could function, and equally looking to develop identified and prioritized issues, opportunities, and initiatives. From an earlier chapter, Figure 3.6 shows two examples of a design thinking workshop board used to extract what a change may "feel" like and how various stakeholders empathize with the current scenario and the need for change.

The ISO defines quality as "the totality of features and characteristics of a product or service that bear on its ability to satisfy stated or implied needs." In simpler terms, "the degree to which a set of inherent characteristics fulfills specified requirements."

ISO 9001 is widely regarded as an essential quality management system worldwide. It brings numerous benefits to an enterprise, and most importantly, these benefits get passed across business units and onto customers. Attributes such as

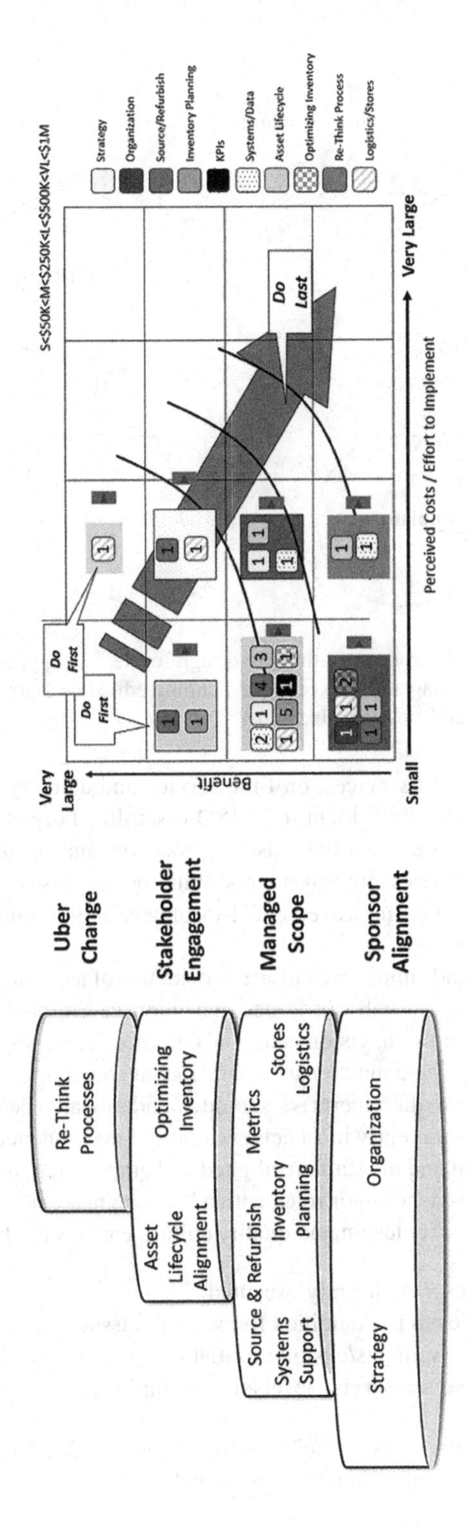


**FIGURE 13.1** Tools of the Maintenance Parts Management Maturity Assessment process. (Copyright MPE ©2022 from Maintenance Parts Management Excellence Training edited by Barry. Reproduced by permission of Asset Acumen Consulting Inc.)

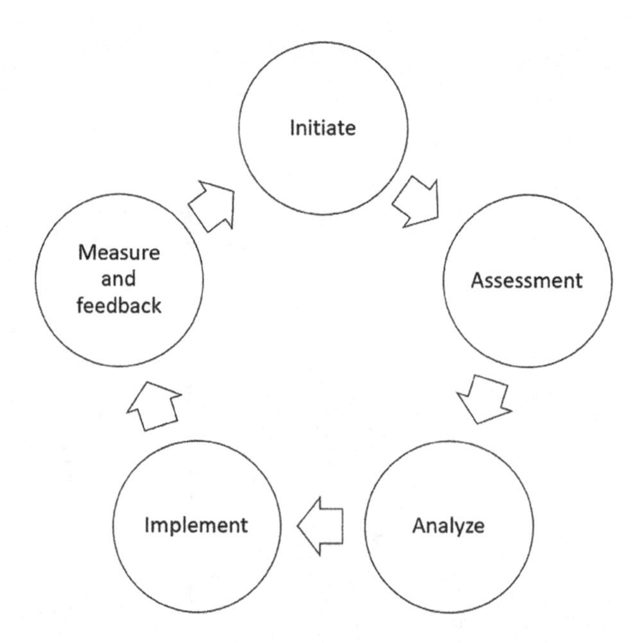


**FIGURE 13.2** High-level process re-think/re-design cycle. (Copyright MPE ©2022 from Maintenance Parts Management Excellence Training edited by Barry. Reproduced by permission of Asset Acumen Consulting Inc.)

consistency in products and services, on-time service and delivery, getting it right the first time, and more can be evident in an ISO-disciplined organization.

An ISO 9001 focus documents the existing processes and, in that effort, challenges many to look for process improvements. With the processes documented, the enterprise creates "continuous improvement" by doing what you said you would do, measuring it, and improving it.

A highly productive and supportive culture is evidence of a culture that empowers all enterprise stakeholders to call out issues and improve the business when they understand the business mission, vision, and goals. It goes a long way in bridging the complex issues in optimizing a maintenance parts management operation. The previous chapters describe how the enterprise should foundationally be organized. The chapters suggest that a community must achieve optimal maintenance parts management. When management and the staff are aligned and empowered, this collaborative group of employees approaches their roles with a "we can make it happen" perspective. A process of re-think/re-design and quality culture enables the team to:

- Improve how things are currently executed
- Look for root causes to fix identified foundational issues
- Drive process improvements/customer satisfaction
- Manage the costs versus service levels to a mutually agreed optimal level

Such a quality culture acts as the "glue" to keep all the stakeholders motivated and feeling that their participation and contributions make a difference to the enterprise.

## CLASSIC PROCESS RE-ENGINEERING PROCESSES AND DISCIPLINES

Process management (an approach for assessing, analyzing, and improving selected key processes) and process improvement can lead to incremental gains in the maintenance parts processes. Process re-think/re-design includes the action of process management and improvement. However, it is functionally looking to break the processes to find and implement quantum improvements to the value that is the output from the maintenance parts management set of responsibilities.

Like the two previous elements discussed (asset lifecycle integration and inventory optimization), the process re-think/re-design focus requires more than a single stakeholder group to achieve significant change and benefit. This focus will re-write the definitions of foundational processes, methods, systems, tools, and execution standards. In addition, this focus will create quantum change and quantum benefits in the maintenance parts operation and its contribution to the enterprise's value and financial bottom line.

A process is a series of definable, repeatable, and measurable tasks leading to a valued result for an internal or external client (customer).

Effective and efficient processes share a standard set of attributes (Table 13.1):

For maintenance parts management, processes come in levels.

At the highest level (Level 1), maintenance parts management may be a declared process family, given its critical role in supporting the enterprise's mission to create value.

At Level 2, processes described in this book (i.e., asset lifecycle planning, inventory planning, inventory optimization, procurement, parts repair, parts sales, logistics and warehousing, inventory financial GAAPs) should be considered topics. This level would represent the principal activities that begin and end with the maintenance parts management client.

Level 3 would support subprocesses of each topic declared in level 2. For example, this group could represent activities between departments within maintenance parts management or critical external business units.

Level 4 could be subprocesses of level 3 or process steps (desk procedures). This level typically involves activities accomplished in a single department or individual.

Depending on the complexity of the processes and subprocesses, there could be additional levels.

**TABLE 13.1**
**Process Attributes Summary**

| Process Attribute | Description |
|---|---|
| Definable | Target process requirements and metrics are documented |
| Repeatable | Processes are a set of sequences of repeatable activities. They are well communicated, understood, and followed consistently. |
| Predictable | Process stability is achieved, and desired results are met from the process activities being consistently followed |

Foundational process re-think/re-design involves starting or taking the initiative and picking a process area to focus on. Then the high-level approach involves three primary stages:

- Initiate a need for action
- Process assessment
- Process analysis
- Process improvement implementation
- Measure and feedback on the initiative influences, outcomes, and issues

Program initiation, measuring, and feedback also need to happen but are assumed tasks in this description.

These three stages should be a closed loop, and the progress of the improvements implemented should be measured.

- Process assessments can include:
  - Initiating and organizing
  - Determining the process owner
  - Interviewing with the process client/customer and confirming client requirements
  - Documenting, confirming, and walking the current focused process (if needed)
  - Rating the quality, success, and value of the process
  - Establishing process priorities
- Process analysis can include:
  - Baseline and benchmarking (and at a minimum, baselining) the selected process
  - Solutions development
  - Reviewing the potential solutions with the impacted stakeholders
  - Lobbying buy-in from the process owner and users impacted
  - Creating a strawman implementation plan and finalizing the improvement plan
- Process improvement Implementation can include:
  - Piloting the proposed solution (if needed)
  - Implementing the process improvement plan
  - Measuring results, obtaining process client feedback on the new process performance
  - Rolling out the solution to the enterprise
  - Looking for a new process focus

Success should be expected when the participating stakeholders start by thinking about why the processes exist in the first place. For example, suppose the asset is part of the enterprise value chain. Maintenance supports the asset's ability to deliver this value – maintenance parts management exists to support the assets' operation and the maintenance mission with the necessary parts. A successful process re-design team will need to consider the following:

- Corporate mission values and its value proposition to the enterprise customer (market)
- How best-of-breed organizations (direct competitors or subprocess support companies) lead in the area of focus
- How to ensure cross-functional participation by including all stakeholder groups
- Assuming the process owner role for the process owner
- Identifying and securing the approval and participation of the area of focus process owner in the initiative
- Promoting individual participation
- Taking a process view of the execution of the work
- Implementing a typical measurement standard that would work across the enterprise where practical
- Promoting ongoing continuous improvement (quantum improvements now waiting until perfection is secured typically not practical)

A high-level map will help guide the team through the re-think/re-design process (Figure 13.3).

In every initiative, a check should be made to confirm the health of the area focus re-think. For example:

- A better understanding of the corporate market value in the eyes of the customer will influence improving how customer needs and expectations now and in the future are being met
- Improving service levels and accuracy in meeting the customer needs will drive a focus on defect elimination
- Reducing the time it takes to work a customer value process will drive a focus on process cycle time reductions
- Measuring the quality and performance and recognizing staff for key-value metric achievements will come from a focus on measurements
- Involving crucial staff from each of the impacted Stakeholder groups in the re-think/re-design efforts to promote employee participation

Many organizations become overly focused on the execution of their domain. For example, in maintenance parts management, no single stakeholder group can be held accountable for the overall output or performance of the inventory policy and delivery process.

The maintenance parts management organization's perception often guides the rest of the enterprise and its customers in forming an opinion of how they are supported. Understanding how the parts operation "actually interacts" with other stakeholders is key to understanding current perceptions and areas for re-design. Mapping the interfaces with the understood process customer (or client) expectations will go

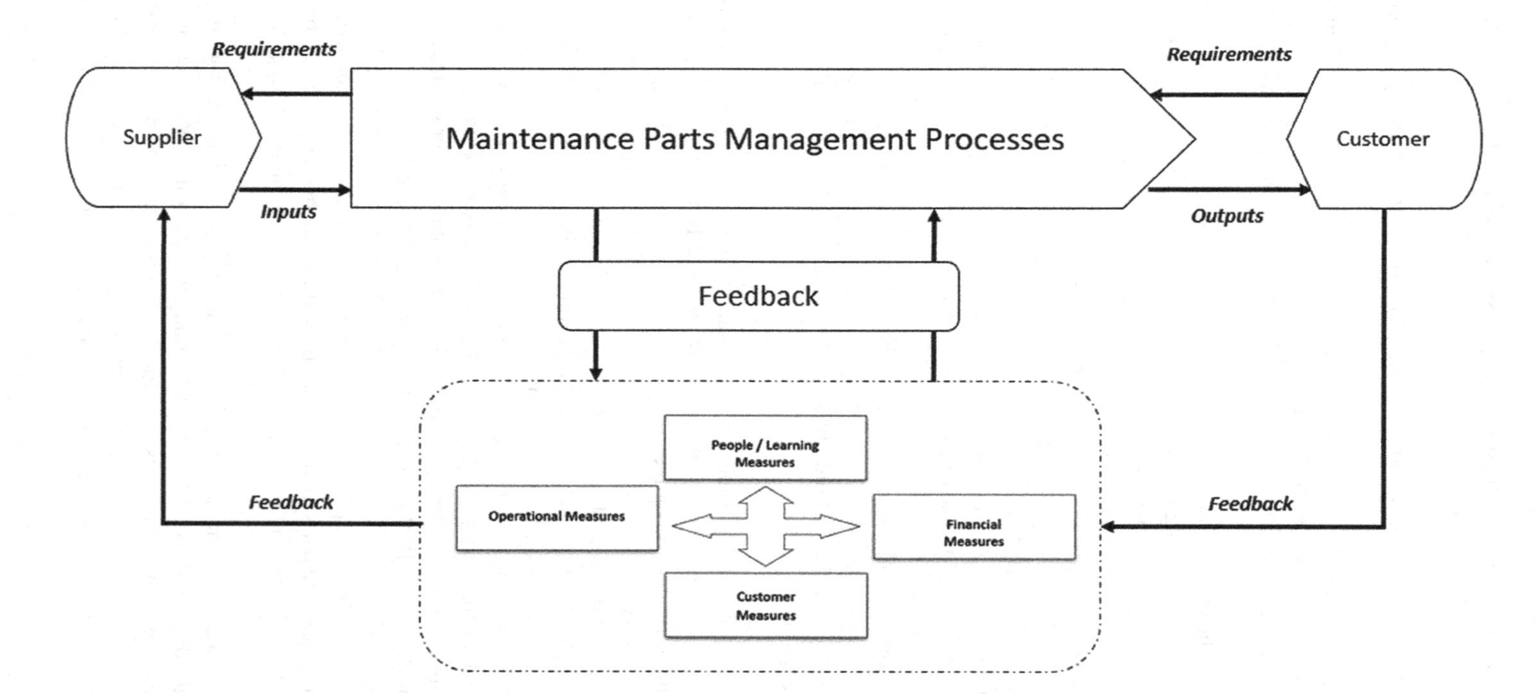


**FIGURE 13.3** High-level re-think/re-design target process map example. (Copyright MPE ©2022 from Maintenance Parts Management Excellence Training edited by Barry. Reproduced by permission of Asset Acumen Consulting Inc.)

**TABLE 13.2**
**Why Re-visiting Parts Processes are Important**

| Areas to Understand | Description | Maintenance Parts Management Examples |
|---|---|---|
| Parts client touchpoints | Client interface experiences – Where maintenance parts meet their affected stakeholders | Parts management Interfaces and process execution<br>• Inventory policy delivery<br>• Inventory service level delivery<br>• Inventory costs management<br>• Distribution quality delivery<br>• Systems support delivery |
| Parts process client needs | Client process expectations – What do clients use to develop a performance opinion? | Understanding the parts client "value" needs<br>• Inventory policy requirements negotiations<br>• Inventory service level commitments<br>• Inventory costs commitments<br>• Distribution Quality commitments<br>• Systems support commitment |
| Parts client perceptions | Client process satisfaction – Perceived achievement of requirements | Being recognized for the "value" provided<br>• Ease of working environment<br>• Inventory service levels<br>• Inventory costs to unit budgets<br>• Distribution quality versus time lost from unresolved issues<br>• Systems ease of use and availability |

a long way in understanding how the current value of a maintenance parts management operation is executing against the process client expectations. The maintenance parts operation should expect a fully supportive and cooperative stakeholder group (e.g., maintenance technician, finance manager, or design engineer). In turn, these stakeholders should suggest that they are very satisfied (if not delighted) with the level of support received when requirements are delivered.

When reviewing existing maintenance parts management processes, it is helpful to understand the client process touchpoints, needs, and perceptions (Table 13.2).

Early indications of poor maintenance parts performance can come from not performing the existing processes effectively or efficiently. Some maintenance parts management external symptoms of poor performance may manifest as:

- Maintenance technician complaints about parts availability, order fulfillment cycle time, or picking quality
- The maintenance planner has concerns about inventory visibility in the system and the ability to stage parts for planned essential work orders
- Financial team concerned about inventory effectiveness and turnover
- Engineering concerns on facilitating a track asset lifecycle support for parts policy
- Supplier concerns about receiving accuracy

Within the maintenance parts management group, internal indications of process performance issues can manifest as:

- Internal turf battles
- A sense of helplessness, employee morale, or recognition issues
- A sense that many internal processes do not add "value" to the process client they serve
- Frustration that the business unit cannot service their process clients well within the existing culture or process set

An effective way to facilitate the documentation of a process is to use sticky notes on a wall. Assuming the process to be documented is agreed to, the steps to do this would be to:

1. Identify the process participants (one per sticky note) and place them vertically on the wall as a column on the left
2. Identify the process client and place them at the top of the participant's list (also on the wall, top of the participant column)
3. Determine who the client first interacts with within this process, and describe the activity on a sticky note. Then place this note next to the client
4. Determine the participant (or client) that is the output of this step and what activity they perform
5. Repeat this facilitation until the documentation of the current process is complete
6. Add dependencies such as system inputs (Figure 13.4)

With the process displayed on the wall, record what is agreed to, and compile a list of lessons learned from this exercise.

Notice that the client can typically only see their interface with the process. The line between the client and the process is how the client perceives the process, the cycle time, and the quality output. This client swim lane is called the process "line of visibility." Every interaction with the process and the client is a "moment of truth" that will contribute to the client's perception of the process's value and performance.

Rating a process can be subjective at times. Equally, rating a new process will be subjective. Often picking a new direction or process will require understanding a process's quality, cycle time, and value (health) before it is changed and the health of a process after the change has been implemented. High-level guidance for rating process health can be a simple maturity matrix (innocence to excellence). For example, Table 13.3 can be a subjective declaration from critical to best in class.

Working the three stages of a classic re-design program supports assessing and analyzing before you jump to implement and avoid the "ready, fire, aim" syndrome. Each stage has steps that may be followed to confirm that the stage has been thoroughly thought through. These stages and steps can easily be managed and executed by the enterprise stakeholders without the hand-holding of a third-party consultant.

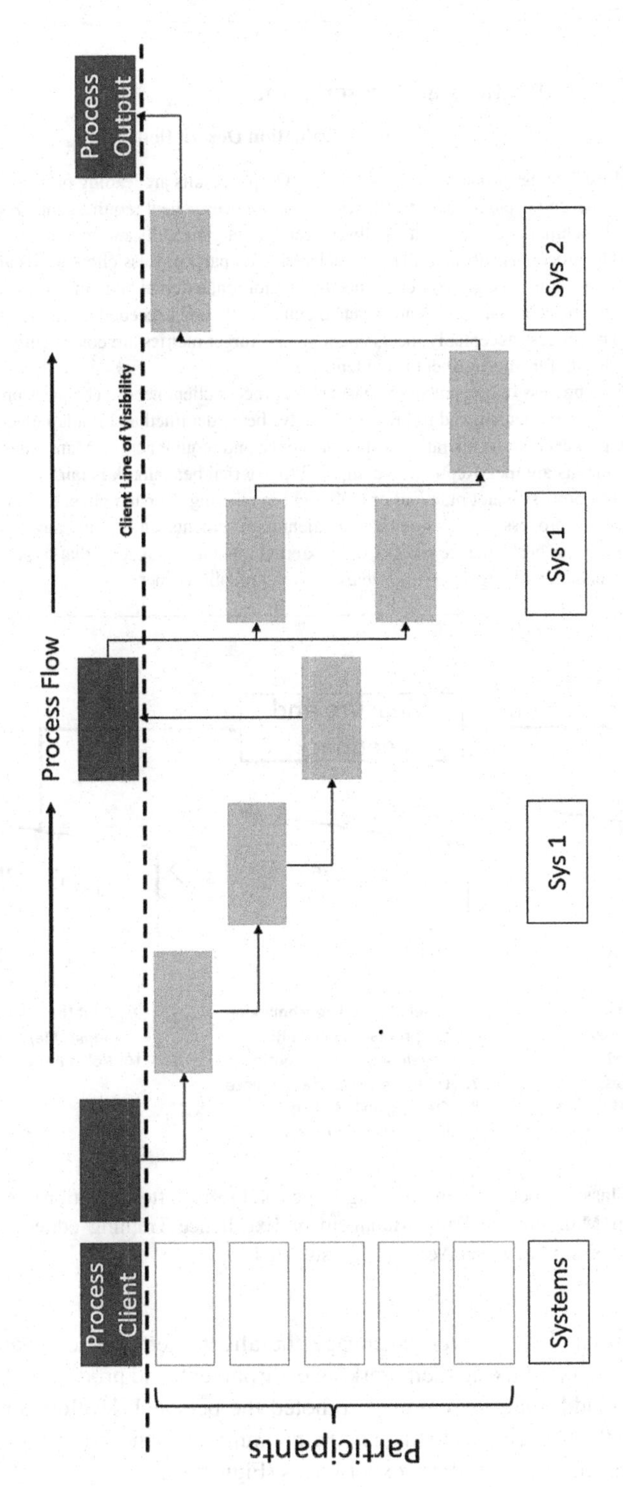


**FIGURE 13.4** Process mapping by line of visibility or swim-lane charts. (Copyright MPE ©2022 from Maintenance Parts Management Excellence Training edited by Barry. Reproduced by permission of Asset Acumen Consulting Inc.)

**TABLE 13.3**
**Process Rating Considerations and Descriptions**

| Process Rating | Process Evaluation Description |
|---|---|
| Best in Class | Best-in-class processes achieve results. Output results are leading or nearly defect-free, and cycle times are well within parts process client requirements. External benchmarks confirm outstanding/exceeding target results and reputation. |
| Strong | The process is substantially over-delivering on parts process client needs and is agile to respond to changing client needs. In addition, external benchmarks and internal metrics confirm targets are reached consistently and exceeded at times. |
| Stable | The process acceptably meets client needs. Target metrics are consistently met. Room for improvement is evident. |
| Fair | The process outputs meet the basic parts process client needs. The parts processes are documented, and improvements have been identified and implemented. However, some operational shortfalls exist and require action. Many internal targets are met. Reputation compared to external benchmark is fair. |
| Critical | The process is not practical or inefficient in meeting the parts client's requirements. Parts process outputs do not meet client requirements, and significant parts-related process performance issues exist. External reputation suggests that targets are often missed, and target service outcomes are unreliability met. |

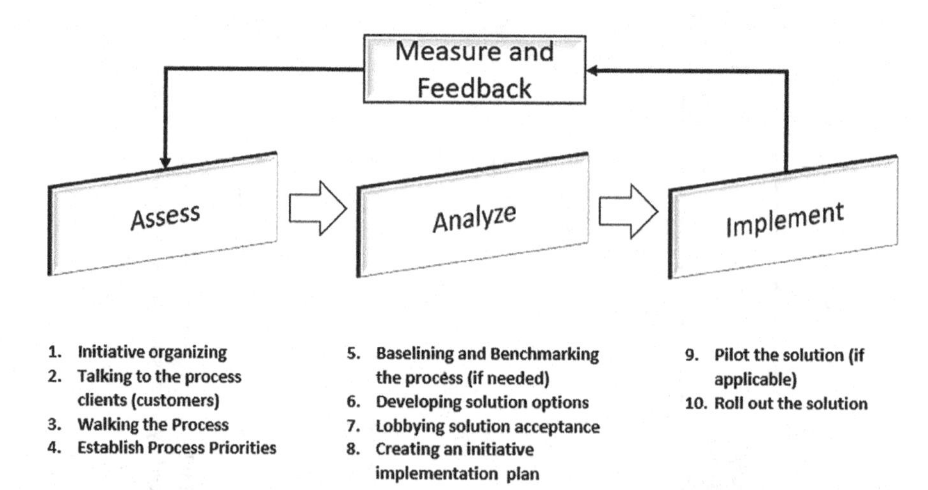


**FIGURE 13.5** Classic process re-engineering stages and steps (after initiation). (Copyright MPE ©2022 from Maintenance Parts Management Excellence Training edited by Barry. Reproduced by permission of Asset Acumen Consulting Inc.)

This approach has been proven to engage the affected enterprise stakeholders, demonstrate ownership of fixing their working environment, and produce a culture of empowerment. In addition, this culture promoted the personal development of the staff and allowed the participants to display their commitment and skills. Often future enterprise leaders emerge from these experiences (Figure 13.5).

## ASSESSING CURRENT (AS IS) PROCESSES

The expected outputs of the Process Assessment steps are shown in Table 13.4:
Getting Organized involves:

- Developing a process overview by:
  - Agreeing on a process name
  - Drafting a brief process description including:
    - Purpose of the process
    - Defining a "starts with" and "ends with" declaration
    - Key activities performed in the process
    - How it relates to other processes
  - Creating a preliminary, high-level process map
  - Identifying the process owner, process client, and participants and supporting systems
- Create a data collection plan by:
  - Identifying data sources for the target process
  - Collecting an early assessment of data sources, quality, and results
  - Brainstorm a potential list of process participants
- Draft a process assessment plan and schedule listing:
  - Include tasks, dates, participants, and expected deliverables from each step

**TABLE 13.4**
**Process Assessment Steps and Expected Outputs**

| Step | Activity | Expected Outputs |
|---|---|---|
| 1 | Initiative organizing | • Target program/process to assess selected<br>• Initiative objectives established, and charter defined (if needed)<br>• A process overview is performed<br>• A high-level "As Is" process map (if valued) is created<br>• Relevant process data collected<br>• The list of process owners, stakeholders, and participants is documented and agreed to |
| 2 | Interview the process clients (customers) and users | • The scope of the processes affected is defined<br>• The process stakeholders defined (e.g., client, users, systems)<br>• The client requirements, standards, and recent experience is collected via interviews<br>• Opportunities for improvement established |
| 3 | Walk and confirm the process | • A detailed "As Is" process map is created or audited and confirmed<br>• "As Is" performance metrics and history are collected<br>• The process with stakeholders affected is confirmed<br>• Process pain points identified/quantified |
| 4 | Establish the Parts Process Priorities against mission and goals | • Process mission, vision, and goals are set as a baseline<br>• A process health rating is established<br>• Prioritize initiatives for improvement are developed/accepted |

Talking to the process clients involves leveraging existing client data and additional collected data to answer essential elements of the process value, such as:

- The process owner and formal client
- The critical outputs expected and at their standards
- The essential performance attributes of these expected outputs (e.g., accuracy, cycle time)
- The importance of each attribute
- The current performance of each attribute
- The performance metric to be used for each performance attribute
- The target level of performance client's expectations by process attribute
- Who should be considered if benchmarking the process is practical?
- The priority process areas for improvement

Customer data may be collected using surveys, interviews, or focus group brainstorming as three relatively simple options. Documenting what the client has shared is key to explaining why a process attribute became a priority or was lowered on the complete list of potential initiatives.

A straightforward approach would be to list the process attributes and ask for a 1–10 rating on how these are performed today, later confirming how important each attribute is on a similar scale.

For example, Figure 13.6 suggests that the process for Parts order fulfillment had five attributes, and when mapped out on an Importance/Performance graph, it would be displayed as shown.

Lines are added to the importance/performance graph to point out the average rating for all five on the importance and the average for performance.

With the attributes mapped and graphed, it becomes easier to suggest to the client and the stakeholder team that the early focus should be on the perceived attributes to be the most critical, but other attributes perform the poorest. For example, displaying this matrix helps prioritize the cycle time and shipping quality as the first two of five areas to look for improvements or re-design.

Walking to confirm the process involves:

- Confirming the processes affected and how each process works versus stakeholder perception and updating the process map to reflect reality
- Confirming the health of the current process and identifying current pain points
- Collecting relevant data missed, including:
  - Typical client issues
  - Suggestions for improvement from both the stakeholders and the client
  - Collecting in progress or planned process changes
  - Defining any reasons why process performance improvements may be blocked or obstructed

As before, surveys, customer interviews, and collaborative brainstorming facilitation sessions may help complete this step.

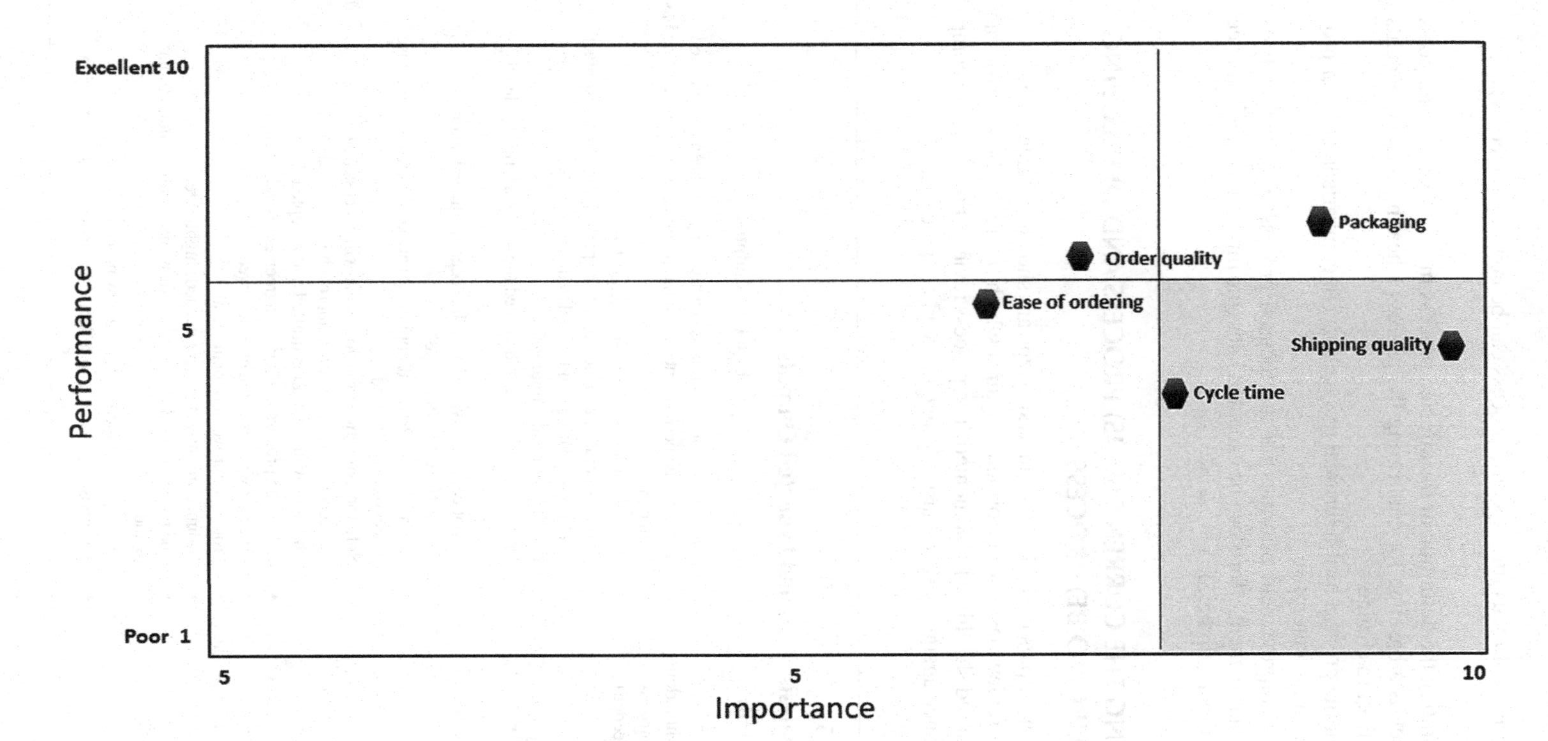


**FIGURE 13.6** Example- process client importance/performance matrix. (Copyright MPE ©2022 from Maintenance Parts Management Excellence Training edited by Barry. Reproduced by permission of Asset Acumen Consulting Inc.)

Setting priorities of focus for a specific process can be now completed by:

- Calibrating the definition of the affected process mission, vision, and goals and set as a baseline for confirming the process's health potential benefits and effort to change
- Rating the process health (critical to excellent) in comparison to other processes in the enterprise
- Setting improvement priorities for the focused process such that analysis and improvement efforts can be executed later (the importance/performance matrix can help here)

## ANALYZING THE CURRENT (AS IS) PROCESS AND DEVELOPING THE FUTURE (TO BE) PROCESS

The expected outputs of the Process analysis steps are shown in Table 13.5:

A plan to review baseline performance against which future attribute performance can be measured should be implemented. The process review should plan to compare its performance against best-in-class organizations where practical.

**TABLE 13.5**
**Process Analysis Steps and Expected Outputs**

| Step | Activity | Expected Outputs |
|---|---|---|
| 5 | Baselining/ benchmarking the process (if needed) | • The baseline "as is" process functionality has been documented<br>• The baseline performance from the previous step's health rating has been confirmed<br>• A benchmark candidate list is generated<br>• Process/operational benchmarks for performance and best practices have been identified and completed<br>• Benchmark results/findings |
| 6 | Develop solution(s) (options) | • Desired outcomes gaps identified (desired performance levels targeted)<br>• A root cause analysis (RCA) has been facilitated against identified poor performance attributes<br>• Alternative solutions (if applicable) are considered, and a revised process map is created<br>• Solution evaluation criterion (what will success look like?) is agreed to<br>• The recommended solution/approach is agreed to<br>• The improvement plan is drafted and accepted |
| 7 | Lobby solution acceptance | • Lobby the process participant agreement is achieved<br>• Process owner agreement is achieved |
| 8 | Create a strawman implementation plan | • Process improvement plan created<br>• Grouped initiatives (where needed) highlighted<br>• The recommended changes in a strawman implementation plan included<br>• The suggested metrics for the new process are accepted<br>• The revised process map reflects the planned "To Be" version |

Benchmarking is a target to shoot for (desired "to be" state) based on process client requirements, the operating context of the process, and the best comparable models available locally and globally. Benchmark data can be gathered internally or externally. The internet is often a good source for gaining insights on which corporate candidates to consider. Understand that often competing corporations will not want to disclose key internal performance insights unless it is to their advantage.

Baselining/Benchmarking the current process involves:

- Reviewing the assessment work already created in earlier steps:
  - Client importance/performance matrix
  - Process rating
  - Process issues
  - Best internal practices
  - Establish a baseline for the internal affected processes
- Looking for external organizations that have similar missions and their performance data is available
  - Best in company
  - Best in industry
  - Best in class
- Selecting benchmarking organizations (internal/external) that:
  - Are accessible to the team
  - Have a like process with a well-performing method and transferable to the process reviewed (Figure 13.7)

The analysis will provide the causes of poor performance and suggest alternative methods to meet our target client requirements. Developing solutions involves:

- Establishing performance targets for essential process metrics
- Reviewing the current process map for opportunities to remove redundant steps, improve execution, and shorten timelines

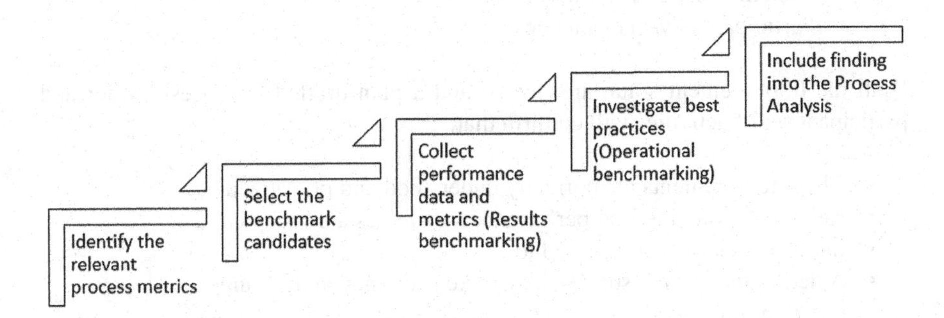


**FIGURE 13.7** The benchmark process steps in process analysis. (Copyright MPE ©2022 from Maintenance Parts Management Excellence Training edited by Barry. Reproduced by permission of Asset Acumen Consulting Inc.)

- Using root cause analysis to brainstorm underlying reasons for the current poor performance
  - (e.g., Brainstorming, fishbone diagrams (root cause diagrams), flowcharts, Pareto charts, scatter diagrams data)
- Identifying alternative methods to achieve the desired client requirements through:
  - Process improvements
  - System improvements
  - Delivery paradigm change (Figure 13.8)

Re-thinking suggests that target new solutions should have more than an incremental improvement. Cycle times should significantly improve. Errors should be reduced by a factor of 10 and ultimately by a factor of 100. The process should be automated to remove the human error factor. Analysts should leverage Artificial Intelligence, Business Intelligence tools, mobile devices, simplified user applications, and other practical and available technical advantages.

Initiative selection (or identifying alternative methods) can be facilitated by a simple financial benefit versus ease of implementing grid (see Figure 2.5 matrix) or a hybrid set of criteria such as:

- The increased value created:
  - Client satisfaction
  - Cost savings/rework levels
  - Market growth potential
  - Cycle time improvements
  - Error rate reductions
  - Employee morale
- Ease to implement criteria such as:
  - Implementation costs
  - Barriers/prerequisites to implementation
  - Implementation timeline
  - Impacts on other business units
  - System constraints/requirements
  - Participant buy-in challenges

With the Improvement solution selected and a plan drafted, a process owner and participant set of activities will confirm that:

- Client requirements are correctly understood and prioritized
- Key inputs, outputs, and performance attributes are accepted
- Initiative priorities are agreed to
- A definition of what success looks like is defined and documented:
  - Client impacts
  - Process impacts
  - Process metrics and expected standards

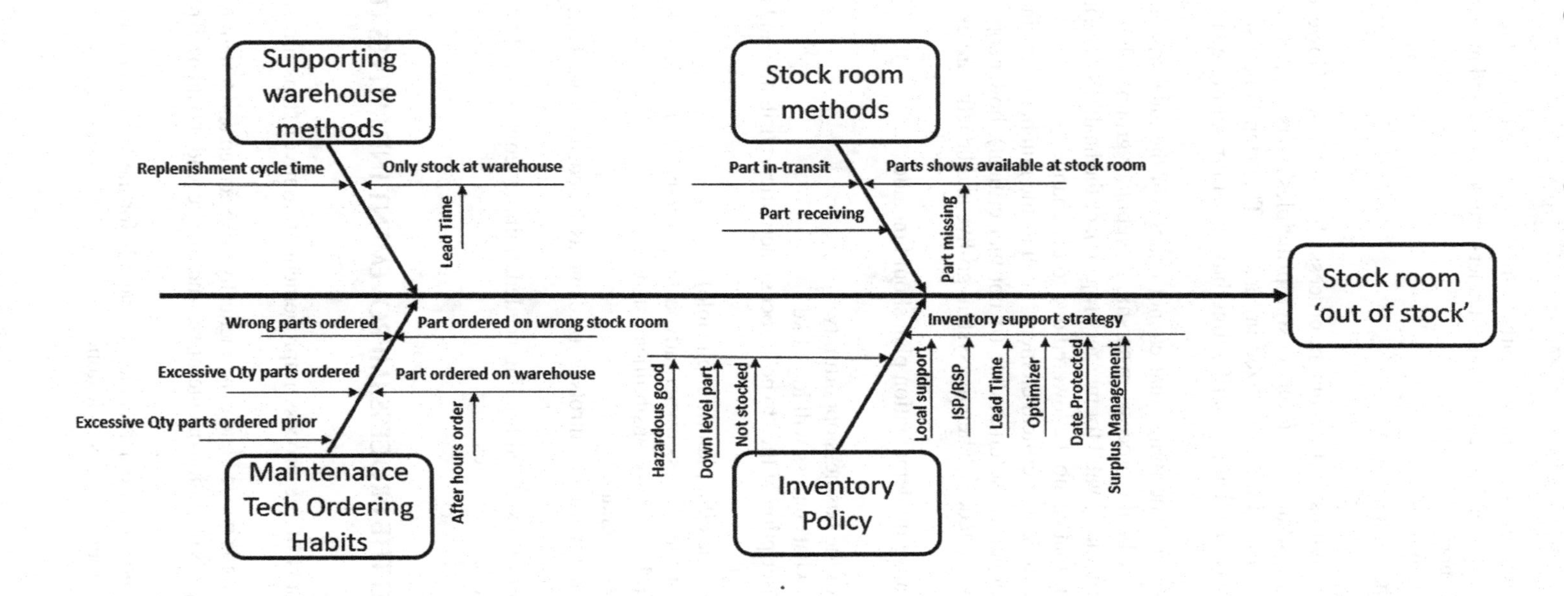


**FIGURE 13.8** Example of a root cause (Fishbone) diagram. (Copyright MPE ©2022 from Maintenance Parts Management Excellence Training edited by Barry. Reproduced by permission of Asset Acumen Consulting Inc.)

- Short- and long-term performance targets are declared and agreed to
- Process ownership and process change leaders are declared and agreed to:
    - Process owner
    - Process champions
    - Process change lead

The above list of activities will document process change buy-in. However, there may be a need to facilitate a solution "design principles" versus "design points" discussion to wrap around the prioritized initiatives. This step is beneficial when senior executives have difficulty coordinating buy-in within multiple stakeholder groups.

Design principles include things that cannot change in the enterprise initiative. Examples of this may be the process ownership stakeholder group, solution implementation date (particularly valid for merger projects), or the health, safety, and environmental standards that cannot be lowered due to the change.

Design points include less critical elements such as the number of business units involved in the change, the solution date (if not that critical), how people change activities will be facilitated, or when the system go-live and benefits are expected to materialize.

The formal initiative implementation plan should include:

- A statement of the business opportunity
- Definition of what success will look like
- A detailed description of the "to be" process, activities, attributes, and management system
- Description of the pilot plan (if applicable)
- Declaration of the process owner, advocate, participants, and lead for the change with roles and responsibilities listed
- A timeline for the change
- New process metrics and controls defined and assigned, and data definitions for related reports
- A summary of the expected funding needed, training required, and systems or equipment dependencies

## IMPLEMENTING THE PROCESS IMPROVEMENT INITIATIVES (TO BE) PROCESS

The expected outputs of the process improvement (Initiative) steps are shown in Table 13.6:

If an initiative pilot is needed and can practically be executed and tested (e.g., for a small subset of the parts operation or enterprise), the change lead should be expected to:

- Create and follow the implementation schedule for the agreed to pilot
- Track key pilot and process milestones

**TABLE 13.6**
**Process Improvement Steps and Expected Outputs**

| Step | Activity | Expected Outputs |
|---|---|---|
| 9 | Solution pilot (if applicable) | • Pilot implemented and lessons learned collected<br>• The improved performance (benefits) measured<br>• Success stories shared |
| 10 | Solution rollout | • Process documentation completed<br>• Rollout plan created and approved<br>• Process initiatives implemented with desired outcomes achieved<br>• The planned change is confirmed, embraced, and executed at all required stakeholder levels |

- Document any initiative challenges and lessons learned so that they can be addressed in the larger rollout
- Revise the initial solution to reflect the learnings from the pilot
- Lobby the new changes to the Executive team that is sponsoring/owning the process impacted

Rolling out (implementing) the initiative solution involves:

- Creating the initial implementation plan to guide the effort, which can include:
  - Describing the implementation approach and locations sequence (if applicable)
  - Implementation schedule
  - Resource requirements (people, systems, etc.)
  - Roles and responsibilities matrix
  - Expected vital milestones and definitions of process success
- Identifying any changes to the process maps, supporting documentation, and metrics addressing the documentation requirements and getting them finalized
- Executing the implementation plan
- Ensure the change is embraced and executed at all required stakeholder levels
  - Supporting the change with change agents
  - Leveraging change enablement disciplines (see Chapter 5) (Figure 13.9)

With the initiative implemented and the stakeholder acceptance observed well in hand, it is time to look for the next initiative to drive significant benefits.

Change is constant. Client expectations and markets will evolve. Competing enterprises will get more competitive. Supporting technology will change the way many processes can be performed. Therefore, it is critical to recognize that re-thinking needs and opportunities are a constant activity.

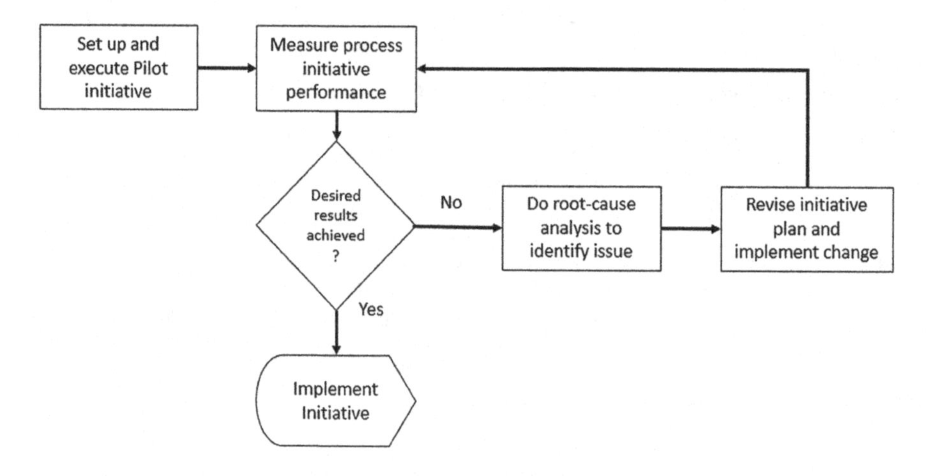


**FIGURE 13.9** Suggested pilot program process flow. (Copyright MPE ©2022 from Maintenance Parts Management Excellence Training edited by Barry. Reproduced by permission of Asset Acumen Consulting Inc.)

## EXAMPLES OF RE-THINKING MAINTENANCE PARTS MANAGEMENT EXCELLENCE

Multiple cases have been executed, siting the fundamental process challenge and re-engineering the status quo when looking at an existing parts operation. Inventory challenges are a typical constant in a Parts operation and an excellent place to apply these change disciplines. Below are some examples where the re-thinking challenge was applied to the parts operation beyond an inventory challenge.

An organization took the opportunity to "re-think" their relationship with their client base and suppliers to reset their staffing roles and numbers. Along with improved service levels, they were to decrease their country-wide (100+ base) staff levels by 25%. The re-thinking and rebalancing of their defined staffing levels allowed many good folks in their operation to move to better opportunities within the same enterprise and create a career path that was not limited to the maintenance or their parts operation.

Applying the re-engineering mindset across all the staff of a large maintenance parts operation, one organization used this mindset to get all their staff engaged in re-thinking how the current operation was executing. Process re-thinking training was provided across the affected staff to calibrate the approach and terms. Assignments were given to the trained group, and sub-group teams were created and asked to apply their newly learned disciplines and approaches competitively. The focus assignment was aligned to their parts management macro metrics (i.e., logistics quality, parts availability, parts costs, systems availability). The approach resulted in competing presentations on potential enhancements to their parts business while they learned the disciplines, felt the executive team listening to them, and experienced the methodology work. This initiative empowered staff at all levels of this parts operation and generated an interest and energy level that other parts staff wanted to experience and engage.

Re-thinking the maintenance parts management is a constant. Encouraging, teaching, calibrating, and promoting this mindset allows every stakeholder to affect change and drive their business not just to the next level but to their future success.

# 14 Opportunities for Outsourcing in Maintenance Parts Management

## INTRODUCTION

As an enterprise expands its market focus, globally or within the same country, it will look for added expertise and resources to keep its business competitive.

Growing organizations continue to look for ways to leverage their core market value and competencies and reduce direct expenses. They will look for ways to consolidate decision-making across traditionally separate corporate business units (functions). Supply chains will become consolidated where it makes sense. Service levels will constantly be called upon to improve at lower costs. A dollar saved in maintenance parts management can directly help an enterprise's bottom line.

Spare parts logistics is often not considered a core competency, although most companies stress the importance of service and reputation in their overall competitive strategies. Original equipment manufacturers (OEMs) like office equipment and automobile manufacturers spend billions developing, producing, and marketing durable goods but do not pay equal attention to servicing these products once they are sold. The parts sales business can be very profitable, but the primary focus for most OEMs is to support the product sales supply chain and treat maintenance parts as a necessary evil. As a result, many OEMs have lagged in their product support supply networks when it comes to sophistication regarding post-sales parts support over the past decades.

To large asset operators like airlines, "service and repair are critical," so they aggressively improve their operations. But for OEMs that make assets, for example, the responsibilities are frequently distributed across a network between the OEM, dealers, parts suppliers, and others. It is also important to note that many parts (e.g., automotive, computer) are foundationally not uniquely "brand sourced" items. Many parts can be sourced through alternative suppliers, from the original component provider to the OEM.

Some significant asset-intensive organizations have outsourced their parts support to a strategic alliance partner Logistics Services support organization. In many cases, this outsourcing was to take advantage of parts already supported by the

DOI: 10.1201/9781003344674-14

alliance partner to improve the availability of their parts by perhaps 5–10% and keep their costs aligned. As a result, alliance partners can boast close to a 100% fill rate (98% of the time) on orders on behalf of their key committed customers. In addition, the asset-intensive company and the alliance parts logistics partner can commit to long-term agreements and team up to manage the integrated logistics processes.

Shrinking margins pressure asset-intensive enterprises to operate lean inventories throughout their networks, including spare parts inventories. At the same time, some OEMs recognize that one way to *increase* margins is to offer very profitable maintenance and repair services. As a result, several high-tech firms make most of their margins with services on repairs and upgrade kits. Similarly, industrial equipment manufacturers earn more than half of their gross margins on parts. They bear almost no sales costs because most equipment users are a captive audience for parts.

Some OEMs have recognized the opportunity for higher profits and a sustained reputation by providing outsourced parts management services to:

- Critical clients for the products they manufacture
- Complementary products not made by them
- Competitive aircraft products.

A personal jet manufacturer provides maintenance and warranty support for their products and markets to maintain the client's full complement of fleet aircraft – parts included.

Going after the spare parts market creates a need to provide perceived added value in the support you provide. Otherwise, a price reduction may be required to match perceived value.

Table 14.1 suggests how some maintenance parts suppliers are going to market.

**TABLE 14.1**
**Market Offerings from Some OEMs and Parts Vendors**

| Spare Parts Services Value Offering | Rationale |
|---|---|
| Marketing a value-added parts service | • High parts availability and quality can be leveraged to drive customer loyalty.<br>• Many service organizations recognize that offering service, repair, and warranty programs across a more significant complement of assets creates account control.<br>• Managing spare parts can be a crucial part of gaining competitive differentiation.<br>• Many OEMs place spare parts inventories with their customers, often with a consignment payment arrangement (e.g., the customer does not own the part until used). |

*(Continued)*

## TABLE 14.1 (Continued)
## Market Offerings from Some OEMs and Parts Vendors

| Spare Parts Services Value Offering | Rationale |
|---|---|
| Bridging client expectations for asset risk/reliability and responsiveness | • Equipment "uptime" and parts availability are attributed to high customer satisfaction.<br>• Just-in-time maintenance parts support mitigates the risk of line shutdowns if the local stock room spare parts are unavailable.<br>• Customers expect equipment to be fixed on the first visit and not be waiting for parts.<br>• Some repair operations (e.g., computer service firms) have long been executing "swap and replace" programs rather than asking customers to wait for their laptops or monitors to return from the repair shop.<br>• Leveraging IoT sensors to predict when a part may be required to complement a service action and injecting a repair mitigation. |
| Leveraging vendor inventories | • Asset-intensive enterprises with large parts support networks are aggressively pursuing direct delivery by part vendors to locations inside their network to minimize duplicate stock:<br>  • not only to warehouses or stock rooms but at times to the asset location itself<br>• Vendor shipments in concert with vendor-managed inventory (VMI) programs. The vendor owns the inventory or the client's consignment. |
| Relationships and Alliances | • Partnerships and joint ventures to enhance the depth and width of their service capabilities are considered when an OEM or parts vendor does not have the resources and competencies to operate their after-sales service business independently.<br>• Achieving geographical parts coverage by partnering with selected local service providers. Examples of comprehensive alliance networks in the maintenance, repair, and overhaul (MRO) business are prevalent in the airline industry. |
| OEM asset lifecycle support and buy-back programs | • OEMs can take steps to increase lifecycle penetration with annual services revenue and product lifetime support by:<br>  • Providing a digital tool that models the client install base and analyzes individual components of assets (e.g., aircraft maintenance)<br>  • Calculating aftermarket lifetime value and applying various improvement levers – there are usually about 40 to 100 possibilities, depending on the industry<br>• OEMs would look to re-market used equipment, re-pricing spare parts more dynamically, or offer to overhaul and modernize a customer's existing equipment through hardware or software upgrades. |

## THIRD-PARTY LOGISTICS PROVIDERS FOR MAINTENANCE PARTS LOGISTICS SERVICES

Third-party logistics organizations can, in many cases, provide improved cost savings, improved service levels, greater flexibility, opportunities to consolidate distribution, reduce overhead and allow for a more focused employee resource set.

Companies who choose a third-party logistics supplier must first look inward and define their business objectives. It could reduce costs and inventory, improve service levels, change reporting, outsource responsibilities, and increase investment viability. Once there is a clear understanding of what is being sought, selecting a solution becomes more apparent if a supplier has the service level attributes and cost-effectiveness. A comfortable partner relationship with the selected vendor is critical, and so is some mutual financial bonus when all enterprise metrics are exceeded.

Selecting a third-party logistics supplier should consider:

- Establishing objectives and selection criteria
- Identifying qualified providers
- Developing a detailed request for bids for the selected third parties
- Developing a plan for integrating the third-party provider into the enterprise systems, processes, and culture
- The responsibility to properly manage a company/supplier relationship that creates a win/win situation

Outsourcing is a significant trend for perceived non-core activities. What needs to be clear is that the enterprise cannot throw out its mandate to keep its critical operational assets running when looking for ways to restructure its business and cash flows. Maintenance and supporting maintenance parts management may be one of their core competencies, particularly when they consider the risk of poor performance from either business unit to their ability to deliver value to the market and the impact on their reputation.

## MAINTENANCE PARTS OUTSOURCING CAN COME IN A MULTITUDE OF TYPES AND COMBINATIONS

When looking to outsource any area around maintenance parts management, the enterprise must establish the selected vendor's contact in their enterprise. It could be finance or procurement or maintenance or maintenance parts. What is important is the behavior and service levels provided by the contact/vendor team. For example, suppose an inventory is perceived as ineffective in one business unit. In that case, it could create a behavior that promotes limiting replenishment to minimize inventory levels, which could, in turn, impact the maintenance and operations groups. The vendor selected is not as important as ensuring that the collective behavior aligns with the enterprise's agreed-upon goals and desired result. The vendor on its own would be a layer distant from the enterprise goals and outcomes and may not be in the best position to be empowered to own this behavior.

## TRANSPORTATION OUTSOURCING

Almost all maintenance parts operations outsource the delivery of parts to their maintenance technician when the part needs to be delivered off-campus. Therefore, on-campus deliveries would be expected to be few or perhaps on a route schedule and, as such, could be handled by the maintenance technician or by an on-campus scheduled delivery route.

The parts delivered to a maintenance parts warehouse are generally accomplished through a "less than truckload" order and, as such, are most often cost-effectively delivered by a third-party vendor.

## SOURCING ALL PARTS FROM A SUPPLIER

The simplest version would be to have no parts operation and depend on a third party for all the parts needs. Full supplier sourcing would allow the client only to have to pay for parts when they need them. Sometimes, a membership fee for access to the parts may apply or at least unrestricted access. However, it leaves the responsibility for all parts costs (before ordering) to the third-party vendor. The parts-related organization, management, warehousing, new defective, and legal responsibilities for the goods shipped are also covered.

An example of effectively sourcing parts from a supplier is fleet maintenance in an urban environment. The maintenance garage may carry some highly used parts (e.g., oils, filters, belts, other fluids), but look to the supplier, or perhaps a series of suppliers, for parts to be delivered within a 30-minute time frame where practical. Suppliers delivering a reasonably high service level will win this business and the client's loyalty. Attributes that are attractive to the (client) maintenance technician would be:

- Parts availability targets met
- Accurate information exchanged on parts available
- Responsiveness to availability inquiries
- Flexibility in delivery support
- Courteous and friendly service/support
- Proactive advice
- Price

## MAINTENANCE VEHICLE OUTSOURCING

For the most part, maintenance vehicles are not an area the maintenance parts management team has to address as part of its scope. Unless the business supported is for transportation vehicles such as a utility van, most organizations outsource the ownership of these vehicles to leasing companies. The maintenance organization would be responsible for outfitting the vehicle to suit the type of parts stocked in the van and the tools needed. The maintenance parts management team and procurement may sometimes be asked to facilitate the maintenance vehicle lease renewals, maintenance schedules, and driver qualification credentials.

## OUTSOURCING WAREHOUSE LOGISTICS

Outsourcing maintenance parts warehousing is a fairly common activity. In cases where the supplier has warehouses in similar locations, economies of scale may allow the third party to operate at a reasonable profit and still provide savings to the maintenance parts client. A transition to a reliable third-party warehouse provider may be relatively transparent to the maintenance technician when the warehouse location is similar. The order picking, systems, and related responsiveness service levels are similar to client expectations. Enterprises requiring maintenance parts support across multiple countries or cities in the same country could benefit by leveraging third-party logistics organizations that already support a similar geographic footprint. Negotiating each location's service levels and expectations must ensure that the current service levels will be achieved or exceeded. As the third-party logistics vendor distributes specific enterprises' parts on their behalf, care must also ensure that customs declarations are correctly executed between country customs. The reputation of the client enterprise is maintained.

The third-party logistics organization will want to consolidate many of their multiple client distribution responsibilities from the same warehouse sites. You should expect the new third-party vendor to manage other items at the same physical site. With the ability to consolidate locations (sites), it is less likely to produce mutual client/vendor financial or service benefits. If the vendor comes on-site with less expensive resources, the enterprise could likely find a way to do the same, assuming no employee relations objections.

## PROCUREMENT SERVICES OUTSOURCING

Third-party procurement vendors are emerging, particularly at the global enterprise level. Specifically, maintenance parts can offer procurement and operational benefits focusing on maintenance practices across its client's enterprise.

Their procurement services solutions can include:

- Consolidating maintenance sourcing strategy standards for a country operation that can be leveraged across the global enterprise
- Leveraging existing vendor procurement infrastructure to facilitate savings in maintenance
- Establishing key performance indicators (KPIs) to drive a total cost of ownership approach for driving out costs
- Additional procurement services support as outlined in the sourcing and refurbishment chapter of this book

The third-party procurement vendors' value approach would declare their combined maintenance parts-related spend leverage to drive parts costs down with selected suppliers. Ideally, they would like clients in similar industries and spend volumes to drive service levels (Table 14.2).

**TABLE 14.2**
**Types of Maintenance Parts Management Procurement Service Offerings**

| Approach Tactic | Description |
| --- | --- |
| Information technology | • Cross-reference and standardized naming conventions for suppliers and items purchased<br>• Leverage standard technologies and inventory policies to optimize inventory<br>• Enhance or standardize online supplier punch-out and hosted catalog interfaces<br>• Consolidate supplier resource access to assist in tracking and managing inventory |
| Materials management | • Standardize parts inventories where applicable to achieve targets across the enterprise<br>• Proactively manage order quantities with the utilization of inventory policy algorithms and tools<br>• Integrate inventory demand planning with parts usage forecasting<br>• Expand VMI with preferred suppliers<br>• Continue to deploy and excess redistribute inventory across the enterprise |
| Performance measures/ benchmarking | • Monitor supplier performance through vendor-specific KPIs<br>• Confirming supplier continuous-improvement goals are worked and achieved |
| Strategy | • Driving procurement policy and supply chain goals<br>• Enhance category sourcing strategy across the enterprise<br>• Expand development and usage of preferred suppliers<br>• Elaborate on contract usage, terms, and conditions |
| Tactics | • Enhance parts refurbishment/repair and warranty programs<br>• Evaluate OEM parts conversion opportunities to alternate sources |

They can boast savings drivers such as:

- Ability to analyze spend patterns to identify opportunities for spend aggregation, supplier rationalization, and product or service alternatives
- Market and benchmark information in both mature and emerging markets
- Supplier relationships that leverage their combined total spend and best practices
- Multi-client agreements in diverse markets, including chemical/industrial manufacturing, gas, and utilities, as well as manufacturing
- Knowledge of local laws and business practices

For example, prominent third-party procurement service vendors can help their clients:

- Facilitate thousands of supplier relationships and past transactions across multiple divisions, business units, and plants across multiple countries

- Manage existing preferred agreements by a sub-category MRO level with multiple suppliers and increase the compliance alignment and reduce the maverick spend in an enterprise
- Measure and manage site-level purchase decisions to influence better compliance with preferred agreements
- Manage the top sub-categories representing closer to 70–80% of the total categorized spend:
  - (Electrical, warranty items, power transmission, safety, etc.)
- Manage the MRO spend over the last three to five years for demand management inconsistencies

As another example, initiatives and benefits created from a procurement outsource vendor can include:

- Negotiations resulting with the preferred suppliers driving additional invoice discounts, specific price structure for everyday products across sites, volume incentive discounts, as well as guaranteed cost savings year over year to drive efficiencies
- Savings averaged out to 13% of spend in the first 12 months with the preferred suppliers
- National agreements are set in place so all facilities can utilize the exact pricing and services to consolidate the supply base
- Improved visibility into the overall spend general MRO supplies enabled better information to be derived from the data that will incorporate additional opportunities in consolidation, standardization, and product substitutes
- Leveraging total spend by increasing purchases with outlined preferred supplier set from 25% to 55% in the first year, which continued to increase driven by a simple communication plan with the plant buyers
- Enhancing core supplier relationships, expanding on the formal cadence of quarterly business reviews, and measuring performance
- Inventory costs are driven down by redeploying inventory, inventory buy-back programs, and vendor managed inventory (VMI) solutions
- Implementing standard products (where practical) and processes across all sites
- Consolidating invoicing (fewer, more efficiencies) while improving and standardizing payment terms
- Reducing and, in some cases, eliminating freight charges
- Establishing a program to monitor MRO warranties (where one did not already exist)
- Implementing lower-cost parts alternatives through substitutions, including OEM part conversions and green alternatives

Procurement outsourcing organizations can provide KPIs to establish baselines and progress throughout their Service tenure (Table 14.3).

**TABLE 14.3**
**Suggested Procurement Outsourced Metrics**

| Support Area | Suggested Procurement-Related KPI |
|---|---|
| Materials | • Maintenance parts inventory turns<br>• Inventory accuracy<br>• Obsolete and excess inventory management |
| Purchases | • Purchasing opportunities (spot buys)<br>• Maverick spend (off contract/system spend)<br>• Item unit price reductions<br>• Overall parts usage<br>• Number of key suppliers |
| Acquisition | • Cost per order/transaction, or cost per line<br>• Automated access versus manual ordering processes<br>• Urgent parts order fulfillment time |
| Assets | • Unplanned asset downtime<br>• Asset productivity metrics |
| Labor productivity | • Maintenance/parts personnel support perceptions<br>• Ease of use, productivity impact on personnel<br>• Expediting impacts in store room from an out-of-stock scenario |

## INVENTORY PLANNING AND INVENTORY POLICY OUTSOURCING

Supply chain planning (SCP) vendors (specific to maintenance parts) recognize that most enterprises have not taken the time or set up the tools for a full complement inventory policy program. An inventory policy would dynamically support all variables needed to be in scope for their enterprise or operating context. This gap has created an opportunity for vendors to provide decision support services to manage a client's maintenance inventories.

Third-party SCP vendors have started to provide this support by reviewing a client's parts-related data, performing analysis in a private cloud, and sending back a revised inventory policy guideline to the client's maintenance parts management Systems.

These maintenance parts SCP-specific vendors offer "intelligent" maintenance parts inventory policy support service. The policy support services can be applied across multiple industries (e.g., oil & gas, mining, manufacturing, and utilities). In addition, it can help each enterprise manage their parts inventories as they manage their capital-intensive assets.

Services from some of these vendors can include:

- Maintenance parts inventory policy analytics
- Supply chain management
- Asset performance management
- Inventory optimization

- Master data services
- Inventory cataloging

The Services are typically cloud-based and scalable while leveraging large data sets and processing flexibility. The data used is vast but limited to what may be available from the supporting client. The SCP vendors market themselves as "quick time to value" (rapid payback cycle) by industry.

Table 14.4 describes how some SCP vendors support maintenance Parts Planning policies.

**TABLE 14.4**
**Examples of Some SCP Vendor Offerings and Benefits**

| Service Offering | Service Description | Suggested Benefits |
|---|---|---|
| Maintenance parts management health Check (Assessment) | • Summary of maintenance parts management holistic assessment to assist in the prioritization of initiatives to improve the client operation<br>• Summary of inventory assessment insights based on an analysis of an inventory location data extract, policy, and the asset's operating context | • Assess the health of the "As Is" parts operation and prioritize potential initiatives the client can do to help themselves.<br>• Assess parts performance and provide initiatives to improve<br>• Identify how your parts operation can be leveraged for a competitive advantage |
| Item-specific inventory assessment | • Leverage existing client data and provide an analysis (analyst to work with client stakeholders) to support a parts management strategy to meet enterprise goals | • Provide fact-based answers and options to a client inventory's effectiveness in supporting their asset availability mission<br>• Create a risk assessment of critical spares, less active spares inventory, potential surplus inventories, and service delivery cycle times |
| Inventory service level analysis | • Create an inventory analysis report to display the balance between service levels and inventory value/ placement | • Optimize inventory based on historical data and known forecast influences<br>• Provide data analysis by item or item support groupings (asset types) |
| Critical inventory/ parts analysis | • Provide flexible scenario testing (simulated analysis) for critical spares | • Test different circumstances and determine a balance between inventory and service level impacts |
| Inventory policy prescriptions | • An asset-specific Optimizer-like service may provide the monthly policy set of updated ordering values, stock levels, and lead times to ensure a client stays on track with their inventory objectives | • Automate available client data collection and discernment to create an inventory policy that works to optimize service levels. (Works to mitigate the stock-out impacts and minimize overstocks.) |

(*Continued*)

**TABLE 14.4 (Continued)**
**Examples of Some SCP Vendor Offerings and Benefits**

| Service Offering | Service Description | Suggested Benefits |
|---|---|---|
| Scenario analysis simulation reporting | • Simple ad hoc query scenario testing and reporting | • Test different circumstances and determine a balance between inventory and service levels by item or inventory set |
| Ad hoc report generation | • Easy access to ad hoc cloud-based reports | • Easy and rapidly generate and display KPI reports |

SCP services is a rapidly growing market segment with a limited number of well-established players.

Third-party inventory policy service providers help their clients work through inventory placement and optimization details so that the enterprise can focus on managing the items that matter most and develop strategies to streamline the rest. Many can provide strategic assessments of your maintenance poperation to help prioritize the next focus. Some provide an assessment and categorization of your item catalog based on engineering performance, reliability factors, safety, and supplier integrity.

Some Inventory third-party outsourcers suggest collecting enterprise data and leveraging their proven algorithms to identify business risk against critical spares policy. They would develop an optimized maintenance parts stocking policy that can be applied for initial spare parts (ISPs) and recommended spare parts (RSPs) and complement the typical Min/Max/EOQ elements of a typical Enterprise Asset Management system. Data analytics can also identify "slow-moving parts" and potential "obsolete parts."

The benefits of a "well executed" service would be:

- Assigned "risk impact" codes, leveraged to determine optimal stocking levels for each item:
  - achieves visibility of the impact stocking levels have on service levels based on criticality and highlights related risks and costs resulting from stock-outs or overstocking
- Aligning stocking levels and reordering recommendations with your materials management strategy:
  - Ensures not too many and not too few spare parts are available when needed
  - Replaces manual effort stocking level approaches with objective analysis to reduce costs, minimize downtime, and improve service levels with this invaluable service
- Objectively predict stock usage for new items:
  - Provide reordering recommendations aligned with materials management strategies
  - Determine reorder levels based on required service levels and economic order quantities
  - Set Min/Max values and ROP/ROQ for each material on a per-item basis

## HIGHLIGHT INVENTORY HOLDING AND PURCHASING COSTS BEFORE AND AFTER A STOCK POLICY RESET

- Analyze and review slow-moving items to avoid excessive reordering and accurately report on opportunities for inventory reduction
- Assist in identifying non-compliance with current materials management strategy for insurance, potential obsolete, and other types of spares
- Generate reports of potential obsolete items by value, quantity, criticality, age, usage, site/equipment groupings, and the net effect on materials value
- Support a review process for potential obsolete items to maximize financial recovery to the business, determine potential risks and costs resulting from an overstocking situation
- Provide visibility of usage trends across equipment and suppliers to enable proactive management of materials

These third-party inventory policy service providers can significantly help maintenance parts operations looking to get a start on proper inventory planner activities and can be a step toward self-management of inventory policy if not currently well managed.

## INVENTORY LIFECYCLE PLANNING OUTSOURCING

Capital spares planning and Initial spares list creations can come from the original equipment manufacturer (OEM) vendors as part of the procurement negotiations process. Ideally, the enterprise design engineer looking to place a new asset solution will look for a solution that mitigates risk through asset design redundancy in the provided solution. The design engineer will be looking for the set of suggested initial spare parts from the OEM in any event. The OEM ISP list takes some of the workloads off of the design role if the vendor can provide the list and the reasons for the part being considered an initial spare part.

This process is described in more detail in the asset lifecycle alignment chapter.

# 15 Maintenance Parts Sales

## INTRODUCTION

Maintenance parts sales has historically been a strategic cash cow for many original equipment manufacturer (OEMs) over the past four-plus decades.

Many industrial OEMs are increasing their focus on aftermarket services. This can include supporting parts, repair, maintenance, and digital services for the equipment they sell.

- OEM parts sales and services provide stable secondary revenue and higher margins than their new equipment sales.
- Across many industries, it is typical that the average earnings-before-interest-and-taxes (EBIT) margin for aftermarket services was 25% or more, compared to an approximated 10% for new equipment.

While product sales are an OEM's priority, helping clients with after-product sales parts support is in the best interest of both the OEM and the client. Many OEMs find that their margins for services can top 40%, whereas margins for finished goods top at around 13% over the past few decades.

OEMs and third-party maintenance operations work to build maintenance and parts sales businesses through loyalty service programs that include access to parts at a competitive value.

For example:

- More aircraft OEM vendors are providing maintenance services as well as parts services
- Airlines receive component MRO services for their passenger aircraft fleet after expanding their aftermarket support with a services component company.
- An aircraft engine manufacturer signed a multi-year fleet management agreement with a continental airline. The agreement will cover the airlines' future fleet of 90+ aircraft powered by their specific engines.
- An airline partnership entered into a long-term component pooling services agreement with a continental-based airline to support its similar passenger aircraft. That contract applies to a total fleet of an estimated 20 wide-body aircraft operated by the partnership

These activities are evolving in the global aircraft after-sales market, while the global commercial aircraft aftermarkets parts market is expected to grow at a CAGR of around 5–7% from 2016 to 2021.

DOI: 10.1201/9781003344674-15

The rule of thumb is that the money spent on after-sales parts and services will outweigh the cost of the original product.

- The accepted rule of thumb is that asset owners spend 5–20 times the initial sales price on subsequent services and consumables.

The global automotive aftermarket industry is seen as a separate business worth more than $700 billion. However, with supply chain shortages and budget constraints at both the consumer and corporate levels, this market is likely to only increase in size.

- The average North American automobile age has grown 15–20% over the past decade. This was likely accelerated during the recent pandemic cycle.
- The vehicle ownership average length for new and used automobiles had increased by 60% over the decade preceding the pandemic. This also is believed to have accelerated when new vehicles were less available due to support constraints.
- 75% of aftermarket auto repair is performed by independent auto repair shops, while 25% of the business lives with dealerships.

When product markets and cash flows fluctuate, maintaining a healthy parts sales market preserves and grows client relationships and future product sales opportunities. There is a battle to retain customers, and OEMs can provide long-term value after a product's initial sale.

If aftermarket services represent approximately 24% of revenue, it often contributes 40–80% of the enterprise profit. Customers are rewarding valued long-term relationships with continued services from their suppliers. Buyers no longer satisfied with the lowest purchase price often factor the total cost of ownership (TCO) into decisions (including product decisions).

A value leveraged from an existing supply chain can help the OEM manage client loyalty without significant new infrastructure costs. Customer satisfaction (with perceived value) is still prioritized across almost all industry sectors.

Some OEMs recently lost market share because their prices were believed to be high without perceived value. Not all OEM components are proprietary. Many parts come from suppliers looking for a piece of the aftermarket parts sales business and related profits depending on the industry.

If the part is purchased in a retail store and not sold as a car would in a car dealership, the gross margins may need to come down to remain attractive and competitive to a discerning client procurement process.

Generally, the gross margin for a part will be expected to be relatively low (competitive) if the procurement client has:

- High product and category knowledge
- High knowledge of where the part is needed as a solution
- Understands the benefits of whether to stock the part or not locally
- An understanding that this part has relatively active usage in their enterprise
- High OEM brand and sub-brand awareness

This scenario works in the OEM's favor to provide value-added services (such as service level commitments or ISP/RSP lists) to drive the relationship and an expected regular transaction volume. Prices may be adjusted for spend volumes. In some cases, the OEM may elect to provide a vendor-managed inventory such that the enterprise does not carry the parts inventories on their books and only pays for the parts when used.

While what has been discussed so far in this chapter may not seem all that surprising, what may be unexpected is the lengths that the OEM (or parts sales vendor) may go to grow the relationship with each of its larger targeted clients.

## STRATEGY DIRECTIONS FOR AFTERMARKET PARTS LOGISTICS SERVICES

Additional services may create or grow loyalty for this profitable market space. In addition, the OEM may elect to expand its after-sales market offering footprint by implementing/deploying new offerings to existing clients or obtaining more clients with existing offerings (Figure 15.1).

With the market development costs suggested in Figure 15.1, OEMs can look to add more value offerings to their existing clients and create new clients.

Complementing parts sales can be:

- Outsourcing the equipment maintenance back to the OEM (as suggested for some aircraft manufacturers)
- Adding a parts exchange program (if the OEM can adequately and cost-effectively repair the part) to drive down the cost of a very active part
- Creating additional parts warranty categories for loyal clients (extending the timeline of warranty and preapproved parts replacement processes for warranty)
- Offering to be the parts supplier of record for selected client locations

New Market Development →

New Tech Development ↓

| | Existing Client Account | New Client Account | New Market |
|---|---|---|---|
| Existing Technology | .10-.15 / Dollar sold | .30-.45 / Dollar sold | .60-.80 / Dollar sold |
| New Technology | .30-.45 / Dollar sold | .60-.80 / Dollar sold | .80-1.00 / Dollar sold |
| New Category Technology | .60-.80 / Dollar sold | .80-1.00 / Dollar sold | 1.20-1.60 / Dollar sold |

**FIGURE 15.1** Potential market development cost considerations. (Copyright MPE ©2022 from Maintenance Parts Management Excellence Training edited by Barry. Reproduced by permission of Asset Acumen Consulting Inc.)

- Offering a tailored inventory set based on the asset (equipment) mix, history, and operating context)
- Offering Parts support consulting services

With these additional services, the part can be expected to be more valued and therefore have a higher gross margin and still be perceived as competitive, particularly when the procurement client has:

- A lower understanding of the product and category dynamics, where the part is needed as a solution
- An unclear appreciation of the benefits on whether to stock the part or not locally
- Some understand that this part has relatively active usage in their enterprise
- A higher perceived value for the OEM brand's additional services.

With parts services aligned to each target client's needs, the likelihood that the procurement client would spend time searching elsewhere for the same part is reduced. Such one-stop shopping (value-added) creates real value for the client. Ultimately the parts supplier may elect to support other manufacturers' products in the aftermarket. This can be a reasonable business approach when the OEM is service-oriented and wants to maintain the relationship to place their products in the following infrastructure re-build cycle. It is also reasonable when competing OEM products have high margins and can be made available from the sub-manufacturer competitively.

The leader in supplier parts, when needed, will have:

- Parts delivered when needed (one hour, same day, overnight, etc.)
- The full scope of parts needed with dependable fill rates and delivery quality
- A reputation for getting it right (right part, correct quantity, to the right place in the committed time)
- Proven parts re-utilization processes
- An easy-to-use interface with their clients' enterprise parts systems
- Easy-to-use parts return program
- High parts quality reputation and
- World-class logistics support

Growing loyalty in a parts support program will have the above attributes and can include:

- Parts warranty support
- Competitive pricing for repairable parts
- Availability in locations close to the expected need in a region
- High reputation for parts quality packaging

OEMs that support parts sales can also provide remote diagnostic support and equipment loaners if a viable market for such services is viable.

## A PARTS SALES CASE STUDY

A computer consumer goods OEM vendor found their customer direct and business partner channels for parts sales were shrinking. The OEM enjoyed an estimated 20% annual revenue growth before this revelation. They offered the sale of a new part (part sale) and the exchange of a defective part for a new part (part exchange) as offerings in their parts sales program. While the direct users were acquiring their parts from the OEM directly, research showed that many business partners had elected to acquire their parts through product dismantling or from the world supplier market. In addition, many of the OEM product line components were not unique to their brand or product.

Research at the time also suggested that the business partner's parts inventories were expanding as they were growing a parts inventory that was from four sources:

- Safety stock supporting their same-day and immediate needs for a critical part
- The OEM, with a next-day delivery service
- Dismantled products that bloated their internal inventories with many parts they could forecast they would never use
- Surplus safety stock of parts acquired at a lower cost from their world supplier sources
- Local parts refurbishment supplier sites (internal and external).

The original OEM had a world-class parts refurbishment infrastructure supporting their maintenance parts inventory sold as new parts. From a financial viewpoint, the value of the parts refurbished and returned to stock as new represented 50–70% of their annual inventory purchases. Statistically, their refurbished parts were of better quality (less likely to fail) than non-refurbished parts from their plant. The OEM typically supported its business partners from a single distribution center in the country with next-day delivery; however, it supported its internal field service technicians from 20-plus stock rooms across the same country with same-day support.

The OEM elected to create an offering that would financially incent the business partners to want to reduce their carried inventories by accessing the inventories the OEM already had available across its distribution network. They offered to provide the business partners access to the 20-plus stock rooms with same-day support and delivery. This offering required a minor "membership fee" to access the parts across the network.

This business partner inventory access program was a great success. All the major business partners embraced the program and eliminated the inventories they recognized they would never use. As a result, the OEM's parts sales revenues increased by four to five times in six to 18 months.

## A PARTS SALES CASE STUDY (2)

An international electrical supply company (OEM) determined that their parts sales were shrinking and that their clientele was becoming dissatisfied with their parts support services' level of service and cost. The OEM asked a third party to investigate the situation, and the following summary of services and client feedback was found:

- OEM parts quality was considered good
- Parts identification, ease of ordering, and parts delivery cycle times were significant issues for their parts customers. (Parts were not shipped same-day and seldom delivered the next day)
- Parts availability from the OEM was considered unreliable
- The biggest issue was the cost of the parts versus alternative sources and local refurbishment costs
- No parts exchange program (at a lower price point) was available
- The cost of parts was considered a deterrent to acquiring future OEM product
- Some clients interviewed were in the process of setting up their parts refurbishment resources to offset the cost of acquiring new parts from the OEM.

The above insights influenced the OEM to change how they supported their parts clients. First, they opened up communication with their primary clients to confirm the issues and get their buy-in on the best mitigations. A focus on improving the parts availability and order cycle time was prioritized. Finally, a local OEM refurbishment solution was sized and added as an offering to take that burden off their clients.

As it happened, a natural climate issue created a unique and immediate need for significant parts sales of their electrical products within a few months of this program. Their parts sales accelerated exponentially because of this unexpected demand, and the programs to hold this client loyalty were already in the works to keep their clients happier. As a result, the clients could focus more on their business and not feel a need to be hunting for a better costing parts solution to support their mission infrastructure.

## THE IMPORTANCE OF OEM PARTS SALES

Parts sales are an important business to the OEM. Product parts sales contribute to the OEM's brand reputation regarding their products and services. Helping clients with after-product sales parts support is in the best interest of both the OEM and the client. It is also very profitable for the OEM. Getting this area of their business right is critical.

Many industrial OEMs consider their focus on aftermarket services essential in today's markets.

# 16 An Anatomy of Maintenance Parts Management Excellence

## INTRODUCTION

Following maintenance parts management trends do not necessarily mean it is a leading practice for a specific enterprise. Furthermore, a "leading practice" may not be an organization's best practice. "Best practice will depend on many factors, including market, employee, quality, and profit focus.

For over three decades, many organizations claim they have found some success in increasing maintenance parts service levels while reducing maintenance parts-related costs and inventories. For example, experience suggests that:

- IT investment and tactical efforts in parts planning have lowered inventory levels by 15–35% while simultaneously improving customer service levels
- Automating inventory management processes can cut costs by 5–15% within specific operations; strategic automation can improve margins by 25% and gain 60% in share.

The simple fact is that increasing service levels and reducing related parts costs improves the enterprise's bottom line. The following figure suggests areas that could benefit from reviewing the maintenance parts anatomy (Figure 16.1).

The key has been to take these tactical actions aligned with the pendulum focus of the enterprise in a given period.

Trends over the past few years have the following attributes:

- Central ownership of maintenance parts inventory and global inventory management
- Global procurement contracts, global processes, global (consistent) metrics
- Consideration of the total asset lifecycle in the maintenance parts inventory planning tactics
- Minimizing parts-related fixed costs
- End-to-end supply chain visibility via tracking and tracing
- Pooling of inventory within an enterprise and with suppliers and partners
- Shared-user distribution infrastructure
- Complete outsourcing of logistics operation (when practical)
- Web-enabled systems to provide visibility and global execution

DOI: 10.1201/9781003344674-16

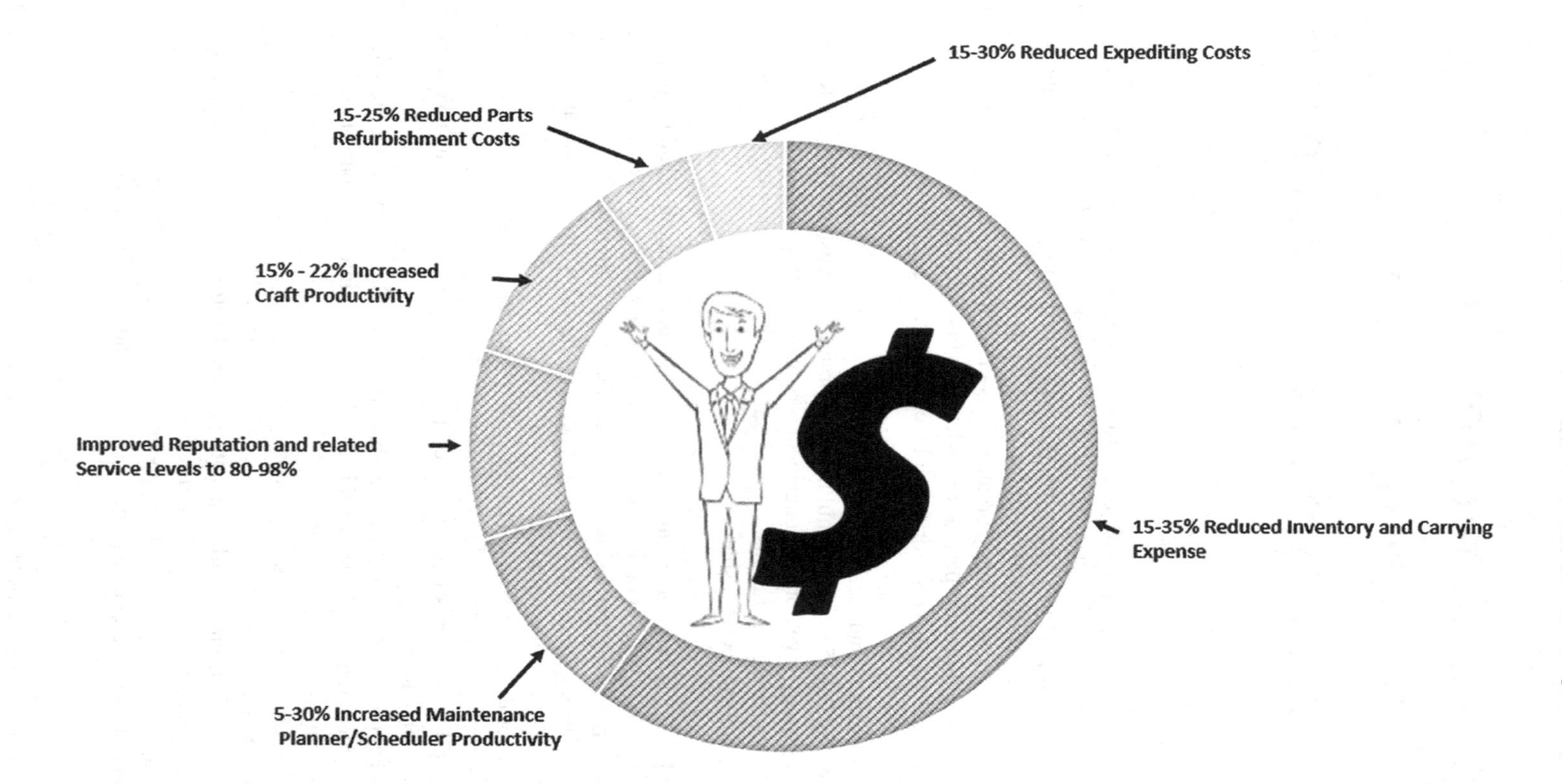


**FIGURE 16.1** Perceived benefits from a focus on maintenance parts management. (Copyright 101 MPE Introduction ©2020 from Maintenance Parts Management Excellence Training edited by Barry. Reproduced by permission of Asset Acumen Consulting Inc.)

- Integrated customer systems to the asset management and maintenance parts systems
- Replacement of proprietary systems with more complete "off the shelf" solutions
- Mobile capabilities integrated with the maintenance parts systems
- Effective parts and rotatable parts repair solutions
- Best-in-class expert applications

Often, many organizations have not implemented a leading set of maintenance parts management best practices for their enterprise. Successful organizations will have:

- Identified the cross-functional community that is actively working together to understand their role in supporting a successful maintenance parts operation
- Defined a workable leading practice set of actions and responsibilities to support maintenance parts excellence
- Accepted the responsibility and put into practice the tactics assigned to create and sustain maintenance parts management excellence

As suggested earlier in this volume, a best practice maintenance parts operation will have active, visible, and aligned sponsorship:

- Think through how they should be organized to achieve their excellence goals
- Assign the execution of maintenance parts activities to manage the scope of each of the assigned areas
- Promote a community of stakeholder engagement for the elements of the business that require cross-functional units working in concert to create and sustain success
- Encourage a culture of quality and drive for uber change improvements for the future (Figure 16.2)

To provoke thought at each level and help an organization challenge its prioritized activities should be against the current and forecasted market challenges. To help, the maintenance parts excellence questionnaire (see appendix A1) was created to assess and list potential new initiatives.

The pyramid above has ten elements of consideration that can assist an organization in self-assessment and then challenging itself to:

- Identify where they are in relation to perceived leading practices
- Quantify the leading practice statement performance level for their organization/enterprise
- Quantify the importance of a given leading practice statement for their organization/enterprise
- Qualify which practice may provide benefit to their enterprise and quantify the benefit potential (one-year versus five-year benefits)
- Estimate the complexity and effort to achieve the desired leading practice statement

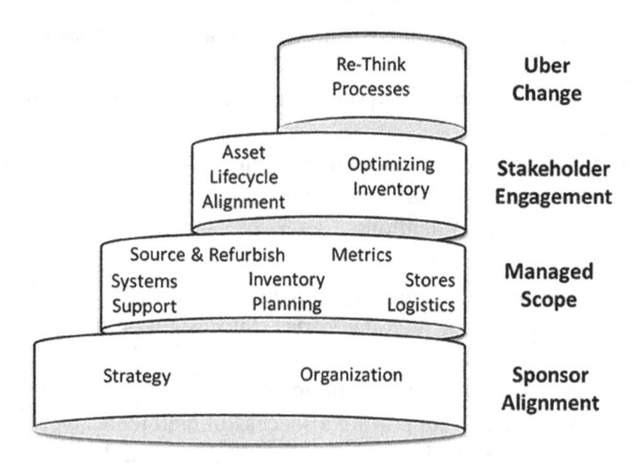


**FIGURE 16.2** Maintenance Parts Excellence Pyramid. (Copyright MPE ©2022 from Maintenance Parts Management Excellence Training edited by Barry. Reproduced by permission of Asset Acumen Consulting Inc.)

Completing this self-assessment will help the enterprise organize the journey to their maintenance parts excellence definition by challenging themselves against leading practices that will define the "best practice" for their organization.

When tackling maintenance parts management, leadership must include the community of stakeholders contributing to an enterprise's maintenance parts management operation.

## ELEMENTS OF LEADING PRACTICES IN MAINTENANCE PARTS EXCELLENCE

### STRATEGY

The strategy must align with the enterprise's prescribed mission, vision, and goals (short term and long term).

Understanding how the maintenance parts operation supports the enterprise, production, and maintenance is key to discussing their existence.

Key elements that should be considered in a leading maintenance parts strategy:

- Confirming that the objectives of the inventory management department support the overall objectives of the organization
- The maintenance parts strategy has an improvement plan with accountabilities clearly defined
- Agreed to improvement plans have actions that are linked to design engineering, maintenance, and inventory management best practices
- Effectively tracking and managing the company's assets and associated maintenance inventories through their entire lifecycle, from the time assets are acquired to the time they are retired

- The inventory management group has a long-term strategic plan to guide maintenance and production business units through asset lifecycle, opportunities, threats, development, and improvement efforts
- Maintenance parts and inventory budgets are related to the expected performance of the assets they support and can expedite delivery of requested inventory if the original request cannot be delivered locally when needed
- A maintenance parts process and budget account for project work on an exceptional or seasonal basis – this activity is tracked outside regular maintenance.
- The maintenance parts strategy is focused upon the total cost of ownership of the production assets, not just inventory turnover, inventory purchase price, or inventory provision costs
- A set of documented policies or guiding principles for maintenance parts management organization
- Clear organizational ownership of the maintenance parts inventory management policy process and hand-offs between design engineering, maintenance/operations, finance, and inventory management
- Health and safety, quality, environmental, and regulatory policies are effectively communicated to employees and contractors and effectively enforced

In previous already mature client self-assessments, an element such as "Inventory management group has a long-term strategic plan to guide maintenance through asset lifecycle, opportunities, threats, development and improvement efforts" was identified as an initiative that would be relatively cost-effective to implement and would create the foundation for significant service level improvement or inventory and cost savings to their parts operation of the enterprise. In addition, simple policy changes such as ensuring that all parts distribution will flow through the Maintenance Parts system for tracking purposes to a work order, assigning ownership to the ISP/RSP lists against the asset lifecycle, can come from diving deeper into the benefits of one strategy element.

## Organization

The maintenance parts management will also need to support the enterprise's prescribed mission, vision, and goals (short term and long term) while aligning with the enterprise management and employee relations culture.

Understanding how the maintenance parts management supports the enterprise, production, and maintenance is key to discussing their importance and role in supporting their enterprise value proposition.

Key elements that should be considered in a leading maintenance parts management are:

- Confirming that the roles between design engineering, maintenance and operations, finance, and inventory management are effective in managing the balance between inventory cost and services levels of critical assets

- Confirming that new employees and contractors have access to an orientation program that covers all areas of health and safety, warehouse/plant operation, maintenance parts, and management procedures
- Confirming that maintenance parts employees understand what may be expected of them
- Confirming that maintenance parts staffing levels are adequate
- Confirming that engineering, maintenance, and maintenance parts personnel have the proper experience, training, and education to support part management policy across an asset lifecycle
- Confirming that the maintenance parts organization is mostly centralized for inventory policy management and procurement and organized by location or product line where appropriate, with centralized support functions as required
- Providing regular training – a minimum of five days/year/employee
- Providing Maintenance Parts Management with a training culture to also receive formal supervisory training in counseling, leading, and motivating their people
- Supporting a formally established apprenticeship program is employed to address the maintenance parts department's future needs
- Leveraging contractors/suppliers to augment plant staff during annual inventories audits and for specific projects or specialized jobs
- Ensuring that supplier services are continually under review for performance and cost-effectiveness
- Maintenance parts organizational charts are current and visible to staff

In previous already mature client self-assessments, an element such as "Ensuring that supplier's services are continually under review for performance and cost-effectiveness" was identified as an initiative that would be relatively cost-effective to implement and could create the foundation for significant service level improvement or inventory and cost savings to their parts operation of the enterprise. Other areas were identified as potential new initiatives, but the direct savings were considered the same as the expected implementation costs. This assessment element (management) often confirms where the organization is in alignment with its enterprise policies.

"Supporting a formally established apprenticeship program" for the maintenance parts department's future needs is observed about 50% of the time an assessment is reported.

## SYSTEMS AND DATA SUPPORT (INFORMATION TECHNOLOGY)

The IT infrastructure business unit will be supporting all areas of an enterprise. Specific to maintenance parts management, IT will need to collect the data required for multiple business areas and make it available for the maintenance parts operation to function effectively. Similarly, data from maintenance parts will need to be visible to other business units for the enterprise to function optimally.

Understanding how information technology (IT) supports the enterprise, production, maintenance, and maintenance parts management is key to any discussion on their importance and role in supporting the enterprise value proposition.

Key elements that should be considered in a leading IT support of the maintenance parts operation are:

- Providing a fully functional and integrated maintenance management system or maintenance parts management system that is linked to the plant's financial and enterprise production and engineering systems
- Providing a system that is easy for the end-user to use
- Providing training and training support for critical systems
- Supporting warehouse automation tools where practical
- Ensuring that Parts information is easily accessible and linked to equipment records – finding parts for specific equipment is easy to do, and the stock records are usually accurate
- Ensuring that maintenance planners/schedulers can easily use the maintenance management system to plan jobs and to select and reserve spare parts and materials
- Providing maintenance parts data query capability and supporting regular use of reports from the systems for parts-related decision-making
- Supporting ad-hoc report capability against maintenance parts data
- Providing support for scheduling of major shutdowns if done using a project management system (linked to the maintenance management system (MMS) work orders) that determines critical paths and required level of parts and resources
- Supporting financial needs in maintenance parts systems such as:
  - Multiple currencies, parts classifications, multiple parts source commodity costing (e.g., new, used, repaired)
  - Financial treatment of parts acquisition (i.e., FIFO. LIFO, Average, Standard) costs following IFRS or similar financial disciplines aligned to enterprise GAAP
  - Efficientlyly managing inventory audits and reconciliation processes according to IFRS or similar financial disciplines aligned to enterprise GAAP
- Systems ease of use for planners/crafts/trades/inventory staff to electronically create parts request orders
- Leveraging mobile technologies for remote craft/trade/inventory staff for real-time inquiries/orders on the system
- Supporting scheduled and un-scheduled orders as well as operational demand, with automated inventory forecast algorithms aligned to inventory policy
- Supporting "out-of-stock" emergencies by electronically referring the order to a sister storeroom or a vendor and maintaining order history to capture demand
- Supporting parts management system inventory surplus balancing between enterprise storerooms
- Support procurement requirements, including electronically accessing technical catalog documents, as required by maintenance personnel
- Providing internal or external analysis tools used to declare or reset inventory policy based on enterprise automatically agreed-to data element values (i.e., ISP/RSP, active, highly active, project demands)

This is not a complete or exhaustive list but allows the self-assessment team to think about some of the services information technology can bring to the maintenance parts operation to help with effective and efficient parts management.

In previous already mature client self-assessments, an element such as "parts information is easily accessible and linked to equipment records. Finding parts for specific equipment is easy to do, and the stock records are usually accurate" was identified as an initiative that would be relatively cost-effective to implement and would create the foundation for significant service level improvement or inventory and cost savings to their parts operation of the enterprise. Other areas were identified as potential new initiatives, but the direct savings were considered the same as the expected implementation costs. "mobile technology," "warehouse automation," and "maintenance technician accessible catalogs" were not thought of as a valued initiative at the time of this initial assessment for one client but have been identified as key for others. The enterprise size and where they are in their maturity versus how they measure up against the market demands can influence which initiatives should be prioritized in any assessment.

## PARTS SOURCING AND REFURBISH

The procurement business unit and the parts repair business unit (if in place) will often be aligned with the maintenance parts management business unit. Parts repair will include any part that can be repaired and returned to stock as a valued part. In any event, the maintenance parts operation highly depends on the ability to source parts when needed.

Understanding how "procure" and "repair" support the enterprise, production, maintenance, and maintenance parts management is key to any discussion on how maintenance parts management supports the enterprise value proposition.

Key elements that should be considered in a leading procure and parts repair support of the maintenance parts operation are:

- Aligning parts purchasing strategies with the corporate enterprise culture
- Supporting number conventions in the enterprise system for items and related equipment internally and from suppliers
- Supporting commodity spend analysis to confirm the best sourcing strategy and alignment to corporate culture
- Supporting maintenance and parts management with an organization that has a defined and standard set of RFI/RFQ processes for commodity types or business environment
- Managing supplier targets that are linked to total parts supply chain objectives and asset total cost of ownership
- Providing supplier management support to minimize the number of critical suppliers used to leverage risk/reward partnership benefits
- Monitoring and reviewing supplier performance in regularly held joint supplier/company meetings
- Leveraging technology to automate and, as a result, minimize the need for procurement workload in the procurement transaction process

- Leveraging inter-company communications and technology in an integrated fashion to optimize inventory demands within your enterprise
- Supporting and optimizing the scope of parts that can be repaired based on the forecasted usage of parts
- Supporting a defined accounting algorithm for dealing with the parts repair process that allows for credit to parts expense or other GL entry for repaired parts
- Supporting each part procured as a potential opportunity lost for savings in parts costs from repair
- Supporting metrics that track used parts capture and repair yields, out-of-box failures for repaired parts, cost of repair, and rotatable serialized controls
- Actively tracking and executing warranty returns to the original equipment manufacturer (OEM) for credit when applicable

In previous already mature client self-assessments, an element such as "focus to minimize the number of key suppliers used to leverage risk/reward partnership benefits" was identified as an initiative that would be very cost-effective to implement and would create the foundation for significant service level improvement and inventory and cost savings to their parts operation of the enterprise. Other areas were also identified as potential new initiatives with additional costs, and the savings made the initiative worth pursuing a bit deeper. These initiatives included:

- Optimization of the scope of parts that can be repaired based on the forecasted usage of parts
- Working to treat each part procured as a potential opportunity lost for savings in parts costs from repair and
- Implementing new metrics to track used parts capture and repair yields, out-of-box failures for repaired parts, cost of repair, and rotatable serialized controls

## STORES AND LOGISTICS MANAGEMENT

Warehouse and logistics needs can range from maintaining a secure warehouse to facilitating the receipt, putting away, picking, packing, and delivering a maintenance part.

Understanding how maintenance parts logistics support the maintenance, finance, and maintenance parts management is key to discussing how maintenance parts management supports the enterprise value proposition.

Key elements that should be considered in leading logistics support of the maintenance parts operation are:

- Ensuring parts and materials are readily available for use where and when needed
- Ensuring parts and material issued from stores is traceable to the equipment, tradesman, and time

- Working so that stock-outs represent less than 3% of orders placed at the storeroom
- Supporting distributed (satellite) stores are used throughout the plant for commonly used items (e.g., fasteners, fittings, common electrical parts)
- Ensuring Parts and materials are restocked automatically before the inventory on-hand runs out and without prompting by the maintenance crews
- Supporting a central tool crib is for special tools
- Ensuring control procedures are followed for all company-owned tools and supplies such as drills, special saws, ladders, gloves, etc.
- Ensuring inventory cycle counts are conducted either annually or stratified across each year
- Reviewing inventory regularly to delete obsolete or very infrequently used items
- Ensuring purchasing/stores can source and acquire rush emergency parts that are not stocked quickly and with sufficient time to avoid plant downtime
- Ensuring order points and quantities are based on lead time, safety stock, and economic order quantities (inventory policy)
- Ensuring material storage facilities are appropriate and secure (24 hours a day) for the type of item stored in terms of its deterioration or risk of theft
- Supporting blanket and system contracts and orders are effectively utilized to minimize redundant paperwork and administrative effort
- Ensuring procedures are in place to alert maintenance personnel regarding receipt of their materials
- Ensuring procedures are in place to alert maintenance personnel regarding receipt of their materials
- Supporting a stores catalog that is up-to-date and readily available for use
- Supporting vendor performance reviews and analyses are conducted
- Executing warehouse disciplines that ensure that safety is paramount, with all racks securely bolted and gas containers securely chained

Some warehouse-related leading practices and requirements could have been listed here. For example, suppose the warehouse mechanic is the topic of an assessment. In that case, it could be taken as a single initiative to help with environmental, safety, process flow, and growth issues (to name a few).

In previous already mature client self-assessments, elements that came to be opportunity initiatives providing significant value for some investment included:

- Supporting order points and quantities that are based on lead time, safety stock, and economic order quantities to help with inventory costs and material handling efforts
- Having an automated inventory management system that dynamically calculates min/max on a weekly/monthly basis based on factors such as criticality, past usage, cost of expediting, and many more factors
- An automated inventory management system considers all aspects of planned, unplanned, and seasonal demand by calculating min/max strategies for each location (to limit redundant order restocking efforts)

The following initiatives are worth pursuing deeper as there is the potential for additional costs but also savings:

- A central tool crib is used for special tools
- Ensuring that material storage facilities are appropriate and secure (24 hours a day) for the type of item stored in terms of its deterioration or risk of theft
- Control procedures for all company-owned tools and supplies such as drills, special saws, ladders, and gloves.

## INVENTORY PLANNING

Inventory planning and policy management can range from coordinating inventory policy elements and influences to reporting to finance on inventory forecasts for inventory levels and related inventory service levels.

Understanding how Inventory planning supports maintenance, finance, and maintenance parts Management is key to discussing how maintenance parts management supports the enterprise value proposition.

Key elements that should be considered in leading inventory planning and policy management support of the maintenance parts operation are:

- Ensuring that all equipment expected to be supported is listed in the maintenance parts management system or enterprise asset management (EAM) system
- Ensuring that all parts transactions intended to support an enterprise asset are executed through the maintenance parts management system or EAM system
- Targeting an 80% planned maintenance culture that allows for parts confirmation and pre-kitting where appropriate
- Ensuring that exceptional parts demand and non-emergency parts requests are screened so that the parts can be ordered in "Just in time" aligned to a planned work schedule
- Ensuring that all parts supported work orders and direct parts orders are appropriately prioritized to pre-defined criteria
- Ensuring that ISP/RSP lists are in place by the plant design engineer or maintenance who leverage a complete reliability-centered maintenance analysis before the commissioning of the new equipment to confirm stocking requirements and locations
- Supporting a transparent process for hand-off between our design engineering group and maintenance/operations group for the ownership and cost of the ISP/RSP being maintained as part of the enterprise inventory policy
- Supporting an inventory management system to complement the ISP/RSP that tracks parts order activity and generates a min/max stock level based on activity history
- Supporting a system that manages Inventory stocking levels automatically by leveraging inventory optimization logic in the system and takes into consideration ISP/RSP lists, recent and seasonal activity, and resets recommended stocking levels on a weekly/monthly basis

- Supporting a system that reviews policy and related replenishment levels on a daily/weekly basis, and replenishment orders are automatically generated to the designated supplier
- Ensuring that the inventory system can distinguish between planned and unplanned activity as part of its Inventory stocking level review
- Supporting an automated Inventory planning process that considers service level requirements and excess inventory in each of your stock rooms across the enterprise
- Supporting a system that manages excess inventory so that it can be redistributed before order quantities are developed for replenishment orders to suppliers
- Performing inventory reviews at least quarterly, along with finance, on surplus inventory sales and inventory scrap activities

Inventory planning and policy management are at the heart of the balance between inventory and service. The inventory planner keeps communications between finance, maintenance, production, and design engineering. For example, the maintenance technician may come to the parts counter for the part. However, the inventory planner coordinates the inventory policy inputs to ensure a likely used part is stocked (policy) and in-stock (service level) when needed.

In previous already mature client self-assessments, elements that came to be opportunity initiatives that could provide significant value for some investment included:

- Initial spare parts (ISP) lists are developed by plant design engineering, who leverage a complete reliability-centered maintenance analysis before the equipment is commissioned to confirm the stocking requirements and locations. An ISP process put into place that leverages the design engineer's perspective of asset operating context (such as used with RCM) when developing the spare parts requirements by location

The following initiatives are worth pursuing deeper as there is the potential for additional costs but also savings:

- Over 90% of maintenance work activities are covered by standard written work order, standing work order, PM work order, a PM checklist, or routine
- Priorities are assigned to all work orders and applied to parts orders using pre-defined criteria
    - to be managed so that it is not abused to circumvent the parts order process
- Inventory stocking levels are generated automatically by leveraging inventory optimization logic in the system and take into consideration ISP lists, recent and seasonal activity, and resets recommended stocking levels on a weekly/monthly basis
- The inventory system can distinguish between planned and unplanned activity as part of its Inventory stocking level review

## KEY PERFORMANCE METRICS AND BENCHMARKING

Key performance indicators (KPIs) can be the dashboard that communicates to staff, process clients, and manage the health and quality of the services supported in maintenance parts management. Having the correct, balanced metrics in place is vital so that if a "key criterion" is stressed, it is likely to see the counter metric balloon to higher-than-expected levels. Also, having metrics that support the needs of executives, supervisors, and staff is vital to work toward the targets that directly relate to what they can influence.

Understanding how KPIs and benchmarking support the maintenance, finance, and maintenance parts management is key to any discussion on how maintenance parts management performs to contribute to the enterprise value proposition. Successful performance measures result in positive and proactive staff and management behavior. However, they should be sufficiently limited in number to recall the measures and their importance with little or no assistance.

Key elements that should be considered in leading KPI and benchmarking support of the maintenance parts operation are:

- Ensuring that there are clear metrics between business domains (design engineering, maintenance/operations, inventory management, and finance) that drive the needed behavior to "optimize" the maintenance parts stock needed and drive return on asset and asset lifecycle cost management
- Supporting a set of performance indicators that are routinely measured and tracked to monitor results relative to the asset lifecycle cost, maintenance strategy, and improvement process
- Inventory management related KPIs are understood and posted for the materials management, procurement, and maintenance personnel to review
- Facilitating key high-level inventory management metrics that include, at order time, parts availability, ease of use for inventory systems, parts quality, distribution quality, parts cost, and distribution costs
- Ensuring that measures are regularly reviewed to ensure they remain relevant to our goals and objectives
- Supporting a culture that believes their inventory management performance indicators are developed from and support their maintenance and asset lifecycle cost strategy
- Ensuring that all maintenance staff have been proficiently trained on the significance of the measures used such that they can determine whether or not overall performance is improving
- Communicating and posting KPIs regularly and making them available/visible for all business domain department staff and trades
- Supporting finance with an inventory value metric such as "turnover" and "reserves," as well as a holistic inventory value metric that is accepted by all levels of your business
- Occasionally benchmarking against external site data and industry standards to set targets and publish performance comparisons.

Well-executed KPIs can provide a scorecard for the stakeholders in the maintenance parts management process, which monitors the health of the business flow.

It can discover early issues and defend inappropriate demands on the inventory levels and staff.

In previously completed and mature client self-assessments, elements that came to be opportunity initiatives that could provide significant value for some investment included balancing inventory levels with the parts availability levels when inappropriate demands to lower inventory without considering the direct number of forecasted additional stock-outs.

The following initiative is worth pursuing deeper as there is the potential for additional costs, but also:

- Confirming that key high-level inventory management metrics include, at order time, parts availability, ease of use for inventory systems, parts quality, distribution quality, parts cost, and distribution costs

## ASSET LIFECYCLE ALIGNMENT

Asset Lifecycle alignment ensures that the maintenance parts operation has insights into new equipment assets planned to be installed in a given enterprise location. The insights need to include the related parts used throughout the assets lifecycle, the suggested ISP that should complement the inventory stocking policy, the volume of equipment installed, and later scheduled for the supported equipment discontinuance.

A reliability-centered maintenance analysis approach is a leading practice in determining which parts could be needed in an asset's lifecycle. Ownership of the ISP list may end after the first equipment commissioning if the enterprise is set up to transfer this data responsibility to maintenance (Figure 16.3).

Key elements that should be considered in leading asset lifecycle alignment support of the maintenance parts operation are:

- Ensuring that equipment history is maintained for all essential pieces of equipment showing failures, preventive replacement, all proactive maintenance activities, causes of failures, and repair work completed
- Recognizing that asset location and criticality are pivotal in the considered inventory placement strategies during the initial spare parts process
- Ensuring that all equipment has been classified based on its importance to plant operations and safety - this classification is used to help determine work order priorities, priority for asset lifecycle analysis, and direct engineering resources
- Programs are in place to successfully avoid failures on equipment that is subject to periodic restoration or replacement work (overhauls)
- Ensuring that SAE JA1011-compliant reliability-centered maintenance analysis, engineering design team, or other RCM structured approach is used to determine the failure management policies for critical equipment and the critical parts needed to maintain asset value and lower risk
- Leveraging the RCM "effects" descriptions that result from the asset lifecycle analyses to be used to help in troubleshooting efforts for failure diagnosis and support parts requirements

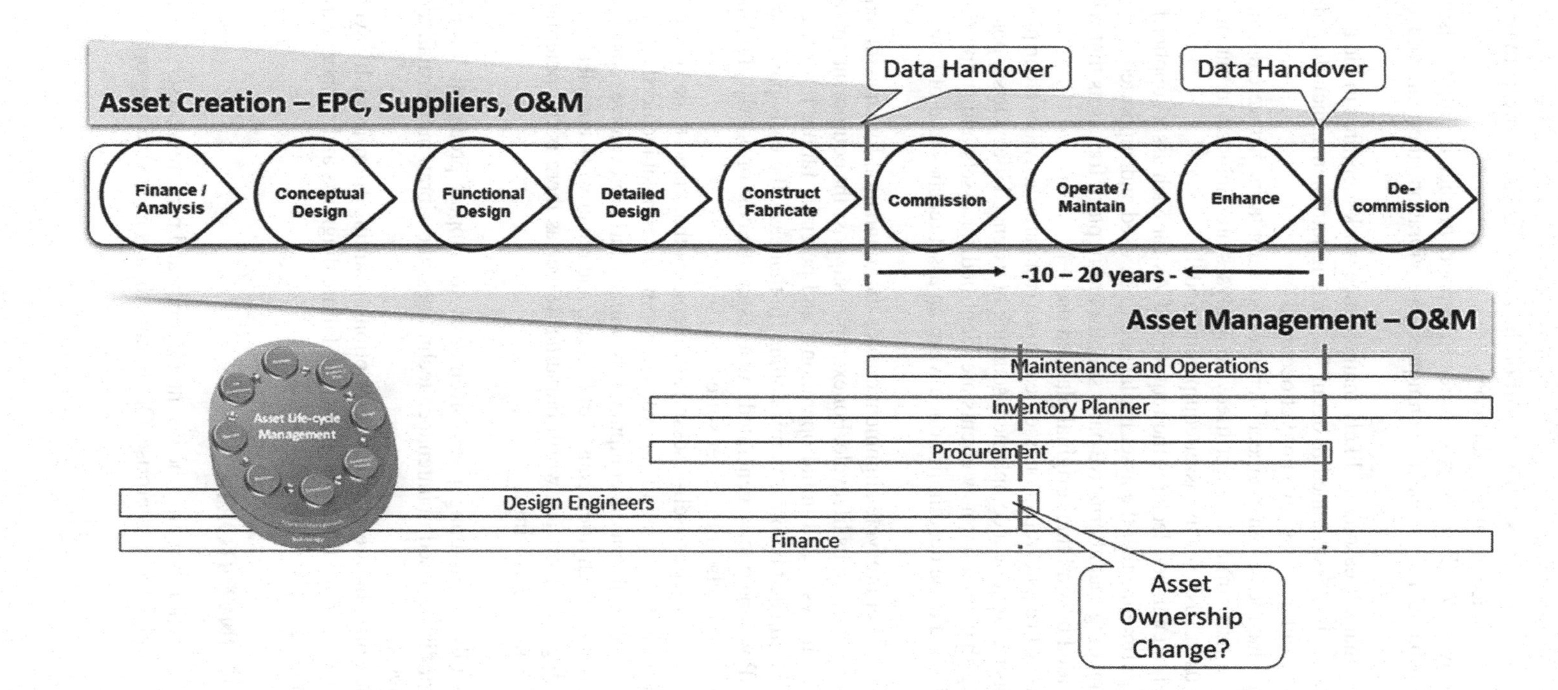


**FIGURE 16.3** An asset lifecycle management focus in inventory management. (Copyright 101 MPE Introduction ©2020 from Maintenance Parts Management Excellence Training edited by Barry. Reproduced by permission of Asset Acumen Consulting Inc.)

- Leveraging value-risk studies as part of the asset lifecycle analysis to optimize maintenance program decisions
- Supporting ISP/RSP parts inventories to support predicted failures using condition-based maintenance techniques like vibration, thermal, and oil analysis
- Ensuring that all reasonably likely maintenance work identified in the asset lifecycle analyses is planned in detail for parts requirements, including the less likely failures that have not yet occurred
- Ensuring that the design engineer identifies operator tasks that were deemed to require parts from the asset lifecycle analyses are included in our parts "where used" lists for the asset/equipment model
- The design engineer declares the required SLAs for each asset supplier for the entire timeline that the asset is planned to be used and supported
- Understanding that alternative parts sources are in place for assets that will be extended beyond their original planned use
- Confirming that parts repair or asset-tear down (for parts) activities are in place for assets that require a complement sourcing or extended lifecycle support
- Ensuring that potential new clients are identified for assets and supporting parts planned to be decommissioned when an asset reaches End of Life

A well-executed asset lifecycle alignment creates the foundation for an asset (equipment) success throughout its lifecycle. For example, suppose the equipment operating context, related analysis, and initial operational and potential failure information is not adequately mitigated before commissioning. In that case, the early parts requirement for an ISP will be poorly generated. As a result, the cost of poorly planning an asset will persist throughout the asset's life.

In previous mature client self-assessments, elements that came to be opportunity initiatives that could provide significant value for some investment included equipment history maintained for all critical equipment showing failures, preventive replacement, proactive maintenance activities, causes of failures, and repair work completed.

The following initiatives are worth pursuing deeper as there is the potential for additional costs but also savings:

- Successfully maintaining parts inventories to support predicted failures using condition-based maintenance techniques like vibration, thermal, and oil analysis
- Documenting "failure effects" descriptions resulting from asset lifecycle analyses to help troubleshoot efforts for failure diagnosis and support parts requirements

## OPTIMIZING INVENTORY

Inventory Optimization is tuning the inventory policy and parts-related service levels to the optimal level for that enterprise. Often more data is needed to support detailed inventory tools such as an inventory optimizer. However, inventory optimization is more than just an additional tool. It can include policy changes, algorithm changes, process tuning, and anything that will improve parts-related service levels while minimizing (reducing) parts-related costs.

Key elements that should be considered in leading inventory optimization support of the maintenance parts operation are:

- Ensuring that all parts activities are managed in an inventory system that can track all orders, movement, and usage between stock locations and assets or work orders
- All reverse logistic activities are tracked, including new parts returns, used parts returns, new defective parts returns, warranty parts returns, etc.
- Executing an automated inventory management system that dynamically calculates min/max on a weekly/monthly basis based on factors such as criticality, past usage, cost of expediting, and many more factors
- Ensuring that the automated inventory management system considers all aspects of planned, unplanned, and seasonal demand by location calculating min/max strategies for each location
- Ensuring that the automated inventory management system considers all aspects of existing orders, reserved activity, and surplus when calculating replenishment requirements
- Ensuring that the ISP process is in place that leverages the design engineer's perspective of asset operating context (such as used with RCM) when developing the spare parts requirements by location
- Ensuring that the inventory policy is constantly reviewed and challenged by inventory planning when the parts complement is not effectively supporting the optimal service levels or costs for a specific asset
- Ensuring that surplus parts are reviewed weekly for potential redistribution to other sites across the enterprise
- Confirming that surplus parts not expected to be used are considered for resale and sold where appropriate
- Ensuring that surplus parts not expected to be used to support assets are regularly removed from inventory (scrapped) to reduce the cost of carrying inventory and establish a financial declaration credit
- Working to share the parts forecast demand with suppliers (including repair suppliers) so that the lead time for delivery can be minimized
- Supporting the option of vendor (supplier) managed inventories in your systems and are used effectively to maintain service levels and minimize inventory carrying costs
- Facilitating the community stakeholder group so that design engineering, maintenance, operations, and inventory management teams work together to optimize the asset and maintenance parts inventory values effectively

In previous already mature client self-assessments, elements that came to be opportunity initiatives that could provide significant value for some investment included:

- Tracking all reverse logistic activities, including new parts returns, used parts returns, new defective parts returns, warranty parts returns, etc. Creating the ability to "return to stock" excess spares from a completed turnaround and creating the ability to recognize inventory that is surplus to emergency repair needs
- Implementing an inventory optimizer tool for equipment where the critical data is viable to complement the inventory policy

The following initiatives are worth pursuing deeper as there is the potential for additional costs but also savings:

- Inclusion of vendor (supplier) managed inventories in systems to effectively maintain the service level and minimize inventory carrying costs
- Sharing of forecast demand with suppliers (including repair suppliers) to minimize the lead time for delivery

## RE-THINKING MAINTENANCE PARTS PROCESSES FOR THE FUTURE

Process re-design of the maintenance parts management processes can be fundamental and continuous. If an organization does not know where to start, a maintenance parts maturity assessment against the potential leading practices listed in this chapter could help. Challenging the status quo against the existing KPIs may be practical if the organization perceives they are already at a reasonable maturity level. Process re-engineering never really stops. Ideally, the initiative's lead should be looking for quantum improvements (e.g., 10X improvements in errors or cycle time, 10% reduction in inventories with improved service levels) (Figure 16.4).

Key elements that should be considered in leading Process re-design support of the Maintenance Parts operation are:

- Ensuring that the EAMS/MMS and other inventory management systems are recognized more for the benefit they provide rather than the effort required administrating them (e.g., data entry, running reports, developing queries)
- Ensuring that the EAMS/maintenance parts system and other management systems are used to automate workflow processes to reduce errors and administrative effort
- Confirming that "as-is" process maps accurately reflect the typically used processes.
- Ensuring that crucial asset design planning, maintenance, and maintenance parts management processes (e.g., planning, corrective maintenance, stock issues) have been identified, and "as-is" processes are mapped
- Supporting a culture where key maintenance processes are re-designed to reduce or eliminate non-value-added activities (e.g., duplicate data entry)
- Supporting a culture where all activities are challenged for effectiveness

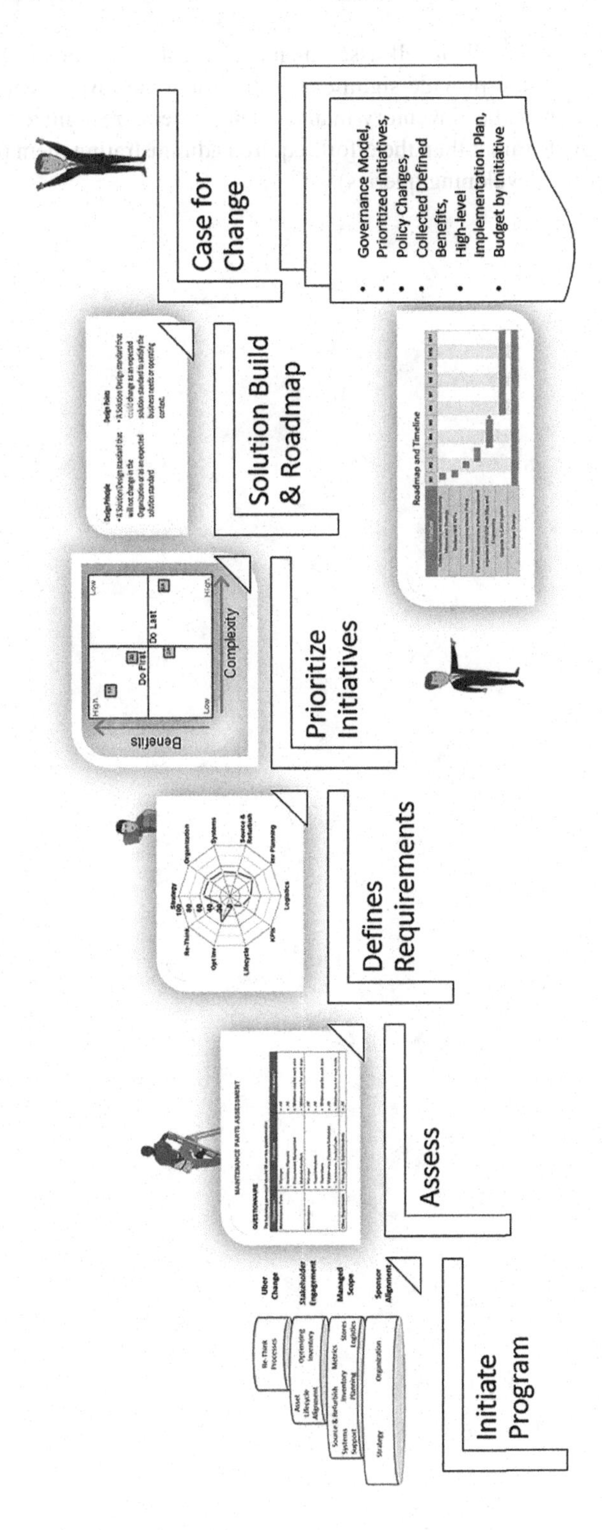


**FIGURE 16.4** Process flow for a Maintenance Parts Management maturity assessment. (Copyright MPE ©2022 from Maintenance Parts Management Excellence Training edited by Barry. Reproduced by permission of Asset Acumen Consulting Inc.)

In previous already mature client self-assessments, elements that came to be opportunity initiatives that could provide significant value for some investment included the EAMS/CMMS, and other inventory management systems recognized more for the benefit they provide rather than the effort required administrating them (e.g., data entry, running reports, developing queries).

# Appendix A1

## *Maintenance Parts Excellence Questionnaire*

### INSTRUCTIONS

This questionnaire is organized into ten sections. Each section has several statements that relate to a specific maintenance topic. For each statement, assign a rating from 0 to 4. The rating definition table below defines how well each statement describes your maintenance organization.

The ratings have the following definition:

| | **Score** |
|---|---|
| Strongly agree | 4 |
| Mostly agree | 3 |
| Partially agree | 2 |
| Disagree | 1 |
| Do not understand | 0 |
| Not applicable | N/A |

A **Glossary** of Maintenance Definitions and a **Comments** section are provided at the end of this questionnaire.

Name (optional): ____________________ Job Title: ____________________

E-mail Address (optional): ____________________

Plant/Site: ____________________ Division: ____________________

| **Primary Responsibility** | **Department** |
|---|---|
| ______ Executive | ______ Maintenance/Tech |
| ______ Management | ______ Inventory Management |
| ______ Supervision | ______ Engineering |
| ______ Trades/Hourly | ______ Purchasing |
| ______ Administrative | ______ Warehousing |
| ______ Other | ______ IT Support |
| | ______ Finance, Parts Repair, Other |

An indicative sample list of assessment questions is in this Appendix.

## 1 Strategy

| Statement | Score (4, 3, 2, 1, 0) |
|---|---|
| 1. The plant's or site's mission statement, vision, and values are communicated and understood at all levels of staff. | |
| 2. The parts/inventory management department's objectives and goals are posted, communicated, and understood by all personnel concerned. | |
| 3. The parts/inventory management group has a long-term strategic plan to guide maintenance, operations, finance, and parts support through asset lifecycle, opportunities, threats, development, and improvement efforts. | |
| 4. The strategy has an improvement plan with responsibilities and accountabilities clearly defined. | |
| 5. The improvement plan is linked to design engineering, maintenance, operations, finance, and inventory management practices. | |
| 6. We effectively track and manage the company's critical assets and associated parts inventories throughout their lifecycle. The asset lifecycle starts with being planned and acquired and ends with decommissioning. | |
| 7. We have established and documented policies or guiding principles for maintenance parts management. | |
| 8. Our approach to maintenance parts management is proactive. We do our best to stock parts to optimize support inventories with asset availability. | |
| 9. The maintenance parts and inventory budgets are monitored to confirm the inventory effectiveness of the assets they support. In addition, outcomes are understood as the likely business impact if the parts inventory policy cannot be delivered when needed. | |
| 10. The parts usage allows any project work to be done on an exceptional or seasonal basis and is accounted for separately (in inventory policy) from regular maintenance. | |
| **Total (max. 40)** | |

*Note*: Strongly agree (4), Mostly agree (3), Partially agree (2), Disagree (1), Do not understand (0).

## 2 Organization

| Statement | Score (4, 3, 2, 1, 0) |
|---|---|
| 1. Maintenance and Parts staffing levels are considered balanced and adequate. | |
| 2. Engineering, maintenance, and parts personnel have the proper experience, training, and education to support part management policy across an asset lifecycle. | |
| 3. Our employees understand what is expected of them. | |
| 4. Maintenance parts organizational charts are current. | |
| 5. Overtime represents less than 5% of the total annual maintenance person-hours. | |
| 6. Overtime is not concentrated in one area (e.g., warehouse or parts planning) but is well distributed. | |
| 7. Regular training is provided to all employees and is more than five days/year/ employee. | |
| 8. A formally established apprenticeship program is employed to address the maintenance parts department's future needs. | |
| 9. A portion of an individual's pay is based on demonstrated skills and knowledge, results, and productivity. | |
| 10. Contractors/suppliers augment plant staff during annual inventory audits, specific projects, or specialized jobs. | |
| 11. Within our organization: The roles between design engineering, maintenance and operations, finance, and inventory management effectively manage the balance between inventory cost and service levels of critical assets. | |
| **Total (max. 44)** | |

*Note*: Strongly agree (4), Mostly agree (3), Partially agree (2), Disagree (1), Do not understand (0).

## 3 Information Technology – Systems Support

| Statement | Score (4, 3, 2, 1, 0) |
|---|---|
| 1. A fully functional and integrated maintenance management system (MMS) is linked to the plant's financial and material management systems. | |
| 2. Our maintenance and materials management information is considered a valuable asset and is used regularly. Thus, the system is not just a "black hole" for information or a burden to use that produces no benefit. | |
| 3. I find our maintenance parts management system easy to use. | |
| 4. All maintenance parts departments have been trained to use their parts system effectively. | |
| 5. Our maintenance planners/schedulers use the maintenance management system to plan jobs and select and reserve spare parts and materials. | |
| 6. Parts information is easily accessible and linked to equipment records. As a result, finding parts for specific equipment is easy, and the stock records are usually accurate. | |
| 7. Scheduling for major shutdowns is done using a project management system (linked to the MMS work orders) that determines critical paths and required levels of parts and resources. | |
| 8. We leverage our system's query capability to create ad hoc reports and analyses supporting parts management. | |
| 9. Our Parts/Maintenance Management systems can manage multiple currencies, parts classifications, and multiple parts source commodity costing (e.g., new, used, repaired). We also treat parts acquisition (i.e., FIFO. LIFO, Average, Standard) costs according to IFRS or similar financial disciplines aligned to enterprise GAAP. | |
| 10. Our Parts/Maintenance Management systems can easily manage inventory audits and reconciliation processes per IFRS or similar Financial disciplines aligned to enterprise GAAP. | |
| 11. Our Parts/Maintenance Management systems can manage scheduled and unscheduled orders and operational demand with automated inventory forecast algorithms aligned to inventory policy. | |
| 12. Inventory surplus balancing between enterprise storerooms is supported through Parts/Maintenance Management systems | |
| 13. Internal or external analysis tools are used to declare or reset inventory policy based on enterprise automatically agreed-to elements (i.e., ISP/RSP, Active, Highly active, project demands). | |
| **Total (max. 52)** | |

*Note*: Strongly agree (4), Mostly agree (3), Partially agree (2), Disagree (1), Do not understand (0).

## 4 Source and Refurbish

| Statement | Score (4, 3, 2, 1, 0) |
|---|---|
| 1. A procurement group manages your purchasing strategies separately and aligns with corporate culture. | |
| 2. Standardized naming/numbering conventions for items purchased and suppliers are used. | |
| 3. Your organization regularly analyzes maintenance parts commodity "spend" to confirm the best sourcing strategy and alignment to corporate culture across the asset's lifecycle. | |
| 4. Your organization has a defined and standard set of RFI/RFQ processes for commodity types, business solution risks, and business environments. | |
| 5. You monitor and manage supplier targets linked to total parts supply chain objectives and asset total cost of ownership. | |
| 6. You focus on minimizing the number of suppliers used to leverage risk/reward partnership benefits as a component of your corporate sourcing philosophy. | |
| 7. Your organization supports the automation of parts-sourcing transaction management. Your automated transaction process minimizes the need for Procurement's direct involvement and enhances the user experience. | |
| 8. You optimize the scope and repair cycle of parts you can refurbish based on the forecasted parts usage and inventory policy needs. | |
| 9. Your accounting process supports an algorithm that facilitates a ledger credit to expense or other GL entry for successfully refurbished parts returned to inventory. | |
| 10. You work to treat each part procured as a potential opportunity lost for savings from parts refurbishment (recycle and repair). | |
| 11. You have a successful parts refurbishment metrics program. It tracks critical elements of the refurbishment process (e.g., used parts capture/returns and repair yields, out-of-box failures from refurbished parts, cost of repair, and rotatable serialized controls). | |
| 12. You have a new defective parts returns program that can actively track and execute warranty returns to the original equipment manufacturer (OEM) for credit when applicable. | |
| **Total (max. 48)** | |

*Note*: Strongly agree (4), Mostly agree (3), Partially agree (2), Disagree (1), Do not understand (0).

## 5 Stores Logistics Management

| Statement | Score (4, 3, 2, 1, 0) |
|---|---|
| 1. Out-of-stock parts requests represent less than 3% of orders placed at the storeroom. | |
| 2. Parts and materials are readily available for work orders where and when needed at an acceptable service level. | |
| 3. Every stocked part has a defined stocking location in the designated stock room. | |
| 4. Parts/Maintenance Management systems can track and trace all parts and materials issued from a stock room to the work order, equipment, tradesman, and time. | |
| 5. Parts and materials are systems monitored and automatically restocked before the inventory on hand runs out without prompting by the maintenance crews. | |
| 6. A central tool crib is used for special tools to manage tool availability and quality. | |
| 7. Parts/Maintenance Management systems support control procedures to track and trace the movement and usage of company-owned tools and supplies such as drills, special saws, ladders, gloves, etc. | |
| 8. Parts inventory cycle counts are conducted and audited within the Parts/ Maintenance Management systems. | |
| 9. The Parts/Maintenance Management systems support the ability to regularly review parts inventories to delete out-of-policy, obsolete, or very infrequently used items (scheduled or ad hoc). | |
| 10. Inventory policy algorithms automatically calculate order points and quantities based on lead time, safety stock, and economic order quantities. | |
| 11. Material storage facilities stewardship supports parts security (24 hours a day) for the type of item stored. Item deterioration or risk of theft is maintained. Parts are only picked from a valid system pick list or work order by Parts personnel. | |
| 12. Procedures are in place to automatically inform maintenance personnel regarding the picking/receipt status of their material orders (e.g., direct order and back-ordered parts to a work order). | |
| 13. Vendor relationship management and vendor performance reviews and analyses are conducted by the supporting Procurement personnel | |
| 14. Warehouse disciplines ensure that safety is paramount, with all racks securely bolted and gas containers securely chained | |
| **Total (max. 56)** | |

*Note*: Strongly agree (4), Mostly agree (3), Partially agree (2), Disagree (1), Do not understand (0).

## 6 Inventory Planning

| Statement | Score (4, 3, 2, 1, 0) |
|---|---|
| 1. An asset register lists all enterprise equipment that requires maintenance, engineering, or parts support during its life. | |
| 2. Over 90% of maintenance work activities are covered by standard written work order, standing work order, PM work order, a PM checklist, or routine. | |
| 3. Over 80% of planned maintenance work (preventive, predictive, and corrective) activities are scheduled to advance execution with sufficient lead time to kit parts when appropriate. | |
| 4. Non-emergency work requests are screened, estimated, and planned (with tasks, materials, and tools identified) by a dedicated planner and parts ordered in time for the scheduled work order. | |
| 5. An agreed-to and predefined criteria are used to assign all work orders and related parts orders. This criterion is respected to circumvent the parts order process. | |
| 6. Initial spare parts (ISP) lists are submitted by Design engineering to confirm the stocking requirements and locations, leveraging a completed Reliability-Centered Maintenance analysis before the equipment is commissioned. This list would include input from the OEM if applicable. | |
| 7. A process for hand-off between our design engineering group and maintenance/ operations group for the ownership and cost of the ISP/RSP (Recommended Spare Parts) is maintained as part of the enterprise inventory policy. | |
| 8. An inventory management "policy-generator" system complements the ISP/RSP inputs tracks parts order activity and generates a Min/Max stock level based on activity history. | |
| 9. Inventory stocking policy levels are generated automatically by leveraging inventory optimization logic in the inventory management system. The algorithms consider ISP/RSP lists and recent and seasonal activity and are run weekly/ monthly to generate the recommended stocking levels. | |
| 10. The inventory system discerns planned, unplanned, and unusual demand activity as part of its automated inventory stocking algorithms regularly run. | |
| 11. Inventory planning strategies and algorithms consider service level requirements and surplus inventory in each stock room when setting inventory policy by location. | |
| 12. Inventory and policy reviews are done at least quarterly to generate surplus inventory sales opportunities and inventory scrap activities. | |
| **Total (max. 48)** | |

*Note*: Strongly agree (4), Mostly agree (3), Partially agree (2), Disagree (1), Do not understand (0).

## 7 Performance Metrics/Benchmarking

| Statement | Score (4, 3, 2, 1, 0) |
|---|---|
| 1. Inventory Management level KPIs are understood and posted for the materials management, procurement, and maintenance personnel to review. | |
| 2. Critical high-level inventory management metrics include at order time, parts availability, ease of use for inventory systems, parts quality, distribution quality, parts cost, and distribution costs. | |
| 3. There are clear metrics between business domains (design engineering, maintenance/operations, inventory management, and finance) that drive the needed behavior to "optimize" the maintenance parts stock needed and drive return on asset and asset lifecycle cost management. | |
| 4. Critical parts metrics are regularly reviewed to ensure they remain relevant to our goals and objectives. | |
| 5. A set of performance indicators are routinely measured and tracked to monitor results relative to the asset lifecycle cost, maintenance strategy, and improvement process. | |
| 6. Our inventory management performance indicators are developed and support our maintenance and asset lifecycle cost strategy. | |
| 7. Our current performance measures result in positive and proactive behavior across the affected stakeholder groups. | |
| 8. The performance measures used for my area or department are sufficiently limited to recall the measures and their importance with little or no assistance. | |
| 9. Performance measures are published or posted regularly and made available/visible for all business domain department staff and trades to read. | |
| 10. An inventory value metric such as "Turnover" and "Reserves" is used, and all levels of our business accept a holistic inventory value metric. | |
| **Total (max. 40)** | |

*Note*: Strongly agree (4), Mostly agree (3), Partially agree (2), Disagree (1), Do not understand (0).

## 8 Asset Lifecycle Alignment

| Statement | Score (4, 3, 2, 1, 0) |
|---|---|
| 1. Equipment history data is available for all critical equipment showing failures, preventive replacement, proactive maintenance activities, causes of failures, and repair work completed. | |
| 2. We successfully avoid equipment failures due to unavailable parts that result from periodic planned restoration or replacement work (overhauls). | |
| 3. We successfully maintain parts inventory placement policies to support potential failures that may arise from condition-based maintenance techniques like vibration, thermal, and oil analysis. | |
| 4. Value-risk studies have been conducted in our Asset Lifecycle analysis to optimize maintenance program decisions. | |
| 5. The "effects" descriptions from our Asset Lifecycle reliability analyses help troubleshoot efforts for failure diagnosis and support parts requirements. | |
| 6. All maintenance work identified in our Asset Lifecycle analyses is reviewed in detail for parts requirements, even for the less likely failures that have not yet occurred. | |
| 7. All parts that have the potential to be used in an installed asset's lifecycle are confirmed to be included in the inventory systems item master database and are identified in the parts "where used" lists for the asset/equipment model. | |
| 8. During the Initial Spare Parts process, asset location and criticality are considered in the inventory policy/placement strategies. | |
| 9. Asset Lifecycle strategies for growth and end of life are factored into the ISP/RSP lists and inventory policy. Active inventory policies are reduced, and surplus inventories are challenged when the volume of assets is forecast to decline. | |
| 10. Alternative parts sources are in place for assets that may be applied to an extended Asset lifecycle scenario. | |
| 11. Future clients are identified for assets and supporting parts planned to be decommissioned when an asset approaches its end of life. | |
| **Total (max. 44)** | |

*Note*: Strongly agree (4), Mostly agree (3), Partially agree (2), Disagree (1), Do not understand (0).

## 9 Optimizing Inventory

| Statement | Score (4, 3, 2, 1, 0) |
|---|---|
| 1. All parts activities (e.g., orders, audits, movement, and usage between stock locations and assets or work orders) are managed in a single enterprise-wide inventory system. | |
| 2. All reverse logistic activities (e.g., including new parts returns, used parts returns, new defective parts returns, and warranty parts returns) are tracked in a single enterprise-wide system. | |
| 3. Your inventory management system is scheduled to automatically and dynamically calculate parts inventory policy factors (e.g., min/max based on inventory policy factors, including asset criticality, past parts usage, cost of expediting, and many more factors) on a weekly/monthly basis. | |
| 4. The Design engineer's perspective of asset operating context (such as used from an RCM analysis) when developing the spare parts support Initial Spare Parts (ISP) requirements by location. | |
| 5. Inventory policy is constantly reviewed and challenged by inventory planning when the parts complement does not effectively support optimal service levels or costs for a specific asset. | |
| 6. All parts outage costs (for out-of-stock and not stocked) are understood and used to influence the inventory policy priorities against the pressures of inventory management. | |
| 7. Surplus parts not expected to be used (outside of inventory policy) to support assets are regularly removed from inventory (scrapped) to reduce the cost of carrying inventory and establish a financial declaration credit. | |
| 8. Forecast demand is shared with suppliers (including refurbishment suppliers) so that the cost of the parts may be optimized and the lead time for delivery can be minimized. | |
| 9. Vendor (supplier) managed inventories are in your systems and are effectively used to maintain the parts inventory service levels and minimize inventory carrying costs. | |
| 10. Finance, Design engineering, Maintenance, Operations, and Inventory Management teams work together to optimize asset and parts inventory value effectively. | |
| **Total (max. 40)** | |

*Note*: Strongly agree (4), Mostly agree (3), Partially agree (2), Disagree (1), Do not understand (0).

## 10 Re-thinking Parts Processes for the Future

| Statement | Score (4, 3, 2, 1, 0) |
|---|---|
| 1. Essential asset design planning, maintenance, and MRO materials management processes (e.g., planning, corrective maintenance, stock issues) have been identified and mapped as "as-is" processes. | |
| 2. The "as-is" process maps accurately reflect the typically used processes. | |
| 3. Essential maintenance and parts processes are re-designed to reduce or eliminate non-value-added activities (e.g., duplicate data entry). | |
| 4. Strategies, tasks, and activities are regularly challenged for the enterprise's mission effectiveness. | |
| 5. Staff are empowered to drive quality program improvements as a regular part of their culture. | |
| 6. The EAMS/MMS or other management systems can automate workflow processes to reduce errors and administrative effort. | |
| 7. The EAMS/MMS or other inventory management-related systems are recognized more for their benefit than the effort required to administrate them (e.g., performing data entry, running inventory policy algorithms, running reports, and developing queries). | |
| 8. A Bill of Materials (BoM) list is used where practical to help with parts ordering, stocking staging, and fulfillment. | |
| 9. Parts order priority system is well-engrained in the corporate culture to distinguish emergency/urgent orders from non-urgent ones. | |
| 10. Process mapping and re-design have been extended to the administration and technical support processes. | |
| **Total (max. 40)** | |

*Note*: Strongly agree (4), Mostly agree (3), Partially agree (2), Disagree (1), Do not understand (0).

### Comment Section

If you could change procedures or processes to improve your site, what are the first two things you would change?

If you could change procedures or processes that would improve things for your department, what are the first two things you would change?

Any other comments?

| Term | Glossary of Definitions for Maintenance Assessment Questionnaire |
|---|---|
| **Asset** | The physical business resources included in a plant, facility, fleet, or parts. |
| **Benchmarking** | Comparing performance with other organizations, identifying comparatively high-performance organizations, and learning what they do allow them to achieve that high level of performance. |
| **CMMS** | Computerized Maintenance Management System. The software is used to manage the maintenance business and is sometimes referred to as MMS (or EAMS). |
| **Condition** | The use of specialist equipment to measure the condition of the equipment. Examples of condition monitoring techniques include listening for abnormal noise, feeling for heat or vibrations, smelling something funny, vibration analysis, tribology, and thermography. |
| **Corrective Maintenance** | Any maintenance activity is required to correct an asset failure that has occurred or is in the process. This activity may consist of repair, restoration, or replacement of components. |
| **Critical Path** | It is a method to determine the minimum time to complete a project based on the activities and their best sequence. |
| **EAMS** | Enterprise Asset Management System. EAMS are broader in functionality in comparison to a maintenance management system. In addition to a maintenance system, EAMS can include inventory, purchasing, finance, human resources systems, etc., but not production material requirements planning (MRP). Enterprise resource planning systems (ERPs) contain MRP systems. |
| **Economic Order Quantity** | A calculation to trigger inventory replenishment. The calculation is based on the cost to carry inventory versus place orders. |
| **Failure** | An item of equipment has suffered a failure when it can no longer perform its required function to a desired or specified standard. |
| **Failure Management Policy** | A general term encompassing on-condition tasks, scheduled restoration, scheduled discard, failure-finding, run-to-failure, and re-design. |
| **Failure** | A failure is a state of being instead of how the equipment got to that state. |
| **Initial Spare Parts (ISP)** | The parts that are determined to be required to be stocked locally to support the early operation of an asset or equipment |
| **Inventory Policy** | Inventory policy is the complement of factors that make up why a specific part is maintained as stocked in a specific location. The factors may include policy decisions such as Min/Max and reorder point, ISP/RSP, and Date influences. |
| **Key Performance Indicators (KPI)** | Several crucial measures enable performance to be monitored against targets. |
| **MMS** | Maintenance Management System. See CMMS. |

| Term | Glossary of Definitions for Maintenance Assessment Questionnaire |
|---|---|
| **Predictive Maintenance (PdM)** | Use measured physical parameters against known engineering limits for detecting, analyzing, and correcting equipment problems before a failure occurs. Examples include vibration analysis, sonic testing, dye testing, infrared testing, thermal testing, coolant analysis, tribology, and equipment history analysis. Predictive maintenance and condition-based maintenance are synonymous. |
| **Preventive Maintenance (PM)** | Maintenance is carried out at predetermined intervals or to a prescribed usage criteria and is intended to reduce the likelihood of a functional failure. The maintenance intervention is a restoration or replacement action. |
| **Recommend Spare Parts (RSP)** | The parts that must be in stock locally to support the asset or equipment when in service. Often an ISP will become an ISP after being accepted and transitioned through the commissioning process. |
| **Reliability-Centered Maintenance (RCM)** | A structured process developed initially in the airline industry that is now commonly used in all industries to determine the maintenance requirements of any physical asset in its operating context |
| **SAE Standard JA1011** | It represents the "Evaluation Criteria for Reliability-Centered Maintenance (RCM) Processes" This standard determines through seven specific questions whether a process qualifies as RCM. |
| **Service-Level Agreements (SLA)** | Service-level agreements can be in place to communicate the supplier's expectations as a baseline. The receiver of the services can then manage these commitments. |
| **Shutdown** | The act and period when the equipment is out of service to perform significant maintenance, repairs, overhauls, or improvements. |
| **Standing Work Order** | A work order that is open either indefinitely or for a predetermined period to collect labor hours and material costs. Standing work orders are used when it has been decided that the cost and history for specific tasks are not required on individual work orders. |
| **Work Order** | The prime document used by the maintenance process to manage maintenance tasks. It may include a description of the work required, the task priority, the job procedure to be followed, and the parts, materials, tools, and equipment required to complete the job. It is also used to collect labor hours, costs, and materials consumed in completing the task and essential information on failure causes, performed work, etc. |

## PERMISSIONS

This Appendix was adapted from copyright Appendix A.22 ©2011, from *Asset Management Excellence, Optimizing Equipment Life-cycle Decisions* edited by Campbell, Jardine, McGlynn. Reproduced by permission of Taylor & Francis Group, LLC, a division of Informa plc.

This Appendix was adapted from copyright MPE ©2022, from Maintenance Parts Management Excellence Training edited by Barry. Reproduced by permission of Asset Acumen Consulting Inc.

# Appendix A2
## *Basic Inventory Planning Replenishment Logic*

### INTRODUCTION

This Appendix explains the basic logic used to support the simple inputs to Inventory Policy (Figure A2.1).

The inventory policy process flow in Figure A2.1 has multiple data inputs for the "weekly policy calculation." The ISP/RSP data creation is explained in Chapter 11, and the Optimizer data creation is explained in Chapter 12. The other data input element logic will be explained in this Appendix.

Local average demand and min/max stock levels can be calculated with a simple set of weeks of stock. In addition, basic inventory data supporting an item's available quantity for a specific location can be supported with data elements such as:

- Quantity on hand
- Quantity on order
- Quantity in transit
- Quantity reserved
- Quantity available and
- Unit cost

These available quantities and an item's unit cost help determine the impact on the inventory. Comparing the available quantity to recent activity can help the inventory planner determine if he statistically has enough inventory available for future demand that may simulate recent activity. Stocking policy by asset/SKU and stock room for recently active parts will use additional data elements to establish the volume of future activity expected based on the recent past activity. Data elements to help with this type of assessment can include:

- Weekly average demand
- Maximum stock level
- Minimum stocking level
- Safety stocking level
- Reorder point and
- Economic ordering point

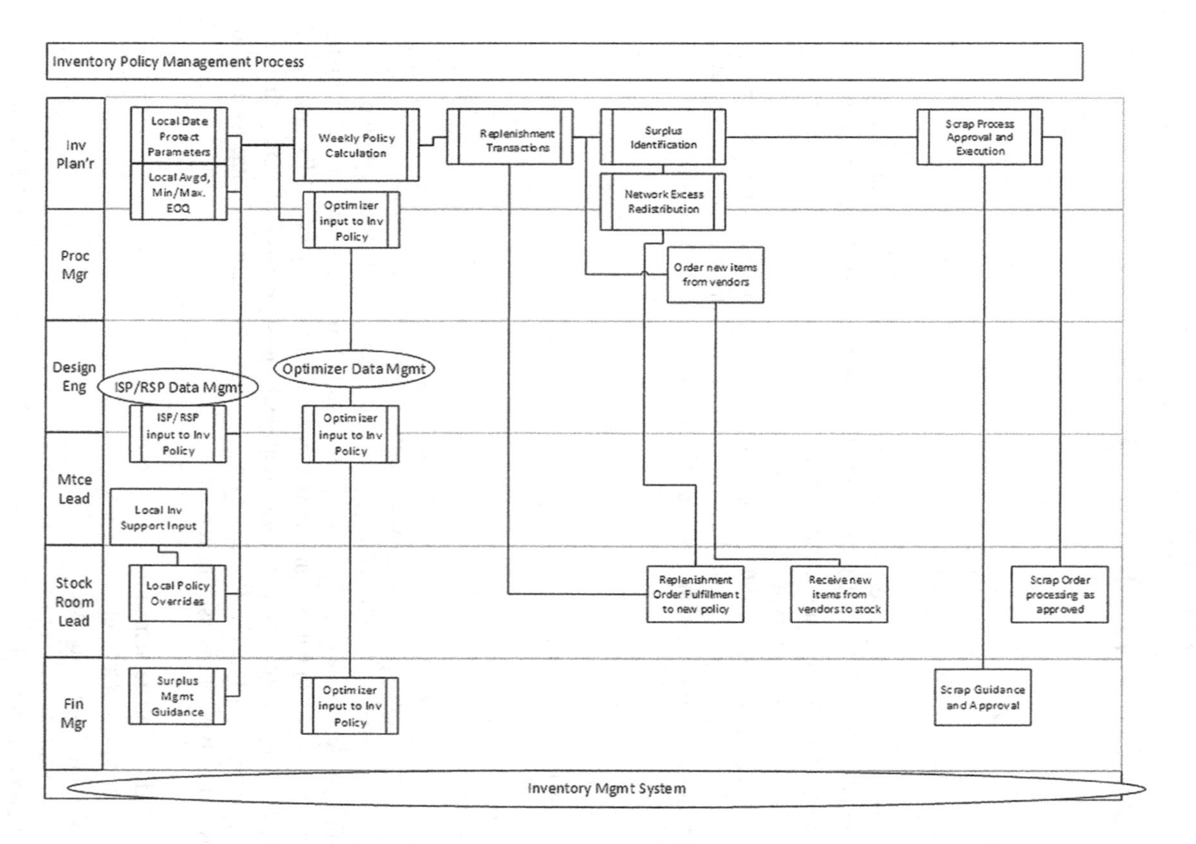


**FIGURE A2.1** Inventory policy management process. (Copyright 104 MPE Inventory Management ©2020 from Maintenance Parts Management Excellence Training edited by Barry. Reproduced by permission of Asset Acumen Consulting Inc.)

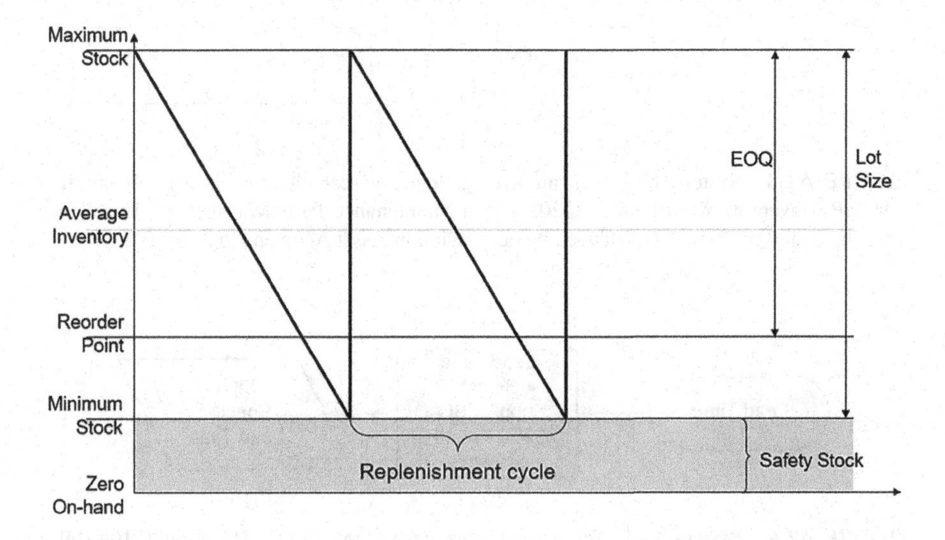


**FIGURE A2.2** Sample minimum/maximum economic order quantity graphic. (Copyright 104 MPE Inventory Management ©2020 from Maintenance Parts Management Excellence Training edited by Barry. Reproduced by permission of Asset Acumen Consulting Inc.)

Graphically the data elements could be plotted as shown in Figure A2.2.

The process would be to determine a minimum stock level for an active part (for example) three weeks of stock, and a reorder point of (perhaps an additional week) four weeks of stock. Depending on the cost of the part, delivery, and put away costs: the reorder quantity could be set to additional weeks of stock. In this example, ten weeks of stock could be the EOQ. If lot quantities are set against the item, then a "lot" quantity could also be leveraged. So, in the example above:

- The minimum stock is three weeks.
- Reorder point is four weeks.
- EOQ is ten weeks and
- The maximum stock level would be (Reorder point + EOQ = Maximum Stock) 14 weeks.

This calculation would only be done in system logic if at least two line disbursements of the same item from the same location in a rolling cycle of 13 weeks (or three months). The weeks of stock would be calculated against a data element called average demand (Avgd).

The Average Demand logic would be as shown (Figure A2.3):

Lead time to get to the Reorder point in a simple system logic set would be as shown (Figure A2.4):

With Lead time known: Reorder point in a simple system logic set would be as shown (Figure A2.5):

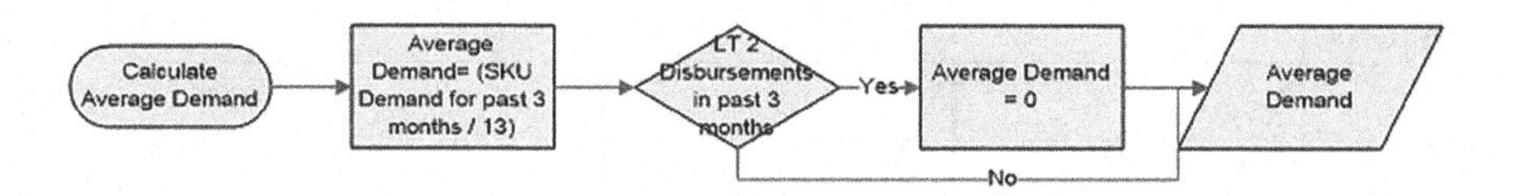


**FIGURE A2.3** System logic for an average demand calculation (Avgd). (Copyright 104 MPE Inventory Management ©2020 from Maintenance Parts Management Excellence Training edited by Barry. Reproduced by permission of Asset Acumen Consulting Inc.)

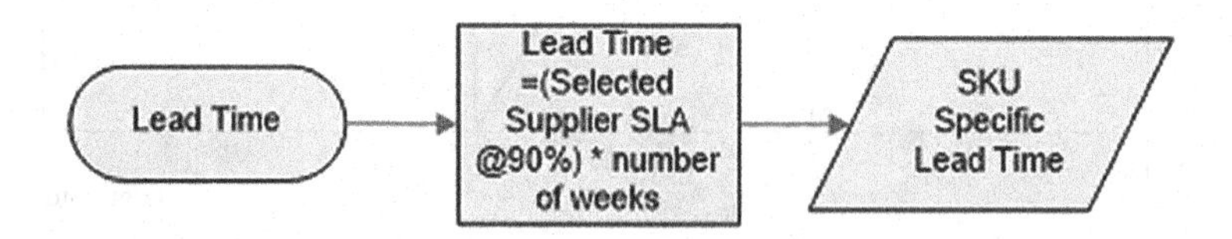


**FIGURE A2.4** System logic for a lead time calculation (LT). (Copyright 104 MPE Inventory Management ©2020 from Maintenance Parts Management Excellence Training edited by Barry. Reproduced by permission of Asset Acumen Consulting Inc.)

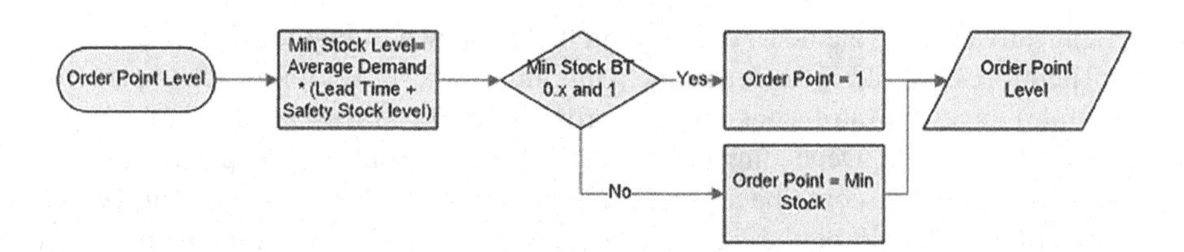


**FIGURE A2.5** System logic for a reorder point calculation (order point). (Copyright 104 MPE Inventory Management ©2020 from Maintenance Parts Management Excellence Training edited by Barry. Reproduced by permission of Asset Acumen Consulting Inc.)

Leveraging cost tables as a weighting factor and system logic, the Economic Order Quantity could be calculated as shown (Figure A2.6):

Maximum stock levels would leverage the above data elements as previously described and as shown (Figure A2.7):

For an inventory stocking policy by Item/Part and Stock room:

- Weekly average demand (Avgd):
  - Calculates the running average demand for the past 13 weeks; and
  - If disbursements are more than 13 weeks apart, then no order point, maximum stock, or EOQ will be calculated

- Examples of this average demand calculation with demand shown by week number over 15 weeks would be as shown (Figure A2.8):

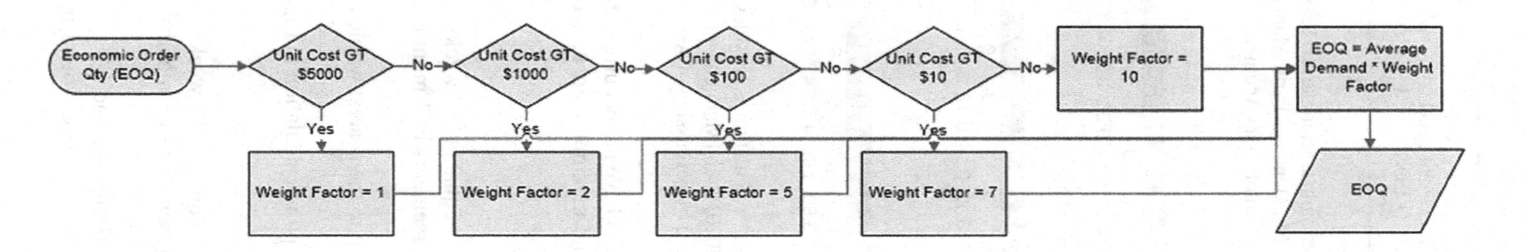


**FIGURE A2.6** System logic for an economic order quantity calculation (EOQ). (Copyright 104 MPE Inventory Management ©2020 from Maintenance Parts Management Excellence Training edited by Barry. Reproduced by permission of Asset Acumen Consulting Inc.)

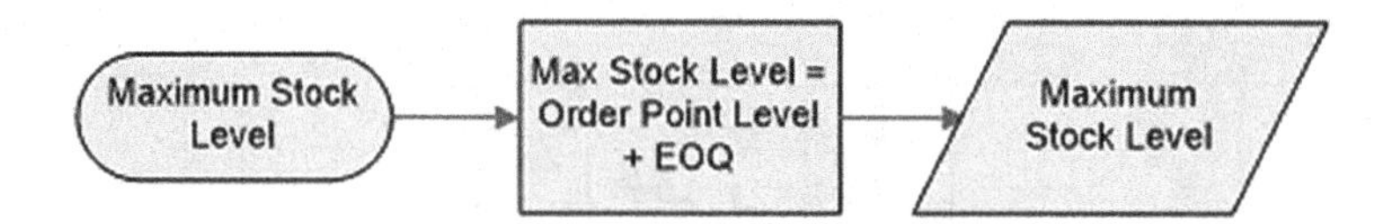


**FIGURE A2.7** System logic for a maximum stocking level (MaxStk). (Copyright 104 MPE Inventory Management ©2020 from Maintenance Parts Management Excellence Training edited by Barry. Reproduced by permission of Asset Acumen Consulting Inc.)

Example 1

| Week | 1 | 2 | 3 | 4 | 5 | 6 | 7 | 8 | 9 | 10 | 11 | 12 | 13 | 14 | 15 |
|---|---|---|---|---|---|---|---|---|---|---|---|---|---|---|---|
| Demand | 0 | 0 | 2 | 0 | 0 | 1 | 0 | 0 | 0 | 2 | 0 | 0 | 0 | 1 | 0 |
| Avgd | 0.00 | 0.00 | 0.67 | 0.50 | 0.40 | 0.50 | 0.43 | 0.38 | 0.33 | 0.50 | 0.45 | 0.42 | 0.38 | 0.43 | 0.40 |

Example 2

| Week | 1 | 2 | 3 | 4 | 5 | 6 | 7 | 8 | 9 | 10 | 11 | 12 | 13 | 14 | 15 |
|---|---|---|---|---|---|---|---|---|---|---|---|---|---|---|---|
| Demand | 1 | 0 | 0 | 0 | 0 | 0 | 0 | 0 | 0 | 0 | 0 | 0 | 0 | 0 | 1 |
| Avgd | 1.00 | 0.50 | 0.33 | 0.25 | 0.20 | 0.17 | 0.14 | 0.13 | 0.11 | 0.10 | 0.09 | 0.08 | 0.08 | 0.00 | 0.07 |

Example 3

| Week | 1 | 2 | 3 | 4 | 5 | 6 | 7 | 8 | 9 | 10 | 11 | 12 | 13 | 14 | 15 |
|---|---|---|---|---|---|---|---|---|---|---|---|---|---|---|---|
| Demand | 1 | 2 | 3 | 4 | 5 | 6 | 7 | 8 | 7 | 8 | 0 | 0 | 0 | 0 | 0 |
| Avgd | 1.00 | 1.50 | 2.00 | 2.50 | 3.00 | 3.50 | 4.00 | 4.50 | 4.78 | 5.10 | 4.64 | 4.25 | 3.92 | 3.57 | 3.20 |

**FIGURE A2.8** Examples of the "Avgd" calculation as it cycles over 15 weeks. (Copyright [104 MPE Inventory Management ©2020] from [Maintenance Parts Management Excellence Training] edited by [Barry]. Reproduced by permission of Asset Acumen Consulting Inc.)

If the two disbursements in 13 weeks have not occurred, then the "Avgd" calculation would continue, but the order point, maximum stock, and EOQ calculations would not be set in the system.

Scenario assumptions:

- Lead Time plus safety stock is equal to "9" weeks.
- Must have two history disbursements in 3 months to calculate order point for active parts and
- Order point always rounds up to the nearest whole number.

This is demonstrated for the order point calculation in Figure A2.9:

Scenario assumptions:

- Lead time plus safety stock is equal to "9" weeks.
- Must have two history disbursements in 3 months to calculate order point and EOQ for active parts.
- Unit cost is $255 => Weight Factor is 5 and
- Order point and maximum stock always round up to the nearest whole number.

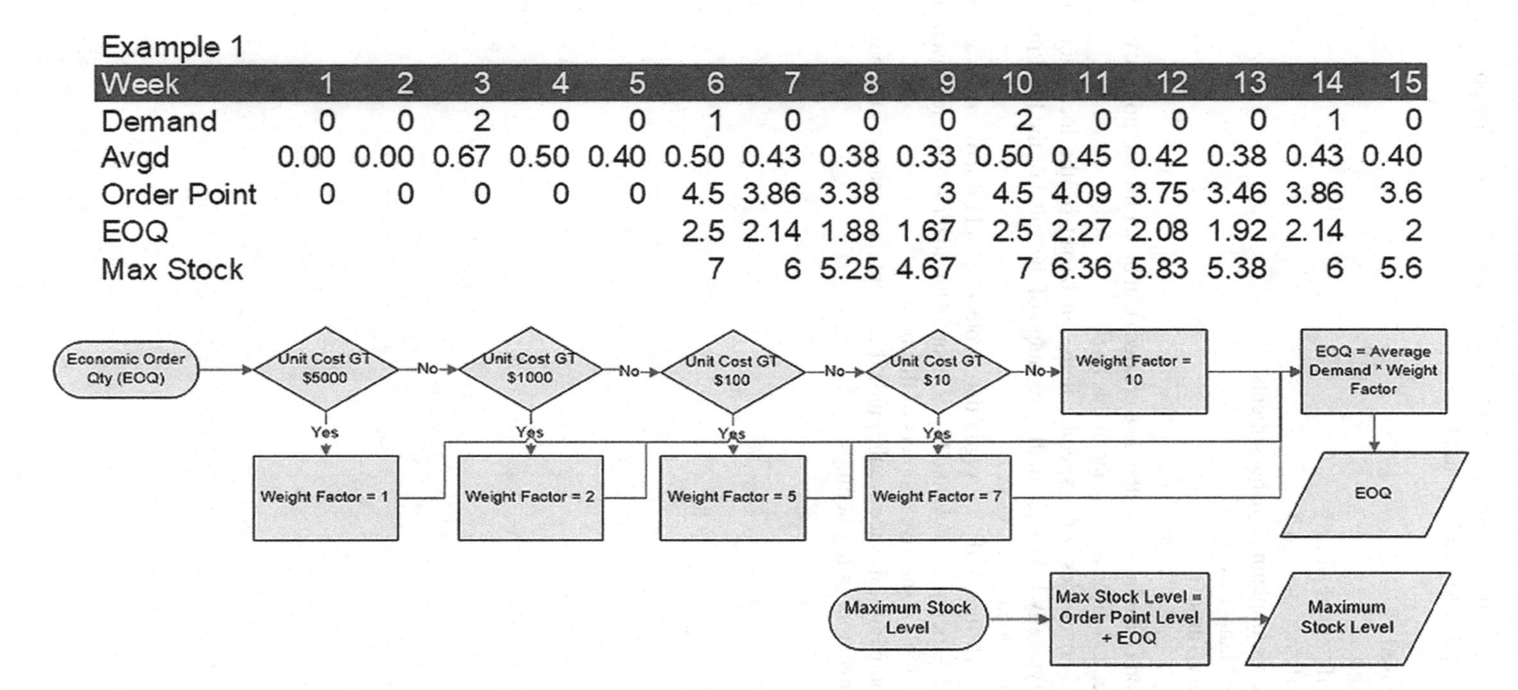

Example 1

| Week | 1 | 2 | 3 | 4 | 5 | 6 | 7 | 8 | 9 | 10 | 11 | 12 | 13 | 14 | 15 |
|---|---|---|---|---|---|---|---|---|---|---|---|---|---|---|---|
| Demand | 0 | 0 | 2 | 0 | 0 | 1 | 0 | 0 | 0 | 2 | 0 | 0 | 0 | 1 | 0 |
| Avgd | 0.00 | 0.00 | 0.67 | 0.50 | 0.40 | 0.50 | 0.43 | 0.38 | 0.33 | 0.50 | 0.45 | 0.42 | 0.38 | 0.43 | 0.40 |
| Order Point | 0 | 0 | 0 | 0 | 0 | 4.5 | 3.86 | 3.38 | 3 | 4.5 | 4.09 | 3.75 | 3.46 | 3.86 | 3.6 |
| EOQ | | | | | | 2.5 | 2.14 | 1.88 | 1.67 | 2.5 | 2.27 | 2.08 | 1.92 | 2.14 | 2 |
| Max Stock | | | | | | 7 | 6 | 5.25 | 4.67 | 7 | 6.36 | 5.83 | 5.38 | 6 | 5.6 |



**FIGURE A2.9** Examples of the "active" part versus the Avgd and order point calculation. (Copyright 104 MPE Inventory Management ©2020 from Maintenance Parts Management Excellence Training edited by Barry. Reproduced by permission of Asset Acumen Consulting Inc.)

Some of the other data elements that could also influence the stocking policy and are not addressed in this Appendix could be:

- Where used
- How used
- Impact if "out of stock"
- Impact if "not stocked"
- Planned vs. unplanned disbursements
- Surplus Management and
- Inventory Turnover

Date protection (Date last used and date created) can be set for six months, one year, two years, etc. and discussed earlier in this book.

Local stock overrides for a specific item can be set by the local stock room lead and are typically held to a small percentage of the value of the local inventory (e.g., 2% of total local stock value).

Active parts (Min/Max/EOQ/Avgd), date protected, and local overrides are inputs into the weekly stocking policy, and ISP/RSP and Optimizer settings are discussed in more detail in their respective chapters of this book.

In most companies, finance would guide the inventory impacts to the business to assist in the inventory policy settings.

# Appendix A3
## *Reliability-Centered Maintenance Process Description*

Reliability leaders and engineers will primarily focus on the "asset performance" business dimension. The expectation is that improved asset performance will drive positive business results across the other three business dimensions (KPIs for the executive, middle management, and staff, etc.). A proper focus on asset reliability leveraging a leading methodology will yield bottom-line financial benefits to the organization and specific benefits such as:

- Greater safety and environmental integrity
- Improved operating performance (uptime, output, product quality, and customer service)
- Greater maintenance cost-effectiveness
- The greater motivation of the team and individuals
- A comprehensive understanding of each asset in its operating context and related constraints and risks
- A comprehensive database (to support long-term asset lifecycle management)

Often a leading risk and reliability disciplines program (also known as Reliability-Centered Maintenance – RCM) will have prioritized its critical processes and assets. The first risk and reliability program step would be defining the asset's operating context for the system it supports, including a business case (justification) for an assessment. The operating context would document the system's and asset's purpose with a plant level, asset level, and business analysis level outline. Next, a trained set of operations and maintenance peers familiar with the asset and RCM methods would be facilitated through an analysis that would identify the primary and secondary functions of the asset and the failed states that could "reasonably" happen within its operating context. Next, a failure modes effects analysis (FMEA) would be completed with identified asset functions and failed states. Next, a "prescriptive" confirmation of the most effective mitigating tactic that should be applied is influenced by whether the failure effect is hidden or will have safety, environmental, and operational or non-operational impact.

A rough definition of a risk and reliability disciplines program is to:

- Determine what disciplines need to be in place; mitigations must ensure that the focused asset does what its users want it to do within its designed operating context.

A summary of the formal risk and reliability disciplines program questions are:

- What is the asset operating context?
- What are its functions (what do the users want it to do)?
- In what ways can it fail (the failed states)?
- What causes it to fail (the failure modes)?
- What happens when it fails (the effects)?
- How does the failure matter (hidden, safety, environment, operational)?
- What can be done to prevent or predict the failure?
- What can we do if we can't prevent or predict the failure?

The strength of a risk and reliability disciplines program is that it provides simple, precise, and easily understood decision criteria. The decision criteria would include predictive or preventive, technically feasible tasks worth doing in any context. When a task is decided, who should do it, and at what frequency? Versions of RCM also consider the probability of failure versus the severity (impact) of the failure and can prescribe additional actions if the calculated outcomes are not considered tolerable (Figure A3.1).

The origins of RCM come from the aviation industry, dating back to the 1960s from an organization called the Maintenance Steering Group (MSG). Their disciplines were later documented due to a United Airlines assessment of the commissioning 747s. In 1978, Nolen and Heap published a "Reliability-Centered Maintenance" report after their exhaustive study of FMEA of the planned 747 aircraft at United.

One of the fundamental truths revealed in the 747 study and is maintained today when you assess a typical complex asset is that the components of an asset fail in six specific failure patterns. Furthermore, the frequency of these patterns is relatively consistent with the original study (within a few percentage points) (Table A3.1).

It is generally accepted that the MSG and United Airlines studies determined that only 11% of failures from this complex asset had failure patterns such as A, B, or C, suggesting that they are more likely to fail over time or a specific usage period.

Conversely, Failure patterns D, E, and F do not increase their likelihood of failure over time and are considered more likely to fail "randomly" over time. The total percentage of assets found to fail randomly over time was 89% versus the 11% for the assets expected to fail over a period of time (Figure A3.2).

Addressing the 89% random failures with asset priorities data, operations data, and maintenance data supported by technology (CBM, PM, IoT, etc.) is key to being a leader in this asset reliability challenge.

If 89% of assets fail randomly, then why do so many organizations:

- Expect asset reliability with only a focus on preventive (time-based) maintenance tactics?
- Fail to identify prioritized assets and their components that need to be addressed with random failure characteristics?
- Ignore opportunities to leverage data to predict random failures?

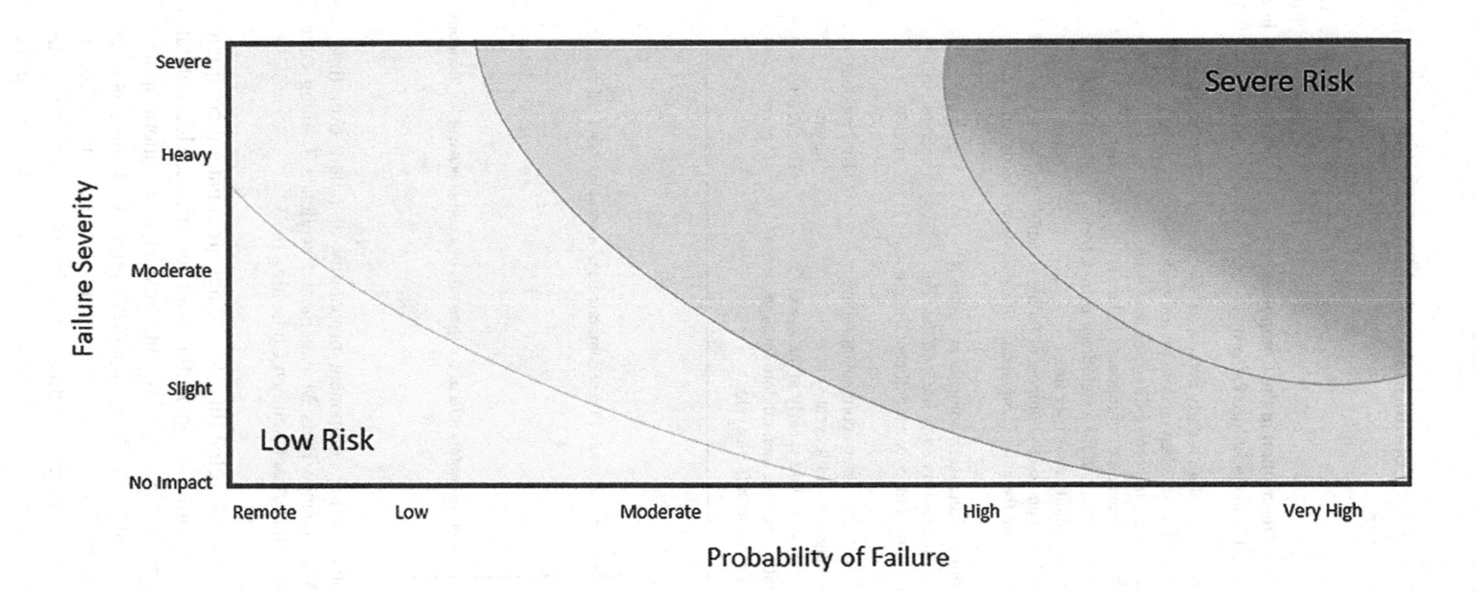


**FIGURE A3.1** Risk grid example- "Severity versus Probability of Failure." (Copyright MPE ©2022 from Maintenance Parts Management Excellence Training edited by Barry. Reproduced by permission of Asset Acumen Consulting Inc.)

**TABLE A3.1**
**Describes a Component's Conditional Probability of Failure over Time within a Complex Asset**

| Failure Pattern | Failure Description in the Component Lifecycle Failure Pattern | Failure Pattern Name | Observed Failure Rate in a Complex Asset (%) |
|---|---|---|---|
| A | The probability of failure is early in the asset component's lifecycle, then random for a significant period, then the failure probability rises over time | The "bathtub curve" | 4 |
| B | The probability of an asset's component failure is random in early life, and for a significant period, then the failure probability rises over time | The "traditional view" | 2 |
| C | The probability of an asset's component failure steadily grows throughout the asset component's lifecycle over time | The "linear increase" | 5 |
| D | The probability of an asset's component failure is virtually zero in its very early life, and then the probability is random over a significant time in its lifecycle | The sharp increase to random" | 7 |
| E | The probability of an asset's component failure is random throughout its lifecycle | The "random failure" | 14 |
| F | The probability of failure is early in the asset component's lifecycle, then random throughout the rest of the component's lifecycle | The "reverse J" curve | 68 |

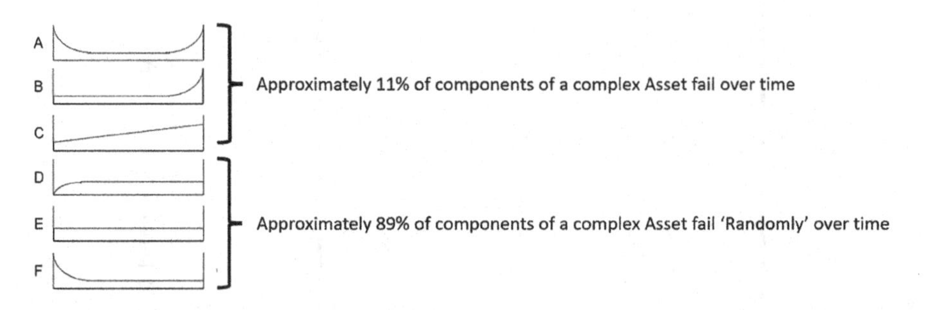


**FIGURE A3.2** Graphs of the component probability of failure over time. (Copyright MPE ©2022 from Maintenance Parts Management Excellence Training edited by Barry. Reproduced by permission of Asset Acumen Consulting Inc.)

Data analytics and risk and reliability disciplines are a natural complement to each other in pursuing asset optimization. At the asset performance level of data collection and analytics, reliability professionals can consider many more options than what may have been previously considered in a traditional RCM analysis of an asset in its operating context. An RCM analysis process for a specific (critical) asset would drive the assessment team to complete the documentation of the asset's Operating context, functions, functional failures (failed states), failure modes, failure effects (FMEA), and associated risks. After completing the FMEA, a decision grid would be reviewed to determine the best mitigating tactic prescriptively, and the first set of tactics considered would be predictive maintenance (PdM).

To confirm if a predictive maintenance tactic is viable in an RCM process: the process expects that the PdM activity would need to be "technically feasible" and financially "worth doing." In other words, the selected predictive tactic must be both (technically feasible and financially worthwhile) if a PdM tactic is to be accepted (prescriptively) as the best mitigating action.

## FOR EXAMPLE

Assume we have a pump supporting a critical operation, and we want 300 liters/ minute of water from this pump, and the pump "bearing" is our focus component. Assume that we know the failing characteristics of the (sealed) bearing is failure pattern "E" and is likely to fail randomly throughout its asset life. In RCM terms, the documented considerations to this point would be:

- Function: To pump 300 l/min of water
- Functional Failure: Unable to pump at all
- Failure Mode: Bearing seized
- Failure Effect: Downstream process stops (Figure A3.3)

If we consider early detection of a failure of a sealed pump bearing as a predictive maintenance mitigating action candidate, we may directly justify inspecting or testing the bearing for early indications of vibration, heat, or noise as "technically feasible" and "worth doing." Based on available skills and tools, we likely would select one of these options and calculate a frequency for the inspection that would provide a tolerable risk-benefit to mitigate the failure of the pump in its operating context. The first signs of trouble, or "potential failed" (P) to "fully failed" (F) state

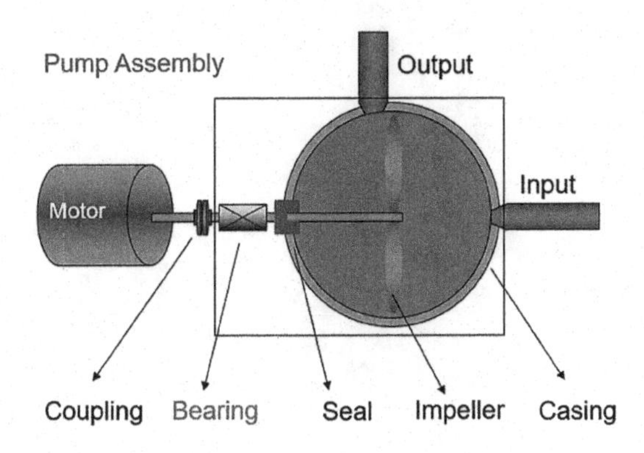


**FIGURE A3.3** Graphic of a pump assembly. (Copyright 104 MPE Inventory Management ©2020 from Maintenance Parts Management Excellence Training edited by Barry. Reproduced by permission of Asset Acumen Consulting Inc.)

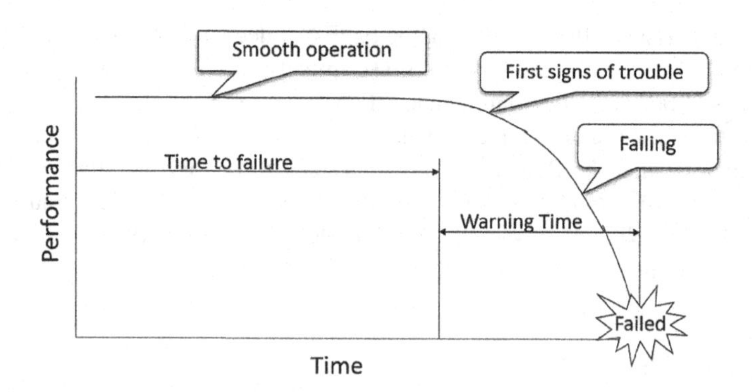


**FIGURE A3.4** Graphic of a "P" to "F" curve. (Copyright 104 MPE Inventory Management ©2020 from Maintenance Parts Management Excellence Training edited by Barry. Reproduced by permission of Asset Acumen Consulting Inc.)

(P to F) timeline, would influence if the early indicator (vibration, heat, or noise) can be detected promptly. A repair action can follow with a minimal outage experienced (Figure A3.4).

*Data analytics can assist and enhance this prescriptive decision grid when other data may be available that will provide further early detection options to be considered.* For example:

- Suppose operations are aware or measure not just the output of 300 l/minute but the state of the flow. What if the flow starts to cavitate due to bearing issues? *or*
- Suppose operations measure the stage differential between voltage and current on the pump's motor. Can this be detected if the bearing starts to fail?

# Index

## O



## P



## Q



## R